AF749400

MATHEMATICAL MODELLING OF WELD PHENOMENA 2

ALSO PUBLISHED BY THE INSTITUTE OF MATERIALS

Mathematical Modelling of Weld Phenomena
Edited by H. Cerjak & K. E. Easterling
Book no. 533 ISBN 0 901716 16 2

Metallurgical Modelling of Welding
Ø. Grong
Book no. 557 ISBN 0 901716 37 5

MATHEMATICAL MODELLING OF WELD PHENOMENA 2

Edited by

Professor H. Cerjak

Department of Materials Science & Welding Technology
Graz University of Technology
Austria

MATHEMATICAL MODELLING SERIES

Editor: Dr H. K. D. H. Bhadeshia
The University Of Cambridge
Department Of Materials Science
and Metallurgy

UNIV DUNELM

THE INSTITUTE OF MATERIALS

14 JUN 1999

Book 594
First Published in 1995 by
The Institute of Materials
1 Carlton House Terrace
London SW1Y 5DB

© The Institute of Materials 1995
All rights reserved

ISBN 0-901716-63-4
ISSN 1352-9692

621.791
MAT

1401026

Typeset by David John Services Ltd
Printed and bound in Great Britain by
Hollen Street Press, Berwick upon Tweed

CONTENTS

INTRODUCTION

The Second International meeting on the 'Numerical Analysis of Welding' was held in May 1993, at Graz-Seggau in Austria. The meeting was a part of a continuing series organised under the auspices of the International Institute of Welding, Commissions IX and IXB, and the Department of Materials Science and Welding at the Graz University of Technology.

The purpose of these meetings was originally defined under the leadership of the late Professor Kenneth Easterling. It was felt at the time that the modelling of welds was a complex problem which required serious documentation and teaching. The contemporary literature seemed inadequate for beginners in the field. Consequently the proceedings of the first meeting held in 1991 were published in the form of a book (edited by H. Cerjak and K. Easterling) containing extensive articles written in a readable style. The book has proved to be extremely useful and has already accumulated numerous citations in the published literature.

The present book is the second in the series and contains similarly detailed articles contributed by experts from many countries and institutions. The subject of weld-modelling is extremely exciting. We believe that the latest addition to the series reflects the enthusiasm and high skill of all the participants, who 'volunteered' the articles. We are very grateful to all these contributors. All of us appreciate the work of Peter Danckwerts at The Institute of Materials; the care with which he has produced the book is satisfying indeed.

H. Cerjak (Austria)
H. K. D. H. Bhadeshia (UK)

List of contributors

Bhadeshia, H.K.D.H. 71
Buchmayr, B. 119–137, 265–277
DebRoy, T. 3–21
Franz, U. 22–39
Gluschenko, O.B. 153–161
Goldak, J. 207–225
Grigorenko, G.M. 172–178, 179–185
Gu, M. 207–225
Hrivnak, I. 162–171
Inoue, H. 54–67
Kotelchuk, A.S. 153–161
Kurz, W. 40–53
Liu, W S. 38–152
Matsuda, F. 138–152
Morimoto, H. 54–67
Ohkita, S. 54–67
Pavlyk, V. 186–203
Radaj, D. 245–262
Roelens, J.B. 226–241
Schmidt, J. 22–39
Taranova, T.G. 172–178
Tayanovskaya, A.V. 172–178, 179–185
Weiss, D. 22–39

SECTION I

MELT POOL PHENOMENA AND SOLIDIFICATION

ROLE OF INTERFACIAL PHENOMENA IN NUMERICAL ANALYSIS OF WELDABILITY

T. DEBROY
Department of Materials Science and Engineering
Penn State University
University Park, PA 16802, USA

ABSTRACT

Since the last decade, our phenomenological understanding of weldability has improved significantly. A combination of advanced numerical methods and detailed description of interfacial phenomena in modelling has led to major progress in several important areas. Quantitative understanding of the role of surface active elements has been achieved by using temperature and composition-dependent surface tension values. Calculations of convective heat transfer in the weld pool have been useful in understanding cooling rates and simple features of the solidification structure. Important practical results are obtained when the solution of the equations of conservation of mass, momentum and energy for the weld pool is coupled with the solution of these equations for the gas phase. Such an approach is useful for reliable weld metal composition control for the welding of alloys containing volatile elements. Furthermore, consideration of the vaporisation phenomena can also provide safe operating limits of variables in pulsed laser spot welding.

INTRODUCTION

The term weldability commonly means a combination of weld design, welding process and material properties that make the welded material perform satisfactorily during service. Early assessment of weldability is usually done with laboratory testing techniques. However, in-service stress distributions and component geometries are often more complex than the simplified loading conditions and sample configurations commonly used during simulation testing.[1] Thus, laboratory testing of a material is often inadequate to fully define weldability, and it is necessary to understand not only the material's sensitivity to external loads, but also the metallurgy that promotes a susceptibility to defect formation.[1] Both material composition and processing conditions are important in determining the weld pool shape, solidification microstructure, compositional segregation, susceptibilities to defect formation, and consequently the performance of the joint during service. Among the various types of materials, the relation between the structure and the properties of crystalline materials has been studied most extensively. Thus, a fundamental issue in welding is to determine a combination of welding process variables and material composition necessary to attain desirable weld metal and heat-affected zone structures.

During fusion welding, the metal in the weld pool undergoes vigorous recirculatory motion driven by buoyancy, electromagnetic, and surface tension forces.[2–9] Buoyancy effects originate from the spatial variation of the liquid metal density mainly because of temperature variations and, to a lesser extent, from local composition variations. Electromagnetic effects are a consequence of the interaction between the divergent current path in the weld pool and the magnetic field it generates. The effect is important when a large electric current passes through the weld pool. The spatial variation of the surface tension owing to temperature and composition gradients at the weld pool surface often provides the main driving force for the convective flow, known as the Marangoni flow. Fluid flow and heat transfer are important in determining the size and shape of the weld pool, the weld macro- and microstructures and the weldability of the material. Although diagnostic techniques for the measurement of pool temperatures are currently being developed and evaluated,[10] direct reliable measurements of velocities, temperatures and species concentrations in the weld pool are very difficult, since the weld pool is small in size and it is often covered by plasma.[11–15] A recourse is to simulate the temperature and velocity fields by mathematical modelling of the essential physical processes of welding. Indeed, the modelling of heat transfer, fluid flow, and mass transfer has already provided detailed insight into the welding processes that could not have been obtained otherwise.

Segregation of surface active elements such as oxygen, sulphur, selenium and tellurium at the weld pool surface has a profound effect on the weldability of steels.[5, 16–19] The spatial variation of the interfacial tension at the weld pool surface determines the direction and magnitude of the Marangoni convection in the weld pool and the structure and properties of the weld metal. The temperature distribution at the weld pool surface determines the rates of vaporisation of the alloying elements. The loss of alloying elements often results in a change in the composition of the weld metal, affects weld properties and is a serious problem in the welding of many important engineering alloys.[20–28] In spot welding with lasers, the escaping vapour can exert a large recoil force and lead to unacceptable metal spatter and weld metal expulsion.[29–30] The weldability of many important engineering alloys are affected by these important interfacial phenomena.

Since the last decade, significant advancements have been made in applying numerical methods to understand several important interfacial phenomena in welding.[31] The role of surface-active elements in affecting weld pool heat transfer and fluid flow, and the development of weld pool geometry, cooling rates and the simple features of the solidification structure can now be predicted from fundamentals of transport phenomena. Considerable progress has been made in understanding the alloying element vaporisation and weld metal composition change because of welding. It is now well recognised that accurate vaporisation rates are important for credible calculations of the weld pool temperature profiles, particularly when high energy density heat sources are used for welding. Improved understanding of alloying element vaporisation has been useful in determining acceptable limits of operating conditions for laser spot welding. In this paper the discussion is focussed on the recent accomplishments, and problems and issues in applying numerical methods to understand weldability with special emphasis on the role of interfacial phenomena in weldability.

NUMERICAL MODELLING OF THE ROLE OF SURFACE ACTIVE ELEMENTS

Most commercial alloys inevitably contain small amounts of surface-active elements such as oxygen or sulphur. Variable depth of penetration during the welding of different batches of a commercial material with composition within a prescribed range has received considerable attention. Previous work has shown that the knowledge of the interfacial phenomena in welding is the key to understanding and controlling weld penetration. Often, the penetration depth is determined by the concentration of the surface active elements in the commercial alloy.[5, 16–19] These elements can affect the temperature coefficient of surface tension, $d\gamma/dT$, and the resulting direction of fluid flow in the weld pool.[5,7,8] The convective heat transfer, in turn, affects the weld penetration. In order to quantitatively determine their role in the transport processes, it is necessary to know how the temperature coefficient of interfacial tension changes with both temperature and composition.

It has been demonstrated[18,19] that the interfacial tension in these systems could be satisfactorily described by a formalism based on the combination of Gibbs and Langmuir adsorption isotherms and is given by:

$$\gamma = \gamma_m^0 - A\,(T - T_m) - \mathrm{R\,T}\ \Gamma_s \ln(1 + k_1 a_i e^{-\Delta H^\circ/RT}) \qquad (1)$$

where γ is the interfacial tension as a function of composition and temperature, γ_m° is the interfacial tension of the pure metal at the melting

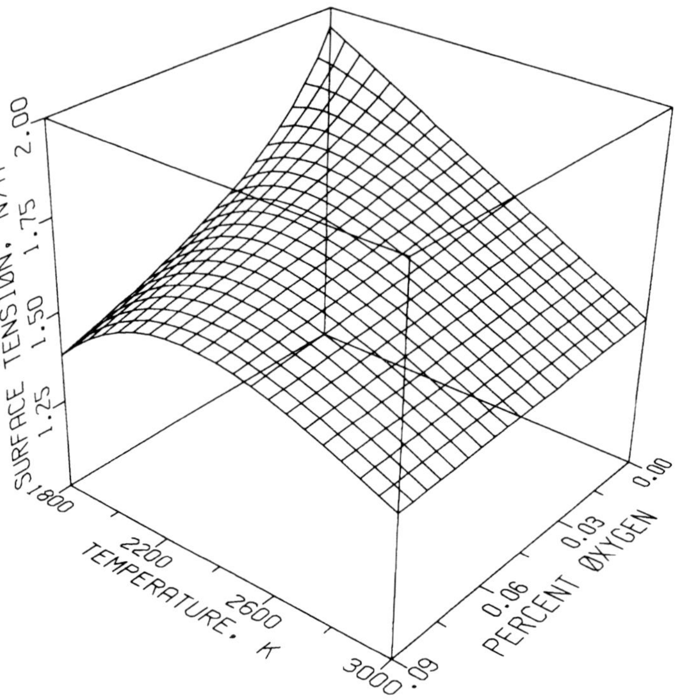

Figure 1. Surface tension of Fe–O alloy as a function of temperature and oxygen concentration.

point, R and R' are the gas constants in appropriate units, T is the absolute temperature, Γ_s is the surface excess of the solute at saturation solubility, k_1 is the entropy factor, a_i is the activity of the solute and ΔH° is the enthalpy of segregation. The calculated values of surface tension[18,32] for Fe–O alloys are presented in Fig. 1. The values of various parameters in equation 1 used for the calculation of surface tension are presented in Table 1. It is observed that for certain concentrations of oxygen, $d\gamma/dT$ can change from a positive value at 'low' temperature to a negative value at 'high' temperature. This implies that in a weld pool containing fairly high oxygen contents, $d\gamma/dT$ can go through an inflection point somewhere on the surface of the pool. Under these conditions, the fluid flow in the weld pool is more complicated than a simple recirculation.

The calculated GTA fusion zone profiles[32] for pure iron, and an iron–0.03 wt% O alloy are presented in Fig. 2. The data used for the calculations are presented in Table 2. The results in Fig. 2 clearly show the significant effect of oxygen concentration on the weld pool shape and the aspect ratio. Near the heat source, where the temperature is very high, the flow is radially outward. However, a short distance away from the heat source, where the temperature drops below the threshold value for the change in the sign of $d\gamma/dT$, the flow reverses in direction. The flow field is not a simple recirculation. The computed values of the aspect ratio are compared with the experimental data

Table 1 Data used for the computation of Fe–O surface tension as a function of temperature and activity of oxygen given in Fig. 1.

Parameter	Value
Surface tension of pure iron, γ_m^o (N m^{-1})	1.943
Temperature coefficient of surface tension for pure iron, A (N/m.K)	-4.3×10^{-4}
Surface excess at saturation, Γ_S, (kg-mole/m$^{2)}$	2.03×10^{-8}
Entropy factor, k_1	0.00318
Enthalpy of segregation, ΔH^o (J/kg-mole)	-1.6×10^8
Gas constant, R (kJ/kg-mole.K)	8314
Gas constant, R'(J/kg-mole.K)	8.314

Table 2 Data used for the computation of velocity and temperature fields given in Fig. 2.

Property/Parameter	Value
Density (kg/m^3)	7200
Melting point (K)	1811
Effective viscosity (kg/ms)	0.1
Thermal diffusivity of solid (m^2/s)	3.9×10^{-6}
Thermal diffusivity of liquid (m^2/s)	4.2×10^{-5}
Specific heat of solid (J/kg K)	710.6
Specific heat of liquid (J/kg K)	836.0
Arc current (Amp)	150.0
Arc voltage (Volts)	15.0
Arc radius (cm)	0.15
Arc efficiency	0.65

of Roper and Heiple[33] in Fig. 3. Good agreement is achieved between the calculated results and the experimental data. A similar behavior was also observed[7,8] for the GTA fusion welding of 304 stainless steel containing 240 ppm sulphur. The results showed the importance of considering composition and temperature dependent $d\gamma/dT$. When a constant $d\gamma/dT$ was used, the predicted depth of penetration was significantly more than what was actually observed. However, when the value of $d\gamma/dT$ was calculated from the works of Sahoo *et al.*,[18] the resulting weld pool shape agreed well with the corresponding experimentally determined value. Apart from the good agreement between the experimental and the calculated weld pool shapes, the

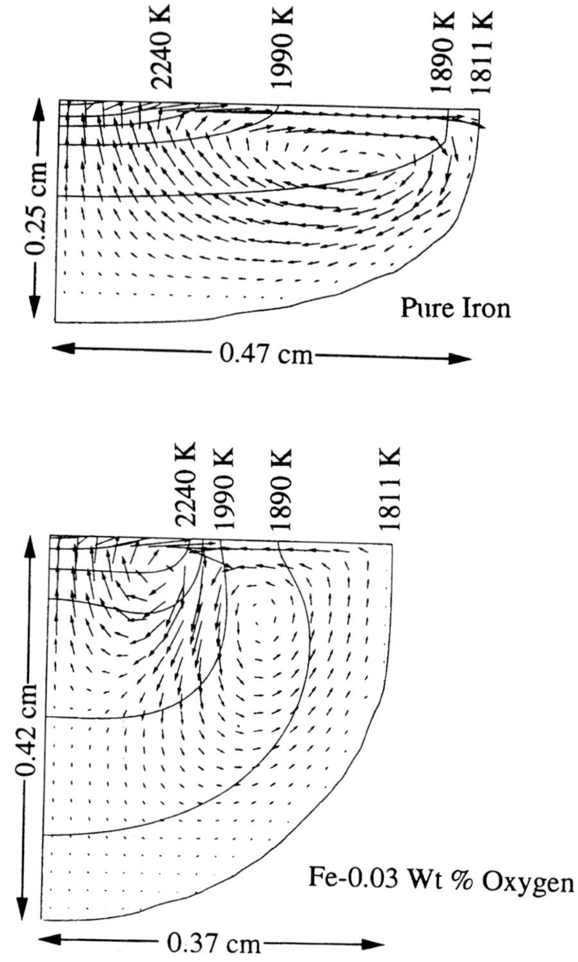

Figure 2. Velocity and temperature fields for two different cases: (a) for pure iron and (b) iron with 0.03 wt% oxygen.

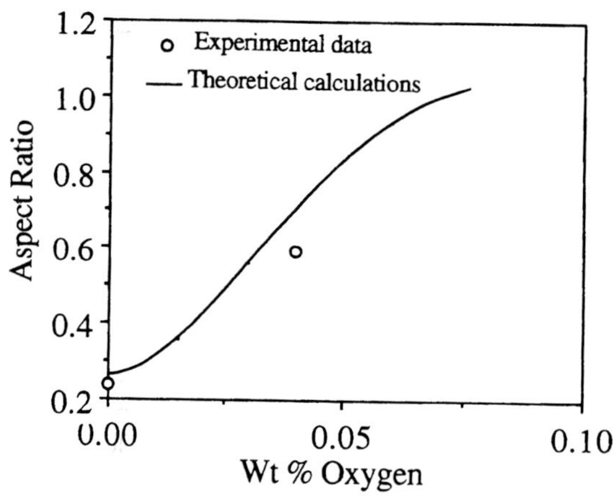

Figure 3. Comparison of the predicted and experimental aspect (depth to width) ratios. The oxygen concentration of 0.04 wt% is obtained from the *Cr–O* equilibrium data for 21 wt% *Cr* given in reference 48.

calculations reveal new insight about the complexity of the flow field. Although the qualitative effects of the role of surface-active elements were known, the numerical calculations provide a basis for quantitative assessment of their role in the development of the weld pool geometry.

SECONDARY DENDRITE ARM SPACING AND THE COOLING RATE

Brower *et al.*[34] experimentally investigated the effect of cooling rate on the secondary dendrite arm spacing for a steel of composition somewhat similar to 201 stainless steel (commercial alloy 440C, 17 wt% Cr, 1 wt% Mn). Experiments[35] in our research group showed that the correlation between secondary dendrite arm spacing and cooling rate for AISI 201 is very close to that of the alloy 440C, as shown in Fig. 4. The cooling rates at the edge of the pool were calculated from the numerically computed temperature field as the product of the temperature gradient and the welding speed, for two welding speeds, i.e. 5 mm/s and 31 mm/s.[5] The secondary dendrite arm spacings corresponding to these cooling rates are 0.9 microns and 0.4 microns respectively. The secondary dendrite arm spacings obtained from the microstructures agreed well with the values predicted by the model. The agreement between the predicted secondary dendrite arm spacing and the values obtained from independent data demonstrates that the calculated values of cooling rates are fairly accurate.

In our previous investigation[9] of the effect of pulsed laser welding on the thermal response of 310 and 316 austenitic stainless steels, the cooling rates were theoretically calculated from fundamental principles of transport phenomena. The secondary dendrite arm spacings were determined from the microstructures. The results indicated agreement between the measured and

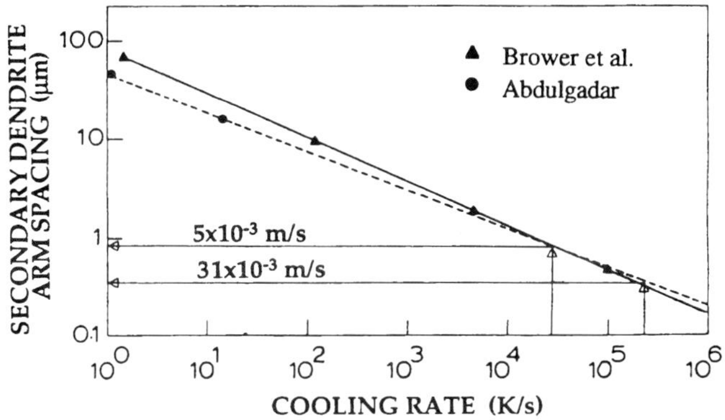

Figure 4. Plot of secondary dendrite arm spacing as a function of cooling rate for conduction mode laser welding of AISI 201 stainless steel.

the expected secondary dendrite arm spacings based on the cooling rates. The investigations with both continuous[5] and pulsed[9] heat sources indicate that simple features of the solidification structure can be determined from the available numerical models of weld pool transport phenomena.

VAPORISATION FROM WELD POOL SURFACE

The weld pool surface temperatures are normally much higher than the melting points of the weld metals. Consequently, pronounced vaporisation of alloying elements takes place, especially when high energy density heat sources are used.[20–25] Such losses often result in a change in the composition of the weld metal, affect weld properties and are a serious problem in the welding of many important engineering alloys. Currently, there is no unified theoretical model to predict, from fundamental principles, laser or electron beam-induced metal vaporisation rates and the resulting weld pool composition changes. A comprehensive theoretical model would allow calculation of the acceptable limits of operating parameters and would be useful for weld metal composition control.

It has been suggested that the weld pool peak temperature is limited by the evaporation of elements from the weld pool. Block-Bolten and Eagar[27] derived the following equation by considering the energy balance to determine the maximum temperature of the weld pool.

$$W(L_e - \Delta\overline{H}) = xP \qquad (2)$$

where W is the temperature-dependent vaporisation rate, L_e is the enthalpy of vaporisation, $\Delta\overline{H}$ is the enthalpy of mixing, P is the power density and x is the fraction of the input power used for vaporisation. Their calculations indicated the peak temperature in arc welding to be about 2773K. Furthermore, they found that the peak temperature in high energy density processes such as the laser and electron beam can be as high as 4273–5273K. It should be noted that the relation between the vaporisation rate W and the temperature significantly affects the value of the temperature calculated from equation 2. In fact, the temperature versus vaporisation rate relation used by Block-Bolten and Eagar,[27] i.e. the Langmuir equation,[36] grossly overpredicts the vaporisation rate at any given temperature, or for a given vaporisation rate, grossly underpredicts the temperature. Kraus[10] has shown that the peak temperature in the GTA weld pool, measured by the laser reflectance method, could reach as high as 2950K. This value is higher than the limit calculated by Block-Bolten and Eagar[27] assuming Langmuir relation between the vaporisation rate and temperature.

The changes in the concentration of manganese as a function of welding speed are presented in Fig. 5 for laser welding of AISI 202 stainless steel.[25] The primary alloying elements in this steel are 7.5 wt% Mn, 17.8 wt% Cr and 4.7 wt% Ni. From mass balance, the decrease in the concentration of manganese, Δ %*Mn*, can be expressed as follows:

$$\Delta\% Mn = \frac{100r}{v\,\rho} \tag{3}$$

where r is the vaporisation rate of manganese, v is the volume of the weld metal melted per unit time and ρ is the density of the weld metal. The weld metal composition change depends on the vaporisation rate and the volume of the metal melted per unit time, often the latter being the dominant factor. The rate of vaporisation increases with laser power. However, the composition change is most pronounced at low powers because of the small weld pool volume at low laser powers.

During some experiments,[20] a portion of the vaporised material was collected as condensation on the inner surface of a hollow, cylindrical, quartz tube opened at both ends which was held stationary and co-axial with the laser beam. Khan and DebRoy[20] analysed the condensed material and showed that the vapour consisted primarily of iron, manganese, chromium and nickel. The vapor composition was about the same as that expected from the base metal maintained isothermally at its boiling point. It should be noted that for the estimation of temperature, the rates of vaporisation of the individual alloying elements are not needed. Since the vaporisation rate is a strong function of temperature, they argued that most of the condensate originated from the centre of the weld pool near the laser beam axis where the local temperature was close to the boiling point of the alloy.

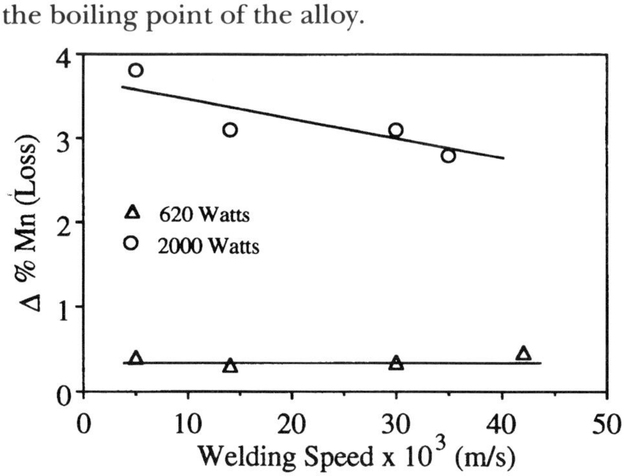

Figure 5. Effect of welding speed on weld metal composition change. Samples of AISI 202 stainless steels were welded with laser power of 620 W and AISI 201 samples were welded with laser power of 2000 W.

DebRoy *et al.* developed a comprehensive mathematical model to understand vaporisation of pure metals[20] and the loss of alloying elements from stainless steel[22,23] weld pools. The fluid flow and heat transfer within the molten pool were stimulated by numerical solution of the Navier–Stokes equations and the equation of conservation of energy to determine the temperature distribution at the weld pool surface. The heat transfer to the shielding gas and the heat loss due to vaporization of the alloying elements were taken into account in the calculations. The computed weld pool temperature distribution was used for the vaporisation rate calculations.

A key feature of the calculations is the consideration of the pressure gradient driven mass transfer.[21–23,37,38] In laser processing of metals and alloys, the peak temperature reached at the surface often exceeds the boiling point of the irradiated material. For example, von Allmen[39] determined molten pool temperatures in excess of boiling point for laser treatment of copper. Batanov *et al.*[40] indicated that temperatures on the surface of a laser-irradiated material can be higher than the boiling point. Paul and DebRoy[5] and Zacharia *et al.*[8] have reported theoretically calculated temperatures close to the boiling point for laser welding. Khan and DebRoy[20] arrived at the same conclusion from an analysis of the experimentally determined values of relative vaporisation rates of various alloying elements. Chan and Mazumder[41] have also reported temperatures greater than the boiling point during laser irradiation of aluminum, titanium and a superalloy. Theoretical calculations of the vaporisation rates by Anisimov and Rakhmatulina[37] and Knight[38] are based on the premise that the liquid pool surface temperatures are higher than the boiling point.

At temperatures higher than the boiling point, the pressures in the vicinity of the pool are greater than the ambient pressure and the excess pressure provides a driving force for the vapour to move away from the surface. The velocity distribution functions were used in the equations of conservation of mass, momentum and translational kinetic energy in the gas phase to calculate the rates of vaporisation of various elements due to a total pressure gradient at the pool surface. In addition, mass transfer rates due to concentration gradients were determined using available correlations amongst various dimensionless numbers.[42]

In Fig. 6, the experimentally determined vaporisation rates are compared[21] with the rates computed from the model and the values calculated from the Langmuir equation[36] for pure iron and titanium. It is observed that the experimentally determined vaporisation rates are closer to the values predicted by the present model than the rates calculated from the Langmuir equation. However, in all cases, the model predictions of the rates are somewhat higher than the corresponding experimentally determined values. The lower experimental rates are consistent with plasma-induced suppression of vaporiastion rates discussed in a previous publication.[43] Similarly, calculated values of vaporisation rates during high-power conduction mode CO_2 laser

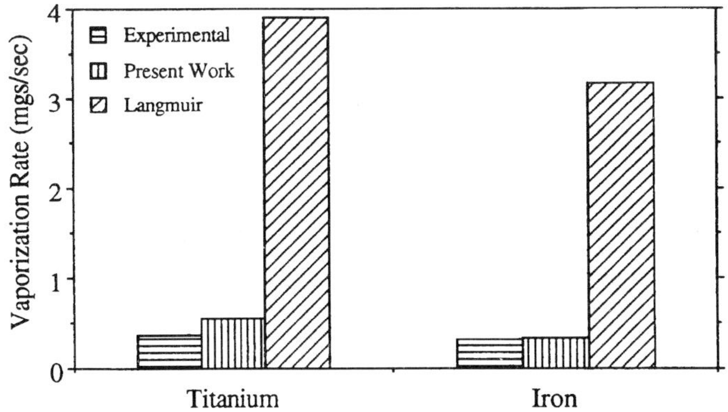

Figure 6. Comparison of the experimental vaporisation rates of pure iron and titanium with the corresponding rates calculated from the Langmuir equation and the present model. Laser power, 500 W; argon flow rate, 1 l/min.

welding of 201 stainless steel agreed very well with the corresponding experimental data as shown in Fig. 7. Although the calculated vaporisation rates are significantly better than the corresponding values obtained from the Langmuir equation, the higher accuracy is achieved by including more realistic and detailed description of the physical processes, and consequently, by doing more complex calculations.

The importance of the evaporative heat loss in the calculation of peak temperatures for different laser powers can be observed from Fig. 8. It is observed that evaporative heat loss significantly reduces the peak temperature and substantial errors in the calculated temperatures result when the heat loss is ignored.[22] The peak temperatures in Fig. 8 decreased slightly with power

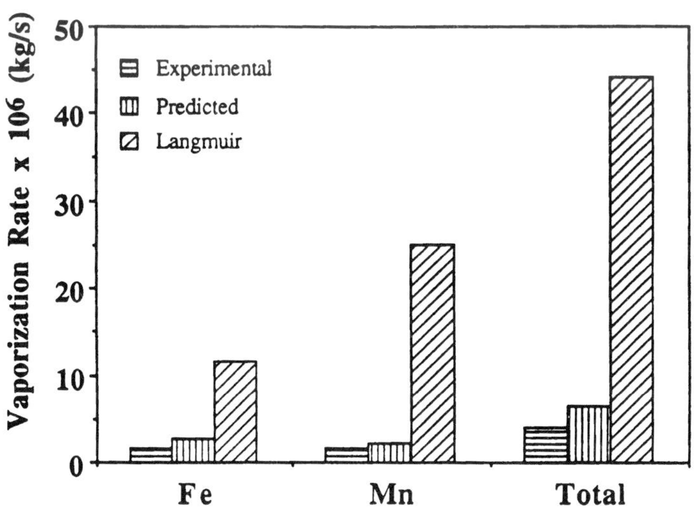

Figure 7. Comparison of the vaporisation rates calculated from the Langmuir equation and from the present model with experimentally determined values for AISI 201 stainless steel. Laser power, 3000 W; scanning speed, 15.24 x 10^{-3} m/s; argon flow rate, 5.5 x 10^{-4} m^3/s.

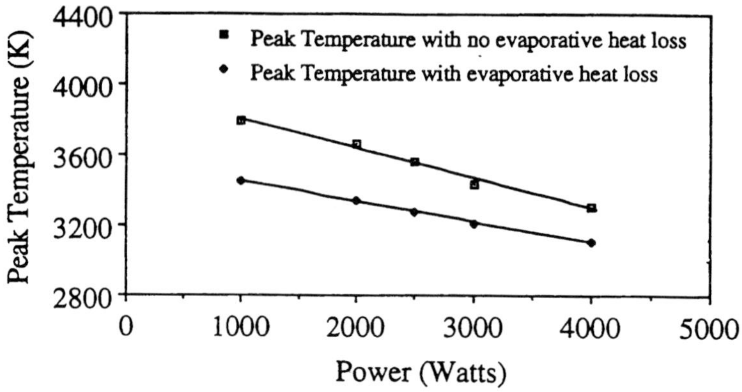

Figure 8. Effect of evaporative heat loss on peak temperature for different laser powers during welding of AISI 201 stainless steel. Laser scanning speed, 15.24 x 10^{-3} m/s; argon flow rate, 5.5 x 10^{-4} m^3/s.

mainly because of the resulting changes in the beam characteristics at high laser powers.

METAL EXPULSION DURING PULSED LASER SPOT WELDING

Significant vaporisation from the weld pool takes place when very high power density energy sources such as lasers and electron beams are used for welding. In many cases, the escaping vapour exerts a large recoil force and, as a result, the molten metal is expelled from the cavity. In welding, the metal spatter and weld metal expulsion are unacceptable. Therefore, it is important to know the role of various factors which affect liquid metal expulsion during pulsed laser welding.

Several investigators have reported experimental results of metal expulsion during laser irradiation. Chun and Rose[44] irradiated an aluminium target with 1–30 J laser pulses of duration 1–500 microseconds and observed that as much as 90% of the material lost was removed from the molten pool as liquid. During laser scribing of ceramics with ruby laser pulses, Wagner[45] observed that a conical molten region was formed and all the liquid was expelled. von Allmen[46] suggested that the vapour pressure acts like a piston on the liquid weld pool and forces liquid metal out of the cavity. He proposed a model to calculate liquid metal expulsion rate. The convective heat transfer and the conduction heat loss in lateral directions were ignored. Furthermore, the spatial distribution of temperature on the weld pool surface was not considered for the calculation of the vapour recoil force. The main difficulty in the application of the melt ejection model[46] to study the critical conditions for the initiation of liquid metal expulsion is that the retarding effect of surface tension is not considered. Thus, a precondition for the application of the model is that the vapour recoil force must be significantly higher than the surface tension force. While such a requirement is not unduly restrictive at very high laser intensities, it precludes application of the model to predict conditions for the initiation of liquid metal expulsion, i.e. when the recoil force just exceeds the surface tension force.

Basu and DebRoy[29,30] examined the conditions for the initiation of liquid metal expulsion during laser irradiation experimentally and theoretically. Micrographs of lead and stainless steel samples irradiated by CO_2 laser pulses are shown in Fig. 9. Melting was observed in each case. However, liquid expulsion did not take place until the pulse length was increased to 0.1 ms for lead samples. On varying the frequency and the pulse length for a total exposure time of 1.0 s, a similar behaviour, i.e. the transition from melting to expulsion, was observed in both stainless steel and titanium samples. A summary of the frequency and pulse length combinations necessary for initiating liquid metal expulsion in lead, titanium and stainless steel is presented in Fig. 10. When irradiated with single pulses, metal expulsion was observed only for large values of pulse length. For multiple pulses, short pulses led to liquid metal expulsion only at high frequencies when the irradiated region could not cool sufficiently between pulses.

The value of the maximum weld pool surface temperature just before the initiation of liquid metal expulsion can be determined from the following criterion. Liquid expulsion takes place when the vapour recoil force overcomes the surface tension force at the periphery of the weld pool. For a given value of peak temperature, T_0, and surface temperature distribution, the recoil force, F, is calculated[30] from the relation:

$$F = 2\pi \int_0^{r_b} r\,\Delta P(r)\,dr \tag{4}$$

where r_b is the radial distance at which the surface temperature is equal to the boiling point, and $\Delta P(r)$ is the difference between the local equilibrium vapor pressure and the atmospheric pressure and is a function of radial distance from the beam axis. The calculated values of the surface tension force at the periphery of the weld pool, $2\pi r_0 \sigma$, where σ is the surface tension at the melting point, and the recoil force for various values of T_0 are shown in Fig. 11. The details of the calculation procedure and the data used for the calculations are given in 29 and 30. It is observed that the recoil force is a strong function of the peak temperature. The value of T_0 for which the recoil and surface tension forces are equal is the critical temperature for liquid expulsion. Liquid expulsion from stainless steel can take place when the temperature at the centre of the pool, T_0, exceeds a critical temperature, T_{cr} which is 3380K. The values of T_{cr} for lead and titanium are 2120 K and 3780 K respectively.[30]

During laser irradiation, the rise and decay of the peak temperature at the surface of a metal can be determined[47] from equations 5 and 6 respectively.

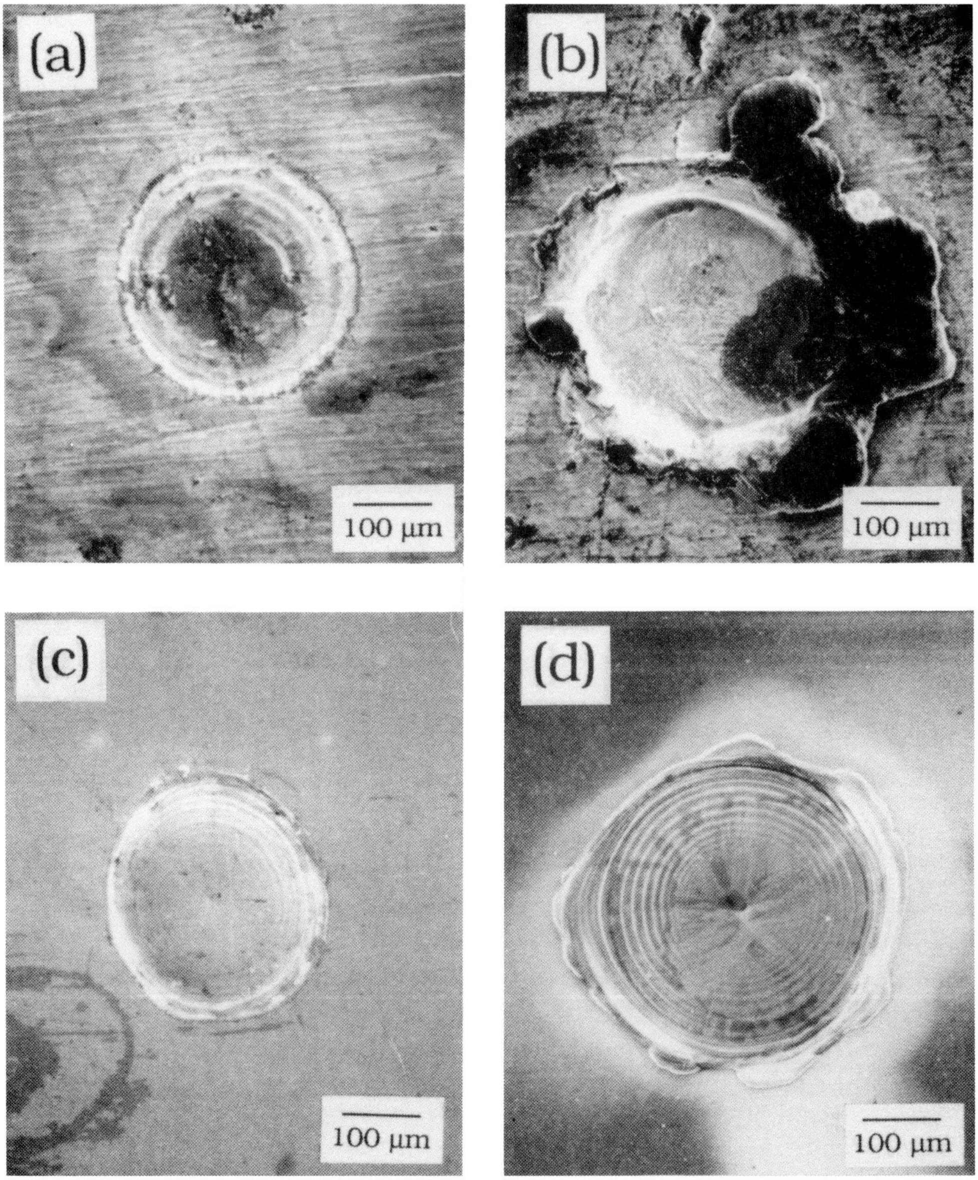

Figure 9. Micrographs of lead and stainless steel samples irradiated by laser pulses of increasing duration leading to liquid expulsion. Lead irradiated by single pulses of (a) 0.05 ms and (b) 0.10 ms durations. Stainless steel irradiated by single pulses of (c) 0.5 ms and (d) 1.0 ms durations. Laser power, 500 W.

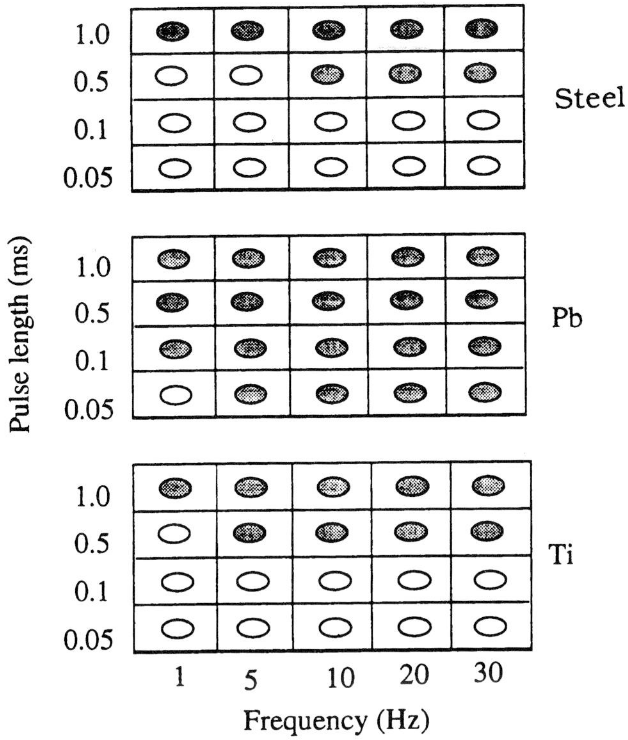

Figure 10. Combination of pulse length and frequency which lead to liquid metal expulsion. Cross-hatched circles indicate liquid expulsion. Laser power, 500 W.

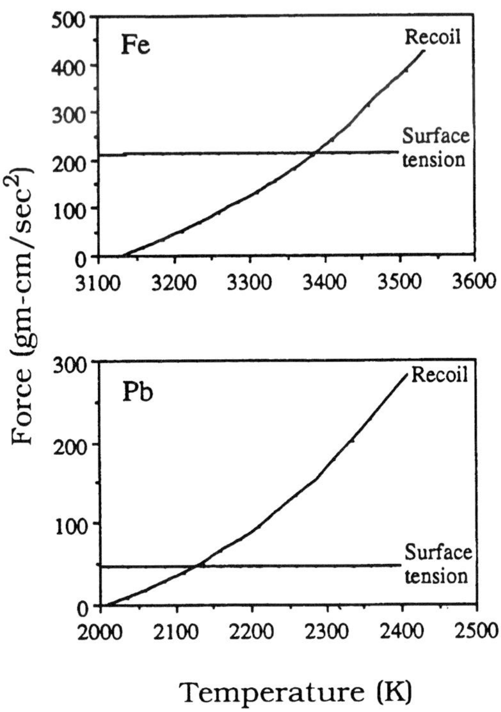

Figure 11. Recoil force as a function of peak surface temperature and the surface tension force at the melting point for iron and lead.

$$T_0 - T_0^0 = 2I\sqrt{\frac{t}{\pi k \rho C_p}} \qquad \text{for } t \leq t_p \tag{5}$$

$$T_0 - T_0^0 = 2I\left[\sqrt{\frac{t}{\pi k \rho C_p}} - \sqrt{\frac{t - t_p}{\pi k \rho C_p}}\right] \qquad \text{for } t \geq tp \tag{6}$$

where T_0 and T_0^0 are the peak temperatures at time t and just prior to irradiation respectively, I is the local intensity of absorbed laser radiation, k, ρ and C_p are the thermal conductivity, density and specific heat of the metal respectively. The temporal temperature profile calculated from equations 5 and 6 neglects the effects of phase change, variable material properties and weld pool convective heat transport but allows rough estimation of the changing peak temperature on the surface.

The peak temperature versus time plots for stainless steel irradiated by single pulses of length 0.5 and 1.0 ms are presented in Fig. 12. It is observed that for a pulse length of 0.5 ms, the peak temperature reached is significantly lower than the minimum peak temperature, 3380 K, required for liquid expulsion to occur. However, the peak temperature exceeds the critical temperature for a pulse length of 1.0 ms. The results are consistent with the experimental data shown in Figs. 9 and 10. Similarly for lead and titanium the peak temperature reached in a single pulse could be used to rationalise the experimental observations presented in Fig. 10. Although there is good

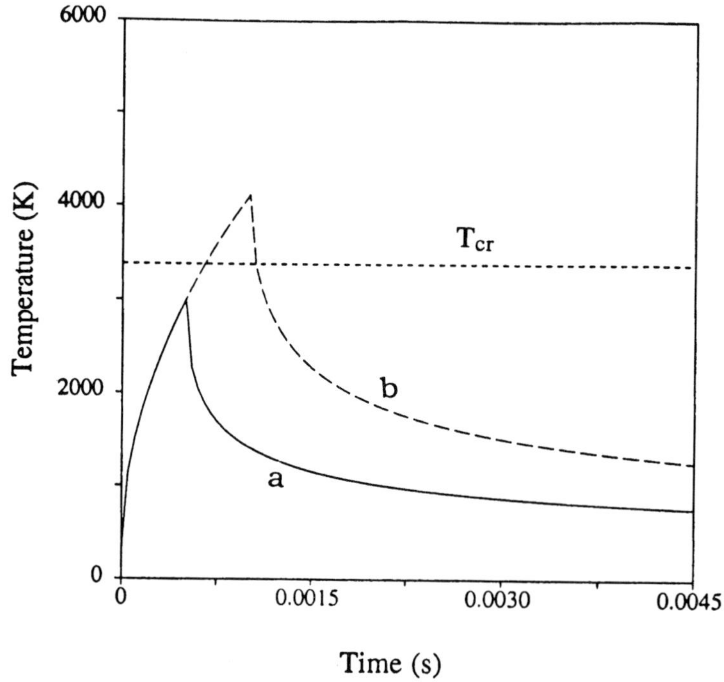

Figure 12. Peak temperature vs time for a stainless steel sample irradiated by single pulses of (a) 0.5 and (b) 1.0 ms durations. Laser power, 500 W.

agreement between the predicted and the observed condition for metal expulsion, the equations employed are not exact, and the computed temperatures provide dependable trends but not rigorously computed temperature data.

When the peak temperature at the centre of the pool exceeds the critical temperature, T_{cr}, liquid expulsion takes place. When molten metal is expelled from the weld pool, the laser beam gets defocussed, resulting in a decrease in the heat flux. As a result, the peak temperature may drop below the critical value, and no further expulsion may take place. Thus, the extent of melting at the time when the critical temperature is reached determines the amount of liquid metal expulsion.

OUTLOOK

In recent years major advances have taken place in quantitative understanding of welding processes and weldability of materials. The interdependence of welding processes and weld metal composition, structure and properties has received increasing attention. Detailed description of several component physical processes are now often included in the mathematical models because of the dramatic increase in the speed of the computers. The examples discussed in this paper are focussed on the advancements in the quantitative understanding of weldability. In particular they show how consideration of important interfacial phenomena in modelling can lead to useful practical results.

However, several key problems and issues related to the role of interfacial phenomena in weldability remain to be addressed. First, inadequate understanding of the plasma–metal interaction, particularly the role of plasma in affecting interfacial tension, and the scarcity of the high temperature thermophysical data affect the quality of the calculations. Second, modelling of the fluid flow and convective heat and mass transfer is undertaken without any special consideration of the rather unique, large recirculating velocities confined in a small cavity. How the small scale of the welding systems and the large recirculating velocities affect the enhancement of the heat, mass and momentum transfer rates above the transport rates encountered in common laminar flows in large-scale systems remain to be examined. Finally, the concentrations of nitrogen, oxygen and hydrogen strongly affect weldability. Serious efforts to understand the origin of these species in the weld metal are just beginning. Understanding the principles of partition of these elements between the weld pool and its environment remain a major challenge. As these problems and issues related to interfacial phenomena in welding are addressed, there will be progress towards a scientifically based control of the weldability of materials. Achievement of such a goal would be an important milestone in the advancement of the welding technology.

ACKNOWLEDGEMENT

This work was supported by the US Department of Energy, Office of Basic Energy Sciences, Division of Materials Sciences, under grant number DE-FG02-84ER45158.

REFERENCES

1. R.A.PATTERSON and K.W. MAHIN (eds): Weldability of Materials, 1990, ASM International, Materials Park, OH.
2. G.M. OREPER and J. SZEKELY: *J. Fluid Mech.*, 1984, **147**, 53–79.
3. C. CHAN, J. MAZUMDER and M.M. CHEN: *Metall. Trans. A*, 1984, **15A**, 2175–2184.
4. S. KOU and Y.H. WANG : *Metall. Trans. A*, 1986, **17A**, 2265–2270.
5. A. PAUL and T. DEBROY: *Metall. Trans. B*, 1988, **19B**, 851–858.
6. T. ZACHARIA, A.H. ERASLAN and D.K. AIDUN: *Welding Journal Research Supplement*, 1988, **67**, 18s–62s.
7. T. ZACHARIA, S.A. DAVID, J.M. VITEK and T. DEBROY: *Welding Journal Research Supplement*, 1989, **68**, 499s–509s.
8. T. ZACHARIA, S.A. DAVID, J.M. VITEK and T. DEBROY: *Welding Journal Research Supplement*, 1989, **68**, 510s–519s.
9. T. ZACHARIA, S.A. DAVID, J.M. VITEK and T. DEBROY: *Metall. Trans. B*, 1990, **21B**, 600–603.
10. H.G. KRAUS: *Welding Journal Research Supplement*, 1989, **68**, 269s–279s.
11. M.M. COLLUR and T. DEBROY: *Metall. Trans. B*, 1989, **20B**, 227–286.
12. R. MILLER and T. DEBROY: *J. Appl. Phys.*, 1990, **68**, 2045–2050.
13. G.J. DUNN and T.W. EAGAR: *Metall. Trans. A*, 1986, **17A**, 1851–1863.
14. G.J. DUNN and T.W. EAGAR: *Metall. Trans. A*, 1986, **17A**, 1865–1871.
15. T.J. ROCKSTROH and J. MAZUMDER: *J. Appl. Phys.*, 1987, **61**, 917–923.
16. C.R. HEIPLE and J.R. ROPER: *Welding Journal Research Supplement*, 1982, **61**, 92s–102s.
17. C.R. HEIPLE, J.R. ROPER, R.T. STAGNER and J.J. ALDEN: *Welding Journal Research Supplement*, 1982, **61**, 72s–17s.
18. P. SAHOO, T. DEBROY and M.J. MCNALLAN: *Metall. Trans.*, 1988, **19B**, 483–491.
19. M.J. MCNALLAN and T. DEBROY: *Metall. Trans.*, 1991, **22B**, 557–560.
20. P.A.A. KHAN and T. DEBROY: *Metall. Trans B*, 1984, **15B**, 641–644.
21. T. DEBROY, S. BASU and K. MUNDRA: *J. Appl. Phys.*, 1991, **70** 1313–1319.
22. K. MUNDRA and T. DEBROY: *Metall. Trans, B.*, 1993, **24B**, 145–155.
23. K. MUNDRA and T. DEBROY: *Welding Journal Research Supplement*, 1993, **.72**, 1s–9s.
24. T. DEBROY: *International Trends in Welding Science and Technology*, S.A. DAVID and J.M. VITEK ed., ASM International. Materials Park, OH, 1993, pp. 17–25.
25. P.A.A. KHAN, T. DEBROY and S.A. DAVID: *Welding Journal Research Supplement*, 1988, **67**, 1s–7s.
26. M.J. CIESLAK and P.W. FUERSCHBACH: *Metall. Trans. B*, 1988, **19B**, 319–329.

27. A. BLOCK-BOLTEN and T.W. EAGER: *Metall. Trans. B*, 1984, **15B**, 461–469.
28. A. BLAKE and J. MAZUMDER: *Journal of Engineering for Industry*, 1985, **107**, 275–280.
29. S. BASU: M.S. Thesis, Department of Materials Science and Engineering, The Pennsylvania State University, 1992.
30. S. BASU and T. DEBROY: *J. Appl. Phys.*, 1992, **.72**, 3317–3322.
31. S.A. DAVID and T. DEBROY: *Science*, 1992, **257**, 487–502.
32. K. MUNDRA and T. DEBROY: Unpublished research, Department of Materials Science and Engineering, The Pennsylvania State University, University Park, PA.
33. R. HEIPLE and J.R. ROPER: *Trends in Welding Research*, (S.A. David ed.), ASM Publications, Metals Park, OH, 1982, pp. 489–515.
34. W.A. BROWER R. STRACHAN and M.C. FLEMINGS: *AFS Cast Metals Research Journal*, Dec. 1970, 176–180.
35. S.A. ABDULGADAR: Ph. D. Thesis, The Pennsylvania State University, University Park, PA.
36. R.D. PEHLKE: *Unit Processes in Extractive Metallurgy*, Elsevier, New York, 1979.
37. S.I. ANISIMOV and A.Kh. RAKHMATULINA: *Soviet Physics – JETP*, 1973, **.37**, 441–444.
38. C.J. KNIGHT: *AIAA Journal*, 1979, **17**, 519–523.
39. M. VON ALLMEN: *Laser-Beam Interactions with Materials*, Springer-Verlag, Berlin, 1987.
40. V.A. BATANOV, F.V. BUNKIN, A.M. PROKHOROV and V.B. FEDORO: *Soviet Physics – JETP*, 1973, **36**, 311–322.
41. C.L. CHAN and J. MAZUMDER: *J. Appl. Phys.*, 1987, **62**, 4579–4586.
42. E.U. SCHLUNDER and V. GNICLINSKI: *Chem. Ing. Tech.*, 1967, **39**, 578–584.
43. P. SAHOO, M.M. COLLUR and T. DEBROY: *Metall. Trans. B*, 1988, **19B**, 441–444.
44. M.K. CHUN and K. ROSE: *J. Appl. Phys.*, 1974, **45**, 614–620.
45. R.E. WAGNER: *J. Appl. Phys.*, 1974, **45**, 4631–4637.
46. M. VON ALLMEN: *J. Appl. Phys.*, 1976, **47**, 5460–5463.
47. H.S. CARSLAW and J.C. JAEGAR: *Conduction of Heat in Solids*, Oxford University Press, 1959.
48. F.D. RICHARDSON: *Physical Chemistry of Melts in Metallurgy*, vol. 2, Academic Press, London, 1974.

A MODEL OF TEMPERATURE DISTRIBUTION AND WELD POOL DEFORMATION DURING ARC WELDING

D. WEISS and J. SCHMIDT
Otto von Guericke University, Magdeburg

U. FRANZ
Köthen College of Technology

ABSTRACT

A mathematical model is presented which allows the calculation of temperature distribution and seam form during arc welding. Special attention is paid to the modelling of the arc intensity distribution in the case of high deformation of the pool surface. The formulation of this distribution is based upon a reduced transitional resistance of the arc. A numerical example is given.

INTRODUCTION

The consideration of the relationships between energy transport, weld pool deformation and cooling conditions is necessary in order to simulate the seam form and the changing of the material structure during arc welding. The rather complex process of the arc effect can be essentially subdivided into five sub-problems (Fig. 1).[9]

Existing mathematical models of welding processes have taken the following into consideration:

- pool convection;[12, 6]
- weld pool deformation;[7, 10, 8, 3]
- energetic parts of arc, radiation, passing drops and vaporisation.[11, 2, 1]

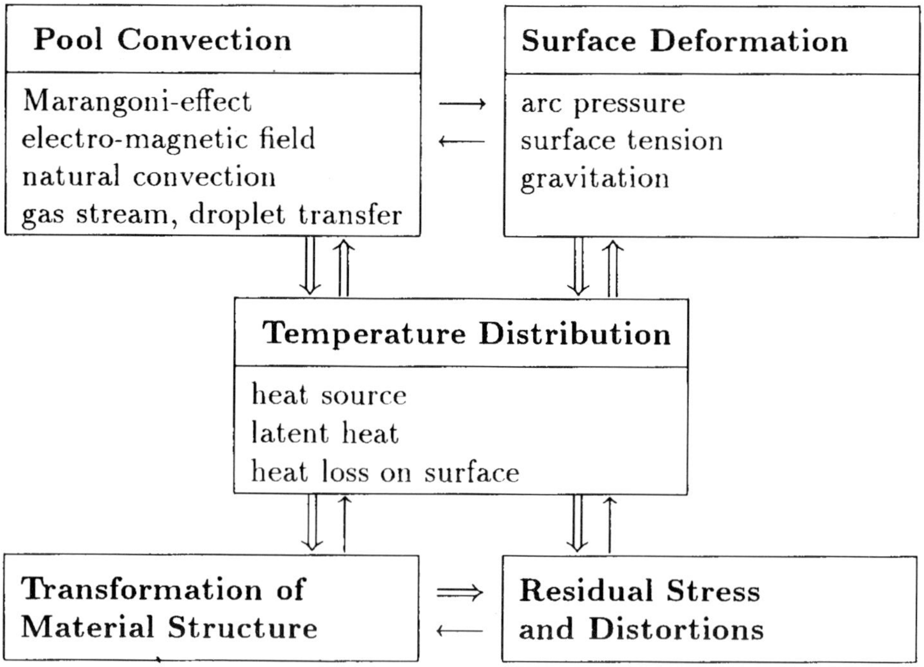

Figure 1. Relations between sub-problems..

The subject of the present investigation is the consideration of the interaction between the temperature field and the deformation of the weld pool. Previous works on simulation of arc welding have shown that the deformation of the weld pool can essentially influence the transport of energy.[10, 6, 8]

In the case of a high deformation of the pool surface (e.g. welding with high current, narrow gap welding) special attention is paid to the description of the three-dimensional arc effect. Effects such as ionisation and arc deflection are considered. This demands a formulation of the source intensity which is adapted to the existing conditions.

MODELLING

In modelling of arc intensity a distribution according to Gauss as a function of the radius r in the $x - y$ plane, independent of z, is often assumed (regarding the co-ordinate system: cf. Fig. 4):

$$\dot{q}(r) = \frac{3\dot{Q}_{eff}}{\pi r_{max}^2} \exp\left(-3\frac{r^2}{r_{max}^2}\right) \quad with\ r^2 = x^2 + y^2 \tag{1}$$

where $\dot{Q}_{eff}$ is the energy input and r_{max} the radius with $\dot{q}(r_{max}) \approx 0.05\dot{q}_{max}$, where the maximum energy input occurs in the intersection of the extended axis of the electrode with the surface of the pool ($r = 0$). In the case of highly deformed weld pools this is not always true. Furthermore the energy is supplied to those places with the lowest transitional resistance, which is clarified by the shaded distribution of heat flow density in Fig. 2. The resistance is determined primarily by the distance between the electrode and pool surface as well as by the electrical conductivity of the plasma.

For the relations in the present model $\dot{q}_{arc}$, $p_{arc} = f(x,y,z,T)$ must be sought which describe the heat flow $\dot{q}_{arc}$ and also the arc pressure p_{arc} as a function of the spatial position and the surface temperature T_{surf}. If a Gauss-distributed surface source is examined for the undeformed surface, a function of the distance between the electrode and pool surface L can be deduced from the radius-dependent distribution [Equation 1]. According to Fig. 3, the lengths are easily convertible.

$$\dot{q}(L) = \frac{3\dot{Q}_{eff}}{\pi(L_{max}^2 - L_{min}^2)} \exp\left(-3\frac{L^2 - L_{min}^2}{L_{max}^2 - L_{min}^2}\right) \tag{2}$$

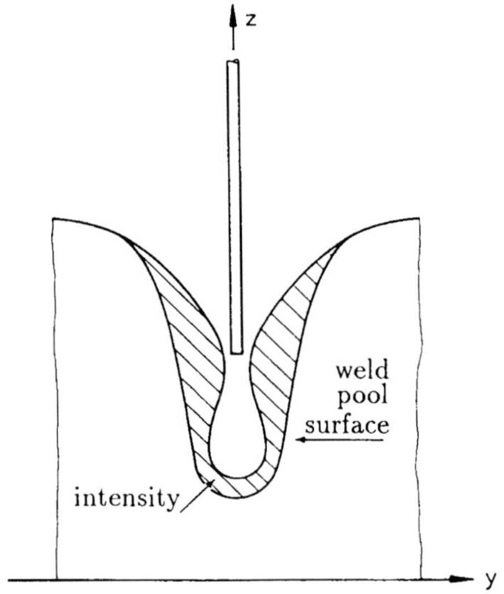

Figure 2. Distribution of heat flow and arc pressure in a cross-section to the direction of movement.

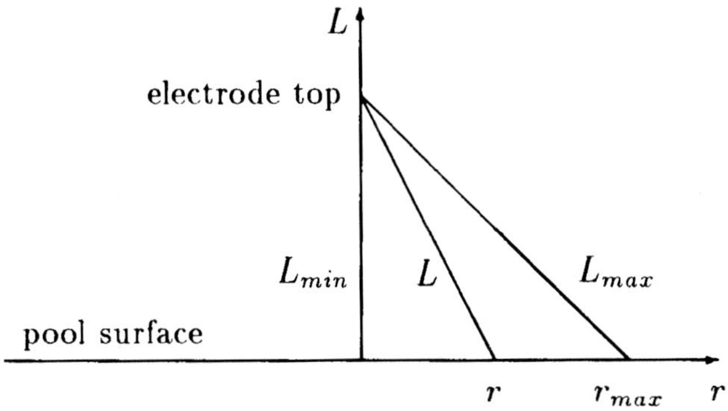

Figure 3. Distances between electrode and an undeformed pool surface.

The ionisation and the electrical conductivity of the plasma are influenced to a high degree by vaporisation of the metal on the weld pool surface. If the improved electrical conductivity is taken into account using a reduced transitional resistance R, the distribution function of the arc intensity ϕ can be defined as follows:

$$\phi(x,y,z,T) = \exp\left(-\frac{R^2 - R_{\min}^2}{R_{\max}^2 - R_{\min}^2}\right) \tag{3}$$

The ratio between intensity at minimum and at maximum resistance should be considered to be smaller than the distribution relation in Equation 1. First of all, a simple relation was used for the dependence of R on pool surface temperature T_{surf}:

$$R(x,y,z,T) = \begin{cases} L & : T_{surf} < T_{liq} \\ L \cdot \left(1 - \vartheta \dfrac{T_{surf} - T_{liq}}{T_{vap} - T_{liq}}\right) & : \text{otherwise} \end{cases} \tag{4}$$

(where T_{liq} is the liquidus temperature, T_{vap} the vaporisation temperature and ϑ the weighting parameter for reduction of resistance by the influence of ionisation).

Maximum values for pressure $p_{arc,max}$ and density of heat flow $\dot{q}_{arc,max}$ are determined in such a way that the corresponding surface integrals on the area of arc activity Γ_2 (cf. Fig. 4) are equal to the total force and the energy input, respectively.

The heat flow density and the arc pressure are as follows:

$$\dot{q}_{arc}(x,y,z,T) = \dot{q}_{arc,max}\, \phi(x,y,z,T) \cdot (\vec{e}_{arc} \cdot \vec{n}_{surf})(x,y,z) \tag{5}$$

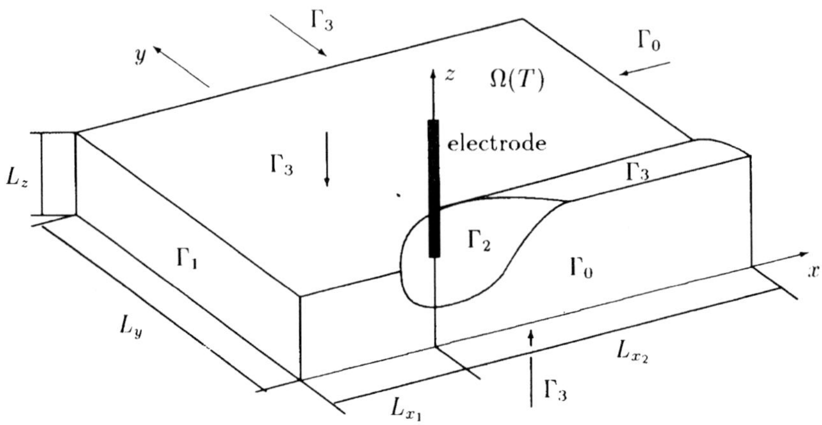

Figure 4. Co-ordinate system and boundary conditions.

$$p_{arc}(x,y,z,T) = p_{arc,max}\,\phi(x,y,z,T)\cdot(\vec{e}_{arc}\cdot\vec{n}_{surf})(x,y,z) \tag{6}$$

where $\vec{e}_{arc}$ is the unit vector (electrode top → pool surface) and $\vec{n}_{surf}$ the unit vector normal to the pool surface. This approach, Equations 5, 6, is applicable to different formed pool geometries, due to its dependence on lengths.

For further modelling the following major assumptions are made:

- the domain is a plate of constant thickness stretching to infinity in the direction of welding;
- stationary conditions exist for energy input and transport;
- restriction of the underside of the pool is through non-molten material or by a pool support;
- de-coupling of the temperature and stress fields is assumed;
- magnetic, electrical and chemical influences are disregarded;
- pool convection is not considered in the calculation of the free surface deformation.

Under these conditions the transport of energy is described by the Fourier–Kirchhoff differential equation. It is appropriate to select a co-ordinate system bound to the electrode (Fig. 4). The non-stationary term can be neglected in this case. From this results:

$$-\nabla \cdot (\lambda(T)\nabla T) + c_p(T)\rho\vec{v}\cdot\nabla T = \dot{q}_{vol} \qquad (7)$$

where λ is the heat conductivity, c_p the mass-specific heat capacity, ρ the density and $\vec{v}$ the velocity relative to the electrode. The source term $\dot{q}_{vol}$ can be neglected according to the assumptions made. The boundary conditions are chosen as follows:

$$-\lambda \left.\frac{\partial T}{\partial \vec{n}_{surf}}\right|_{\Gamma_0} = 0 \qquad (8)$$

$$T|_{\Gamma_1} = T_\infty(x,y,z) \qquad (9)$$

$$-\lambda \left.\frac{\partial T}{\partial \vec{n}_{surf}}\right|_{\Gamma_2} = \dot{q}_{arc}(c,y,z,T) + \dot{q}_{drop}(x,y,z) - \alpha_{eff}(x,y,z,T)(T_{surf} - T_\infty(x,y,z)) - \dot{q}_{vap}(T) \qquad (10)$$

$$-\lambda \left.\frac{\partial T}{\partial \vec{n}_{surf}}\right|_{\Gamma_3} = -\alpha_{eff}(x,y,z,T)(T_{surf} - T_\infty(x,y,z)) \qquad (11)$$

where T_∞ is the ambient temperature, α_{eff} the heat convection coefficient which covers heat transfer by radiation and natural convection, $\dot{q}_{drop}$ the energy input caused by the droplet transfer and $\dot{q}_{vap}$ the heat loss by vaporisation.

The influence of phase transformation is covered by the use of an effective heat capacity. An effective heat conductivity takes the influence of convective energy transport in the molten pool into consideration.

The formation of the free surface can be modelled on the basis of the equilibrium of the mechanical forces caused by arc pressure, gravitation and surface tension (Fig. 5).[10] An approach via minimisation of the surface energy is also possible (with $Z = L_z - z$):[8]

$$\int_\Omega \left\{ \sigma(T)(\sqrt{1 + |\nabla Z|^2} - 1) - p_{arc}Z + \frac{1}{2}\rho g Z^2 + p_0 Z \right\} dx\,dy \Rightarrow \text{Minimum} \qquad (12)$$

where σ is the surface tension of the liquid metal and g the gravitational constant. Term $p_0 Z$ considers all further effects excluding that of surface tension, arc pressure and gravitation.

On the assumption that the molten volume is constant, the problem can be solved using the Euler equation. In the resulting equations the term for local pressure $p_{local} = p(\theta_{plate})$ was adapted for cases where the plate is inclined:

$$p_{surf} - p_{arc} + p_{local} + p_0 = 0 \qquad (13)$$

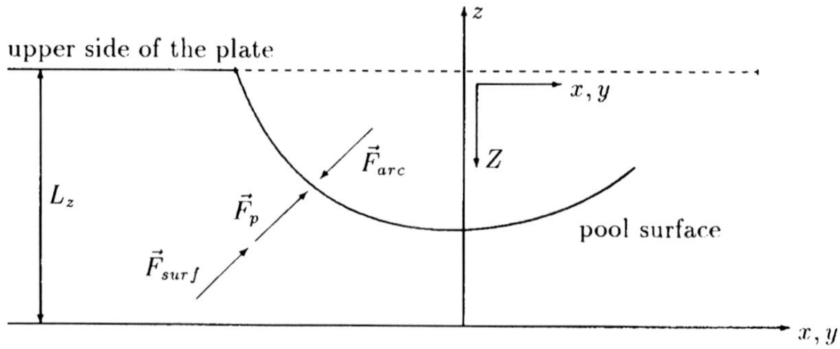

Figure 5. Forces on free weld pool surface.

$$p_{surf} = \nabla \cdot \left(\frac{\sigma(T)\nabla Z}{\sqrt{1 + |\nabla Z|^2}} \right) \tag{14}$$

(15)

where θ_{plate} is the inclination of the plate in the $x - z$ plane.

NUMERICAL SOLUTION

The coupled non-linear system described is linearised and solved iteratively (Fig. 6). The cycle of alternating surface and temperature field computation is carried out using under-relaxation ($0.05 \le \theta \le 0.33$).

The governing equations were solved using the finite element method. When applying the Galerkin method for the solution of the problem, where u is the unknown function T or Z, respectively, L is the linear operator, W is the Sobolev function space and φ is the test function:

$$Lu \equiv -\nabla \cdot (a\nabla u) + \vec{b}\cdot\nabla u + cu = f \text{ in } \Omega \tag{16}$$

with the weak formulation:

$$u \in V^g := \{v \in W^{1,2}(\Omega) : v|_{\Gamma_1} = g\} \tag{17}$$

$$a(u, \varphi) = f(\varphi) \qquad \forall \varphi \in V^0 \tag{18}$$

physically unmotivated oscillation can appear in the numerical solution of the temperature field in the present convection dominant case. One remedy for

$m = 0 \quad T^0 = T^1 = T_{start} \quad Z^0 = L_z$

repeat

$m = m + 1 \qquad Z^m = Z^{m-1} \qquad T^m = \Theta_T T^m + (1 - \Theta_T)\, T^{m-1}$

Pool surface deformation: Find $Z^{m+1} = Z(Z^m, T^m)$ so that

$$-\nabla \cdot \left(\frac{\sigma(T^m) \nabla Z^{m+1}}{\sqrt{1 + |\nabla Z^m|^2}} \right) + c\, Z^{m+1} = f(x, y, Z^{m+1}, T^m)$$

$$c = -\varrho g \cos \theta_{plate}$$

$$f = f(x, y, Z^m, T^m) = -p_{arc,max}(x, y, Z^m) \Phi(x, y, Z^m, T^m) \cdot (\vec{e}_{arc} \cdot \vec{n}_{surf})(x, y, Z^m) + \varrho g x \sin \theta_{plate} + p_0(Z^m)$$

Temperature field: Find $T^{m+1} = T(Z^{m+1}, T^m)$ so that

$$-\nabla \cdot (\lambda(T^m) \nabla T^{m+1}) + c_p(T^m) \varrho \vec{v} \cdot \nabla T^{m+1} = 0 \quad \text{in}\ \Omega(T^m)$$

$$-\lambda(T^m) \left. \frac{\partial T^{m+1}}{\partial \vec{n}_{surf}} \right|_{\Gamma_2} = \dot{q}_{arc}(x, y, Z^{m+1}, T^m) + \dot{q}_{drop}(x, y, Z^{m+1}) - \alpha_{eff}(x, y, Z^{m+1}, T^m)(T^{m+1}_{surf} - T_\infty(x, y, Z^{m+1})) - \dot{q}_{vap}(T^m)$$

$$-\lambda(T^m) \left. \frac{\partial T^{m+1}}{\partial \vec{n}_{surf}} \right|_{\Gamma_3} = -\alpha_{eff}(T^m)(T^{m+1}_{surf} - T_\infty(x, y, Z^{m+1}))$$

until $||T^{m+1} - T^m|| < \varepsilon_T$

Cooling conditions $t_{8/5}(y, z)$

Seam form $Z^{m+1}(y)|_{x = L_{x_2}}$

Figure 6. Coupling of temperature and free pool surface calculation.

solving this is offered by stabilisation through a Galerkin/least-squares formulation:[4]

$$\tilde{\varphi}_h = \varphi_h + \delta L \varphi_h \tag{19}$$

$$u_h \in V_h^g := \{ v \in V^g, v \mid_K \in P_1(K) \ \forall \ K \} \tag{20}$$

$$a(u_h, \varphi_h) + \sum_K \delta_K \int_K (Lu_h - f) L\varphi_h \mathrm{d}K = f(\varphi_h) \qquad \forall \ \varphi_h \in V_h^0 \tag{21}$$

where δ is the damping parameter, u_h the approximation of the solution and K the finite element. Additional testing of the numerical defect $Lu_h - f$ by means of functions of the form $\delta L\varphi$ proves the method to be consistent. Good adaptability to locally changing diffusive and convective influence should be noted. The method is stable for any local Péclet-numbers.[4, 5]

Figure 7 shows the finite element mesh. In the middle of the domain, a kernel cube can be seen. Its volume is calculated in such a way that it can contain the whole molten pool. At present grids are being used with around 60 000 tetrahedra.

The domain for the free surface problem is described by the upper side of the kernel cube of the finite element grid (Fig. 8). During iteration, the constant p_0 should be calculated in such a way that within one unit of time the volume of molten-off filler material is equal to the sum of the gap volume and the volume of the reinforcement (Fig. 9):

$$A_{nominal} = \frac{\dot{V}_{wire}}{v_w} = A_{gap} + A_{reinforcement} \tag{22}$$

$$A_{reinforcement} = \int_{-b_{seam}}^{+b_{seam}} -Z(x = L_{x_2}, y) \mathrm{d}y \tag{23}$$

where b_{seam} is the half of the seam width, $\dot{V}_{wire}$ is the feed of the filler metal, A_{gap} and $A_{reinforcement}$ are the area of the gap and the reinforcement in the cross section (with $x = L_{x_2}$), respectively.

The field of $t_{8/5}$ – time (cooling time from 800°C to 500°C) is determined by the temperature field. With this, and by knowledge of the composition of the basic material direct statements are possible regarding microstructure formation and consequently quality assurance.

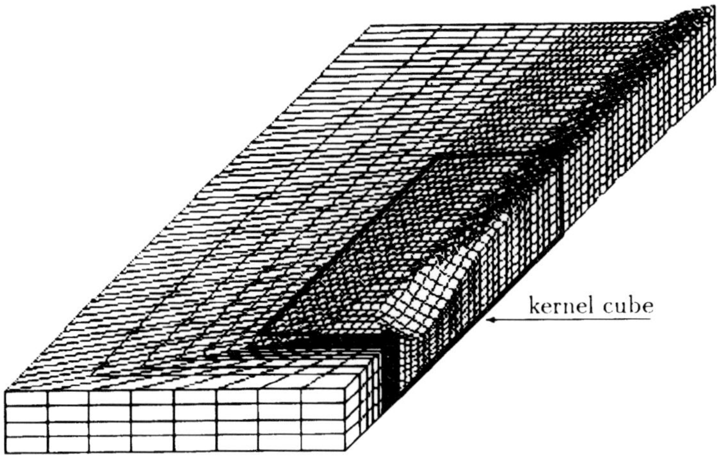

Figure 7. Finite element grid.

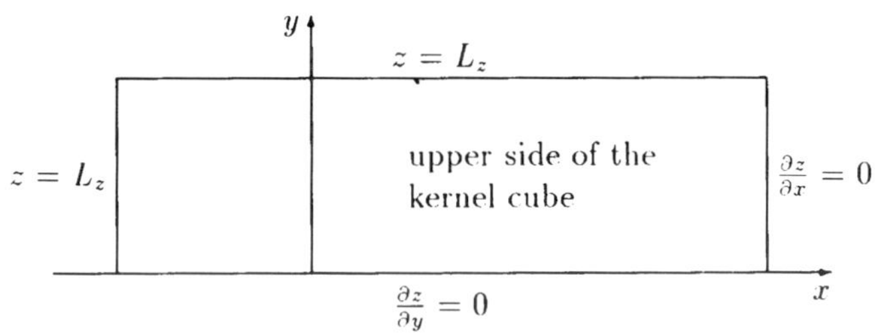

Figure 8. Boundary conditions for free pool surface calculation.

RESULTS

The testing of the model regarding its effectiveness and sensitivity was carried out with the question in mind as to whether it reacts to alterations under ionisation conditions and to the altered effects of forces in the molten pool caused by the arc as well as the position of the workpiece in such a manner that any experimental knowledge may be reproduced.

In this example only one parameter is altered. The others are considered to be constant, so the influence of the respective parameter may be demonstrable. This theoretical test case is not technically realisable in an experiment because of the close dependencies between the parameters. Thus the results are only of a qualitative nature.

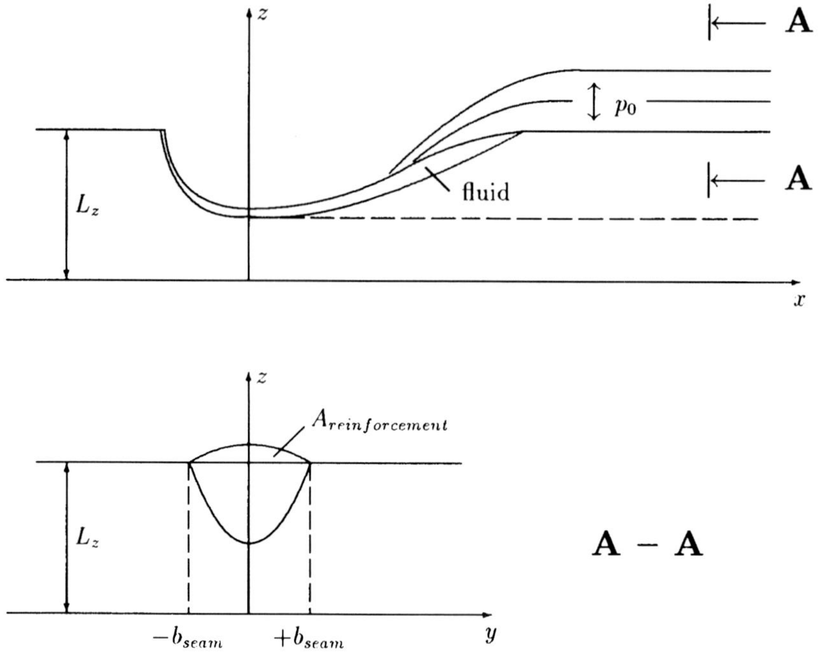

Figure 9. Mass balance for pool surface calculation.

A build-up welding experiment is simulated with the following model parameters:

weighting parameter for ionisation	ϑ	= 0; 0.5; 1
plate inclination	θ_{plate}	= 0°; 30°; 60°
plate thickness	L_z	= $10{\cdot}10^{-3}$ m
plate width	L_y	= 0.1 m
welding velocity	v_w	= 0.0066 m/s
average arc pressure	$\tilde{p}_{arc}$	= 300 Pa
energy input	$\dot{Q}_{eff}$	= 1500 W

The energy input according to Equation 5 also covers the part of energy of the filler metal.

In Figs 10 and 11 transverse section drawings heavily stretched in the direction of z are shown for the plane $y = 0$. For the experiment $\vartheta = 0.5$ and $\theta_{plate} = 0°$ were chosen. The isoline step was $\Delta T = 100$ K.

Figure 10 demonstrates the influence of the weighting parameter ϑ [cf. Equation 4]. The upper part of the figure shows the temperature distribution without taking the improvement of conductivity caused by ionisation into consideration with $\vartheta = 0$. The increase in ϑ moves the preferred area of arc

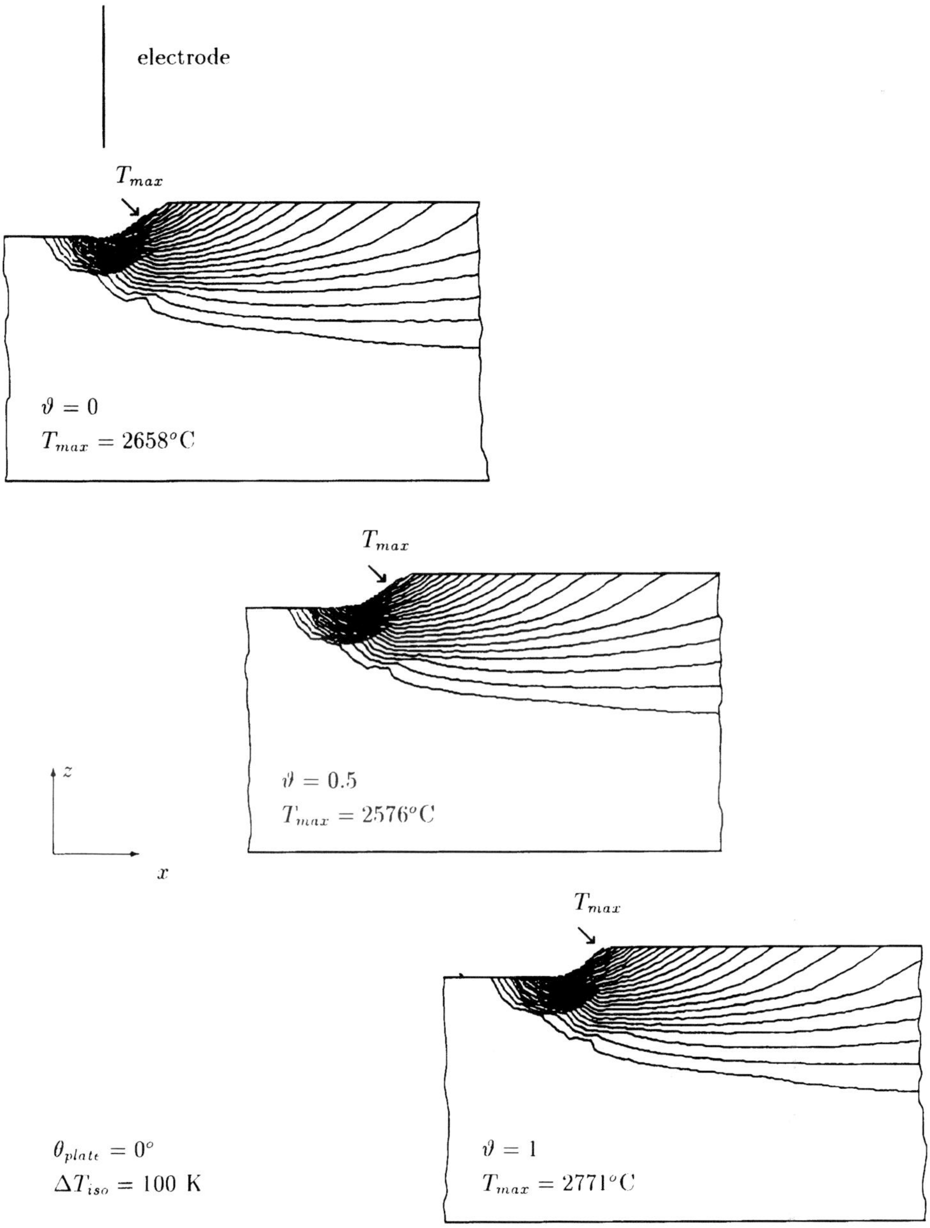

Figure 10. Shifting of maximum temperature on the weld pool surface caused by the influence of ϑ_S.

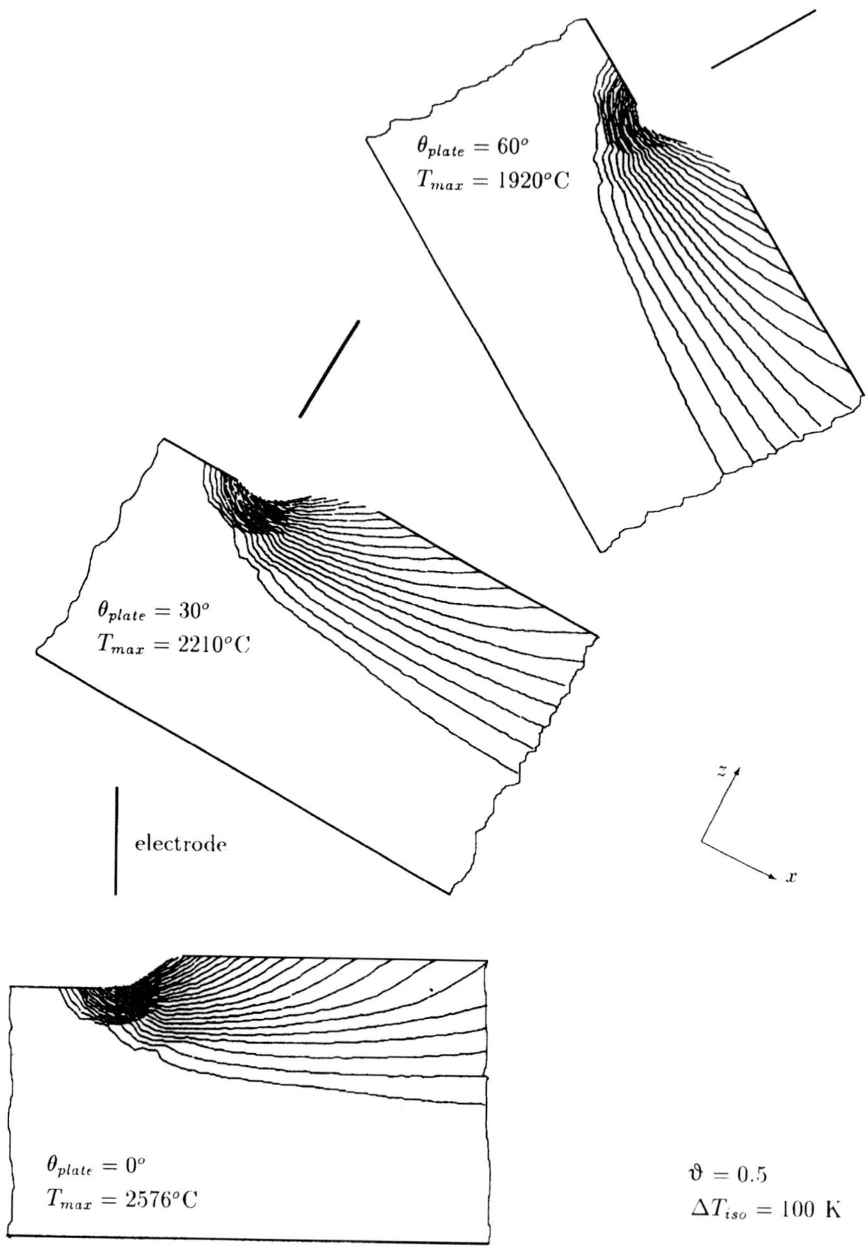

Figure 11. Effect of workpiece inclination on the pool surface deformation.

activity in the direction of the *x*–axis. This is made clear by the shift in the position of the point of the maximum temperature of the weld pool T_{max}.

Change in the workpiece position by variation of the plate inclination θ_{plate} influences the pressure conditions in the weld pool. With the increase in inclination, the depth of welding penetration increases (Fig. 11) and the overheating of the pool decreases (cf. maximum temperature of the weld pool).

The suitability of the presented model for the description of the influence of a high deformed weld pool surface may be proved by simulation computations in comparison to the experiment.

Figure 12 shows the results of an experiment which was carried out with the following data:

welding velocity	v_w	= 0.0027 m/s
voltage	U_w	= 32 V
current	I_w	= 205 A
angle of workpiece inclination	θ_{plate}	= 45°
plate width	L_y	= 0.1 m
thickness of the plate	L_z	$= 20 \cdot 10^{-3}$ m
diameter of wire	d_{wire}	$= 1.2 \cdot 10^{-3}$ m
shielding gas		100% CO_2
steel		RSt37-2

In the melting zone of the weld pool we can see that both are in agreement. In the solidification range a certain deviation can be noted. This can be explained by the fact that one part of the heat transfer caused by convection is not taken into consideration.

CONCLUSIONS

A mathematical model has been developed to describe the temperature distribution in the workpiece during arc welding. The most important finding is that the presented distribution of arc intensity, taking into consideration its three dimensional character during the process, is suitable to describe the temperature distribution and seam forming also in the case of high pool surface deformation being in agreement with the experimental results.

Restrictions of the model at present still consist of a simplified approximation of the convective energy transport in the weld pool and neglect of electro-magnetic influence. Both these effects necessitate further investigation.

The present intention is to use this model for the calculation of seam form and the formation of the microstructure in different welding positions, in

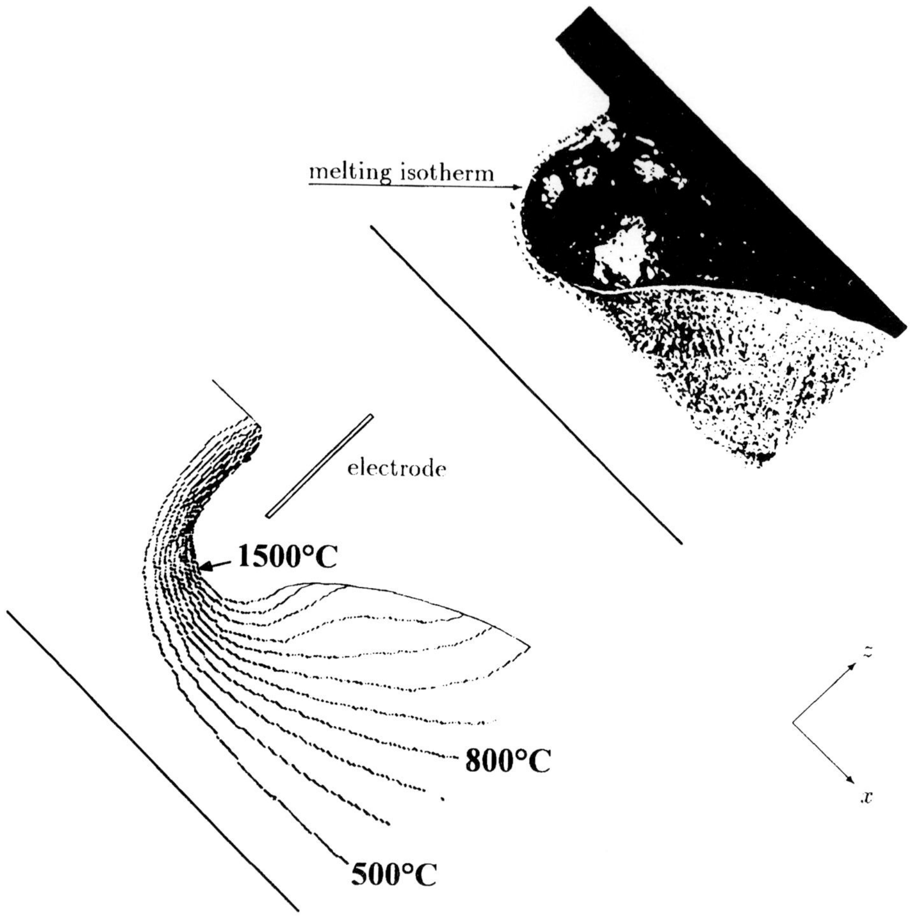

Figure 12. Comparison of the numerical result with the experiment (build-up welding with inclination of plate $\theta_{plate} = 45°$).

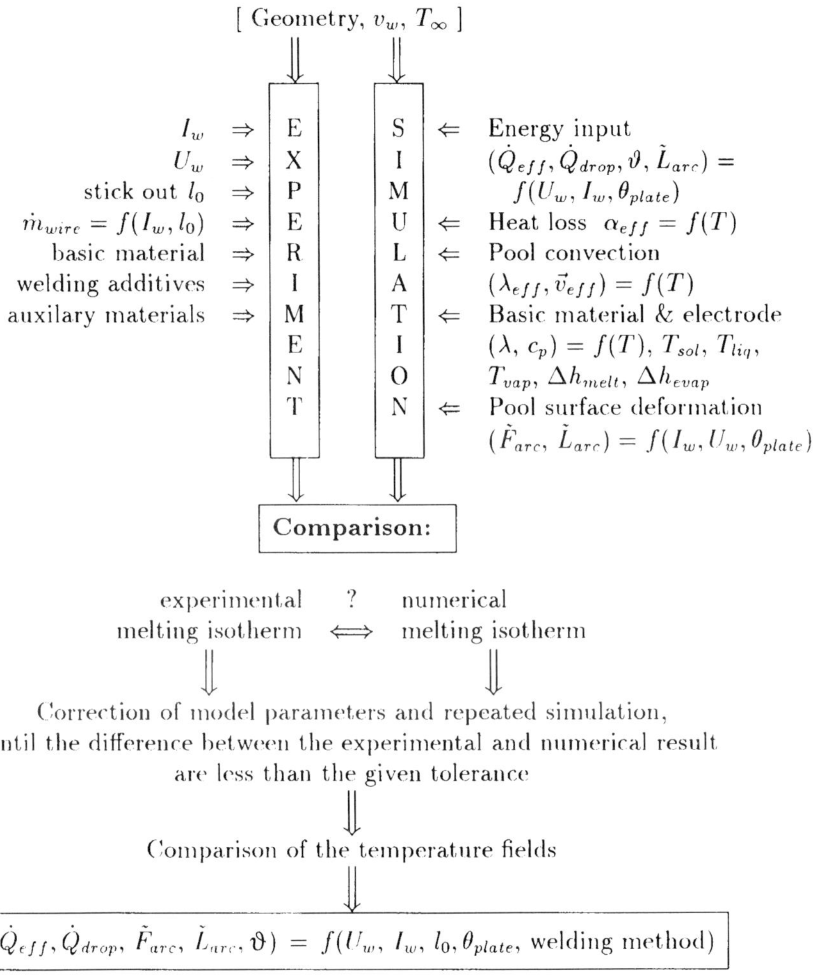

Figure 13. Derivation of model parameters from technological parameters.

order to predict suitable technological parameters for welding. It can be used for quality assurance in production. With the results of further theoretical and experimental work in the field of gas shielded arc welding, adaptation of model parameters to technological parameters will be carried out (Fig. 13).

LIST OF SYMBOLS

A	area [m^2]	c_p	specific heat [J/(kg K)]
$\vec{e}_{arc}$	unit vector (electrode top → pool surface)	g	gravitational constant [m/s^2]
$\vec{n}_{surf}$	normal unit vector on pool surface	Δh_{melt}	latent heat of fusion [J/kg]
p	pressure [N/m^2]	L	distance [m]
$\dot{q}_{vol}$	heat source density [W/m^3]	$\dot{q}$	heat flow density [W/m^2]
r	radius [m]	R	reduced resistance [m]
$t_{8/5}$	cooling time from 800 to 500°C	T	temperature [°C]
v_w	welding velocity [m/s]	V_h	function space
W	Sobolev function space	v_{wire}	velocity of wire [m/s]
		x, y, z	co-ordinates [m]
α	heat transfer coefficient [W/(m^{2}K)]		
ε	emissivity	δ	damping parameter
φ	test function	ϕ	arc intensity distribution
λ	thermal conductivity [~W/(mK)]	Γ	surface of domain Ω
σ	surface tension [N/m]	ρ	density [kg/m^3]
ϑ	weighting factor for reduction of resistance by ionisation	θ_{plate}	angle of inclination in the $x-z$ plane [°]
		Ω	domain

ACKNOWLEDGEMENT

The authors wish to thank the German Research Society and the Land SachsenAnhalt for the support of this investigation.

REFERENCES

1. R.T.C. CHOO, J. SZEKELY and R.C. WESTHOFF: On the calculation of the free surface temperature of gas–tungsten–arc weld pools from first principles: Part I. modelling the welding arc, *Metall. Trans.* B, 1992, 357–369.
2. T. DEBROY, U. BASU and K. MUNDRA: Probing laser induced metal vaporization by gas dynamics and liquid pool transport phenomena, *J. Appl. Phys.* 1991, **70** (3) 1313–1319.
3. U. DILTHEY, G. HABEDANK, T. REICHEL, V.A. SUDNIK and A. IVANOW: Numerical simulation of the metal-arc active gas welding process, *Welding and Cutting*, 1993, E50–E53.
4. L.P. FRANCA, S.L. FREY and T.R.J. HUGHES: Stabilized finite element methods, I. Application to the advective-diffusive model. CMAME 95, 1992, 253–276.
5. G. LUBE: Stabilized Galerkin finite element methods for convection dominated and incompressible flow problems, Banach Center Publications, 29, Polish Academy of Science, Warsaw, in press.
6. S.D. KIM and S.J. NA: Effect of weld pool deformation on weld penetration in stationary gas tungsten arc welding. *Welding Journal*, 1992, 179–193.
7. S. NA and J. RUGE: Temperaturfelder beim Schweißen von Aluminium Druckguß, *Schweißen und Schneiden*, 1990, 327–332.
8. T. OHJI, A. OKHUBO and K. NISHIGUCHI: Mathematical modelling of molten pool in arc welding. IUTAM Symposium: Mechanical Effects of Welding, Lulea, Sweden, Jun 1991, Springer-Verlag, Berlin 1992.
9. RADAJ, D: *Heat Effects of Welding*, Springer-Verlag, Berlin, 1991.
10. V.A. SUDNIK: Research into fusion welding technologies based on physical-mathematical models, *Welding and cutting*, 1991, 216–E217.
11. K.C. TSAO and C.S.WU:Fluid flow and heat transfer in GMA weld pools, *Welding Journal*, 1988, 70–75.
12. T. ZACHARIA, S.A. DAVID, J.M. VITEK and H.G. KRAUS: Computational modeling of stationary gas-tungsten-arc weld pools and comparison to stainless steel 304 experimental results, *Metall. Trans.* B, 1991, 243–257.

DENDRITE GROWTH IN WELDING

W. KURZ
Swiss Federal Institute of Technology Lausanne, CH 1015 Lausanne, Switzerland

ABSTRACT

Understanding of microstructure and phase selection in weld solidification is important for controlling mechanical properties of welds. A selection criterion based on the maximum growth temperature of the competing growth morphologies / phases is presented. Solidification parameters relevant to weld microstructures are given and a model for columnar dendritic growth under conditions of rapid solidification is discussed. The dendrite model together with the selection criterion is applied to the transition of ferrite to austenite in steel with increasing welding rate.

INTRODUCTION

In most metals and alloys dendrites constitute the dominant growth morphology. This is the case in casting, welding and rapid solidification processing. Welding is an intermediate solidification process as far as the rate of solidification is concerned. In certain cases such as high rate laser welding (of the order of 0.1 m/s), mechanisms of rapid solidification are operative and theory should take this fact into account. In recent years there has been much work done on the various aspects of dendritic growth theory, the fundamentals of which have been described for example in reference 1.

The aim of this paper is to present the relevant solidification mechanisms, theories and parameters which allow understanding and prediction of the microstructures formed in the fusion zone of a weld at normal welding rates as well as under rapid welding conditions. A microstructure selection map connecting the expected structures to the solidification conditions is presented for Fe–Ni alloys.

MICROSTRUCTURE SELECTION CRITERION

The problem of microstructure selection under given conditions of transformation is a very complex one. In the absence of a complete quantitative theory of this phenomenon a simple and, according to experimental observations, reasonable criterion will be used. A variant of this criterion has been proposed by Tammann and Botschwar as early as 1926.[2] These authors stated that the structure with the highest growth rate is the one which is observed. This criterion, however, is useful for the case of undercooled growth only and neglects nucleation. In columnar growth such as in welding where growth is epitaxial with respect to the substrate and where a strong positive temperature gradient is imposed at the solid/liquid interface (so called constrained growth) the relevant criterion is:

> The phase and its growth morphology which shows the highest interface temperature for a given growth rate is the one observed.

This criterion is an empirical one and it has been applied successfully to various solidification situations. This approach[3] allows (a semiquantitative) prediction of microstructures when (i) the processing conditions and (ii) the corresponding system parameters are known. These are (i) the velocity of the growth front (i.e. velocity of the isotherms corresponding to the temperature of the growth front) and the temperature gradient, and (ii) the properties of the alloy such as composition, phase equilibria, diffusion coefficients, interface energy, etc.

In this paper the theory will be applied to the transition from ferrite to austenite dendrites in steels. This transition is of great technical importance as, under others, the hot cracking susceptibility of a steel depends strongly on the presence of these phases.

SOLIDIFICATION CONDITIONS

Before discussing microstructure theory one has to deal with the relevant solidification parameters of the welding process. For their estimation the classical Rosenthal solution will be used.[4] But instead of evaluating cooling rates and cycle times, as is usual in the welding literature, we will concentrate on the more relevant, parameters such as growth velocity, V, and temperature gradient at the interface, G. Note that for columnar growth such as in welding where the temperature gradient is positive the absolute value of the cooling rate is the product of the latter two processing parameters,

$$dT/dt = -GV \tag{1}$$

This relationship shows that the cooling rate defines the products of G and V. The theory of solidification shows that generally the effect of G and V is not the same on the formation of microstructures. Often the solidification rate, V, is the more important one. Equation 1 does not specify if for a given cooling rate V is large or small. Therefore:

> Cooling rate is **not** the most relevant parameter for microstructure prediction.

For a rough estimation of the solidification parameters one can use the Rosenthal solution[4] for a point source of effective power, P, moving at constant speed, V_b, in a semi-infinite medium (neglecting latent heat, temperature dependent properties, melt-convection and keyhole formation)

$$T - T_0 = \frac{P}{2\pi\kappa r}\exp\left\{-\frac{V_b(\xi + r)}{2a}\right\} \tag{2}$$

where T_0 is the temperature of the plate far from the heat source, r is the polar coordinate measured in a reference frame moving with the heat source ($r = \sqrt{(\xi^2 + y^2 + z^2)}$), κ is the thermal conductivity and a is the heat diffusivity. The ξ coordinate measures the distance along the translation direction of the heat source in the moving reference frame. As has been shown from this solution, various relationships can be developed which allow the estimation of important solidification parameters in welding or in laser surface treatment.[5] They depend on the location in the weld pool. Typical orders of magnitude can be obtained from the following relationships (all evaluated at the end point of solidification, C in Fig. 1).

The local growth velocity (normal to the solid/liquid interface) $V = V_b \cos\theta$, (where θ is the angle between the interface normal and the translation direction of the heat source). θ is close to 0° at point C, therefore the growth rate is largest there and is given by

$$V \cong V_b \tag{3}$$

The temperature gradient at point C can be approximated as

$$G \cong \frac{2\pi\kappa}{P}(T_m - T_0)^2 \tag{4}$$

with T_m, the melting temperature. The cooling rate and the G/V ratio at point C are therefore

$$\dot{T} \cong \frac{2\pi\kappa V_b}{P}(T_m - T_0)^2 \tag{5}$$

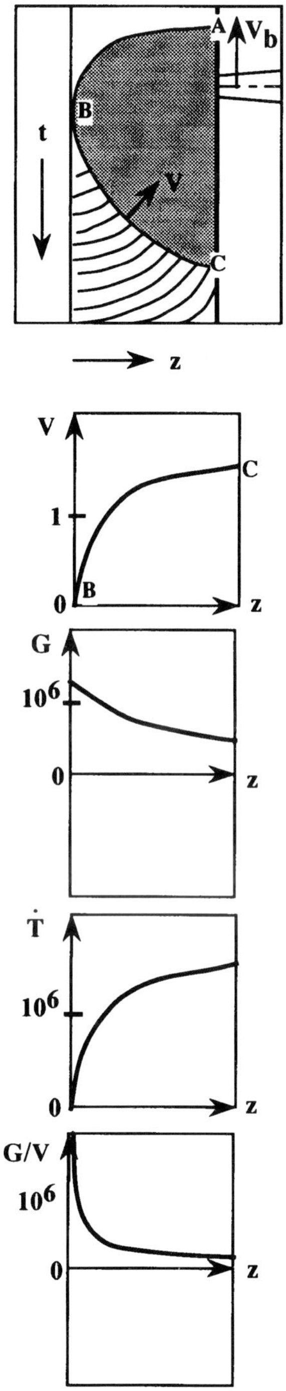

Figure 1. Important solidification parameters in the weld pool varying over the pool depth.

$$\frac{G}{V} \cong \frac{2\pi\kappa(T_m - T_0)^2}{PV_b} \quad (6)$$

Figure 1 shows schematically the evolution over the weld pool distance of these four parameters which control the formation of the microstructures. As, for the melt pool form shown, the velocity increase is larger than the temperature gradient decrease, the cooling rate increases strongly from point B to point C. The G/V ratio controls the onset of constitutional undercooling and therefore the onset of morphological instability. G/V is infinite at point B as θ is 90° and the growth velocity is zero. Therefore in certain cases plane front solidification is observed.

MODELLING DENDRITIC GROWTH

Columnar dendritic growth is controlled essentially by solute flux around the tips. Generally the solidifying crystal has a lower solubility for alloying elements than the melt so that solute is rejected by the growth front. For more effective diffusion the crystal takes the form of a needle, the tip of which is close to a paraboloid of revolution. For this morphology the dimensionless supersaturation, Ω, is related to the Péclet number of the dendrite tip, p, through[6]

$$\Omega = Iv(p) \quad (7)$$

here the Ivantsov solution, $Iv(p) = p\ exp(P)E_1(p)$ and $p = RV/2D$ with R, the dendrite tip radius, V, the tip growth rate, D the solute diffusion coefficient in the liquid and $E_1(p)$ the exponential integral.[1]

Evaluation of the tip concentration gradient and use of marginal stability arguments[7] leads to the following equation for a linearised phase diagram (for more details see reference 1):

$$V^2A + VB + G = 0 \quad (8)$$

where

$$A = \frac{\Gamma}{\sigma^*(2pD)^2}\ ; \qquad B = \theta\ \xi/D$$

and

$$\theta = \Delta T_0^v\, k_v\, (C_l^*/C_0)\ ; \qquad \Delta T_0^v\, k_v = m_l^v C_0(k_v - 1)$$

$$\frac{C_l^*}{C_0} = \frac{1}{1-(1-k_v)Iv(p)} \tag{9}$$

Γ is the Gibbs–Thomson parameter (ratio of solid/liquid interface energy to melting entropy), σ^* is the stability parameter of the order of $1/(4\pi^2)$, C_l^* is the tip composition, C_0 is the alloy composition, k_v is the growth rate dependent distribution coefficient, m_l^v is the kinetic liquidus slope and ΔT_0^v is the growth rate dependant liquidus/solidus interval ($\Delta T_0^v = C_0 m_l^v (k_v - 1)/k_v$).

There are four parameters in the above equation which become important in high speed welding:

1. the solute distribution coefficient at the interface changing at high velocities due to the solute trapping effect, which is represented by k_v [Equation 10];
2. the non-equilibrium liquidus slope, m_l^v [Equation 11];
3. the diffusional part of the stability equation which is given by the stability term ξ [Equation 12];
4. the diffusion coefficient, D, which changes with interface temperature.

According to Aziz[8] the growth rate dependent solute distribution coefficient, for dilute solutions, is given by the relationship:

$$k_v = \frac{k_0 + p_i}{1 + p_i} \tag{10}$$

where k_0 is the equilibrium distribution coefficient, p_i ($= a_0 V/D$) is the interface Péclet number, a_0 is an atomic jump distance. The thermodynamics of non-equilibrium processes have been considered by Boettinger and Coriell,[9] who obtain, for dilute solutions:

$$m_l^v = m\left[1 + \frac{k_0 - k_v(1 - \ln(k_v/k_0))}{1 - k_0}\right] \tag{11}$$

where m is the slope of the equilibrium liquidus which is assumed to be linear. The parameter, ξ, has been given as[7, 10]

$$\xi = 1 - \frac{2k_v}{[1 + (2\pi/p)^2]^{1/2} - 1 + 2k_v} \tag{12}$$

For a certain interface velocity, Equation 8 can be readily solved for different Péclet numbers and the radius obtained through the definition of p:

$$R = 2Dp/V \tag{13}$$

The tip radius is an important parameter for determining the scale of the dendritic microstructure such as primary and secondary dendrite spacings.

The dendrite tip temperature, T_d*, which is needed for the prediction of microstructures, may now be obtained for a linear phase diagram as

$$T_d^* = T_m + m_l^v C_l^* - 2\Gamma / R \tag{14}$$

In the above equation, T_m is the melting point of the pure solvent, $m_l^v C_l^*$ is the solute undercooling (with m_l^v negative) and $2\Gamma/R$ represents the curvature contribution to the undercooling of the tip.

Following Bobadilla *et al.*[11] these equations have been generalized for slow solidification of multicomponent alloys and extended to rapidly solidified ternary Fe–Cr–Ni alloys by Rappaz *et al.*[12] Figure 2 shows the variation of the tip radius of a Fe–15Cr–15Ni alloy and Fig. 3 shows the tip temperature of the same alloy for different temperature gradients which are typical for high energy beam processes such as electron beam (or laser) welding.[12] Note that above a velocity of some 150 mm/s the dendrite tip radius increases suddenly with *V*. This is a sign of the tip approaching the limit of absolute stability. Above this limit no dendrites can grow and a planar interface morphology is restabilized. This limit is obtained as[13]

$$V_a = \frac{\Delta T_0^v D}{k_v \Gamma} \tag{15}$$

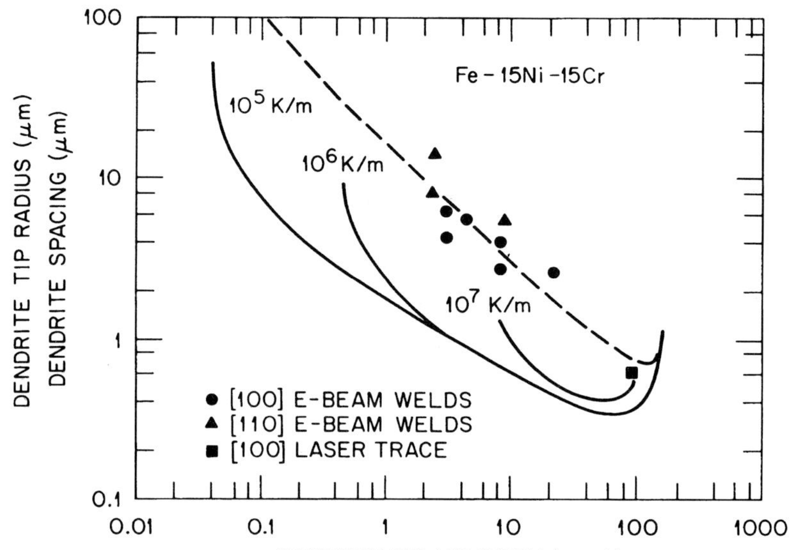

Figure 2. Calculated dendrite tip radii (continuous lines) vs tip velocity for an Fe–15Ni–15Cr alloy with three temperature gradients typical for high energy (EB, laser) beam welding.[12] Points representing experimentally determined dendrite spacings.

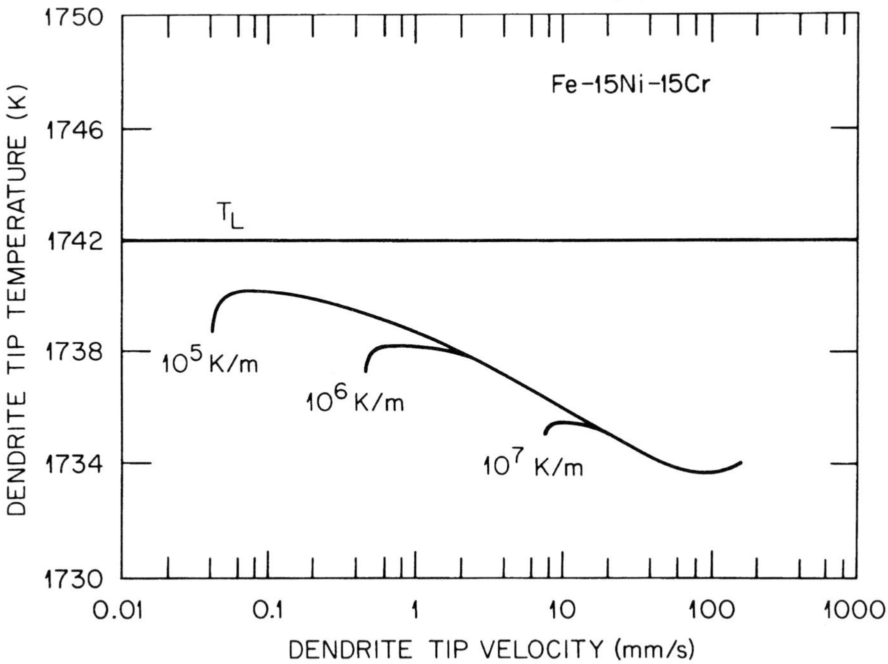

Figure 3. Calculated dendrite tip temperature vs tip velocity for an Fe–15Ni–15Cr alloy with three temperature gradients.[12]

Up to now only the dominant effect of the growth rate on microstructures has been discussed. However, the temperature gradient can also have an important effect at smaller velocities where a decrease in constitutional undercooling may stabilise a planar interface morphology, i.e. when $G/V > \Delta T_0/D$. Therefore, at the bottom of the weld pool where G/V is tending to infinity, a planar front may be observed. However, diffusional transient phenomena will often inhibit formation of this morphology.

DENDRITE GROWTH IN WELDS

Dendritic growth occurs along selected directions. These are generally low index directions of the underlying crystal structure such as ⟨100⟩ in cubic materials. As has been discussed in detail by Nakagawa *et al*[14] and Rappaz *et al.*[15] this fact changes the growth velocity given in Equation 3 to

$$V_{hkl} = V_b \cos\theta / \cos\psi \tag{16}$$

where V_{hkl} is the velocity of the dendrite trunk of crystallographic orientation $[hkl]$ and ψ is the angle between the solidification front normal and $[hkl]$, as

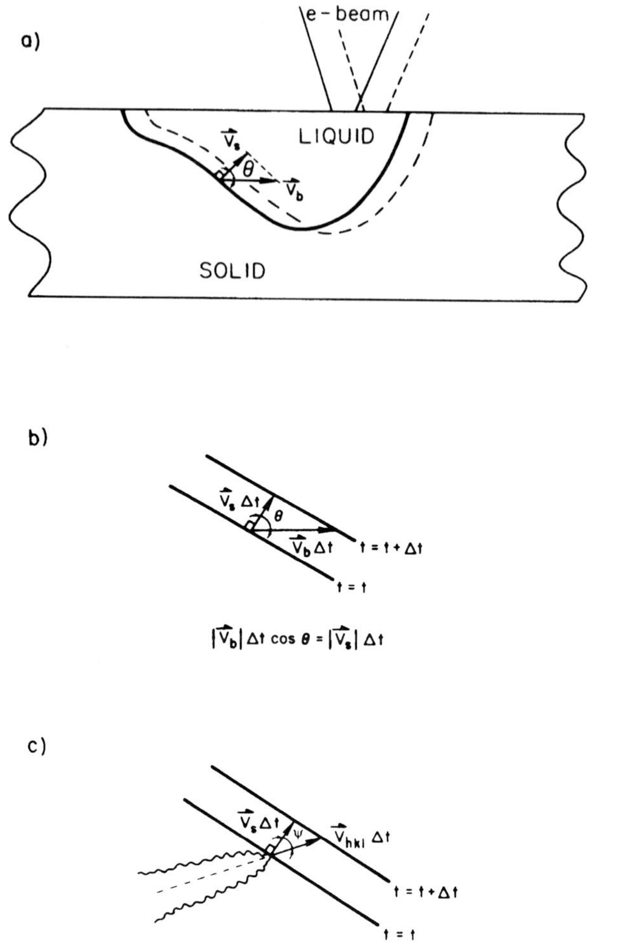

$$|\vec{V}_b|\,\Delta t\cos\theta = |\vec{V}_s|\,\Delta t$$

$$|\vec{V}_{hkl}|\,\Delta t\cos\psi = |\vec{V}_s|\,\Delta t$$

Figure 4. Relationships between beam velocity, solidification front velocity and dendrite growth velocity.

given in Fig. 4. Corresponding dendrite patterns in stainless steel single crystals have been analysed in a series of papers by Rappaz *et al.*[12, 15, 16] These results may be helpful in the interpretation of welding experiments.

Figure 5 shows the effect of ψ on the selection of the dendrite branch which grows at the highest temperature, i.e. at the lowest velocity for one specific orientation of the dendritic grain. Dendrite trunks with [001] orientation are selected at the bottom part of the melt pool, between point B and the point where $\psi = 45°$ while beyond 45° the [100] directions of the secondary branches will be preferred by the trunks. More complicated cases have been discussed by Rappaz *et al.*

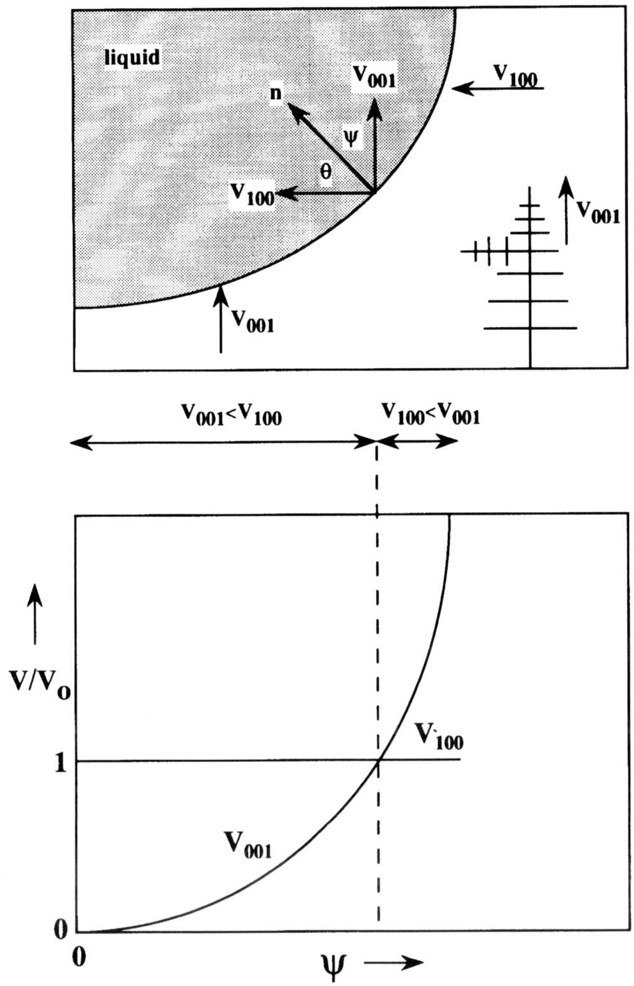

Figure 5. Diagram showing the selection of crystallographic dendrite orientations.

FERRITE/AUSTENITE TRANSITION IN STEEL

For many years it has been established that a sufficient Cr/Ni-equivalent is necessary to avoid solidification cracking induced by heavily segregating trace elements such as P and S. The precise value of the Cr/Ni-equivalent for safe welding depends on the level of these trace elements (Fig. 6).[17] Increased cracking tendency has also been observed with higher welding velocities. This is caused under others by:

(a) higher microsegregation of such trace elements as S and P, due to an increase in the fraction of primary γ-dendrites with respect to δ-ferrite dendrites at high *V*.

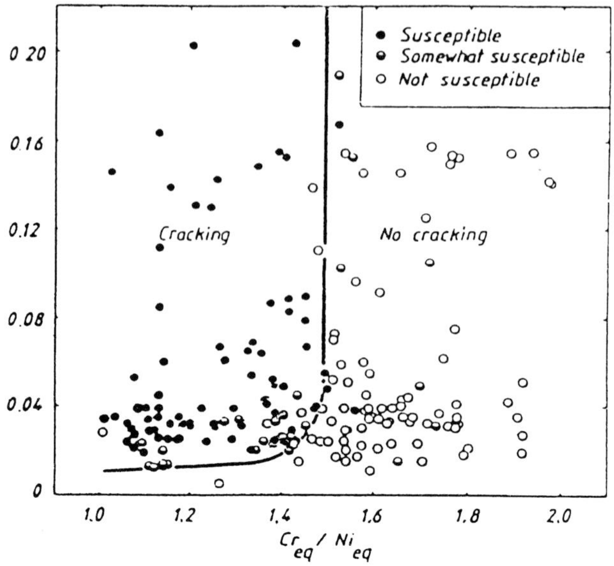

Figure 6. Hot cracking susceptibility of stainless steels according to the Cr/Ni equivalent.[17]

(b) higher stresses due to increased temperature gradients

The reason for the microsegregation behaviour has been treated by the author in the context of continuous casting of low carbon steels.[18] As has been discussed there, distribution coefficients and solid state diffusion coefficients (controlling back diffusion) are much smaller in γ than in δ. Therefore microsegregation is increased and the final solidification temperature is decreased substantially by the presence of γ leading to a larger extent of the brittle mushy zone (where the solid fraction is close to one).

The reason for the increase of the austenite fraction with increasing solidification velocity is worth discussing. This phenomenon has been explained in detail by Bobadilla *et al.*[11] (These authors extended the approach described in the section on dendrite growth to ternary systems; however, the rapid solidification effects were neglected.)

For reasons of simplicity the discussion will be centred around the binary Fe–Ni system. The phenomena are similar to Cr–Ni steels and low carbon steels. A peritectic is formed by the three phases, liquid, δ-ferrite and γ-austenite (Fig. 7a shows the relevant stable and metastable equilibria).[19] Due to the larger distribution coefficient $k_{Ni}(\gamma/l)$ compared to $k_{Ni}(\delta/l)$, the solute rejection and therefore the growth undercooling of γ for a given growth velocity is smaller than the corresponding undercooling for δ. However, for compositions lower than 5.2 at%Ni γ is a metastable phase and therefore its liquidus temperature is lower than that of δ. Both effects decide which phase has the higher growth temperature at the given growth velocity (Fig. 7b). The higher the velocity the greater the advantage for the γ-phase.

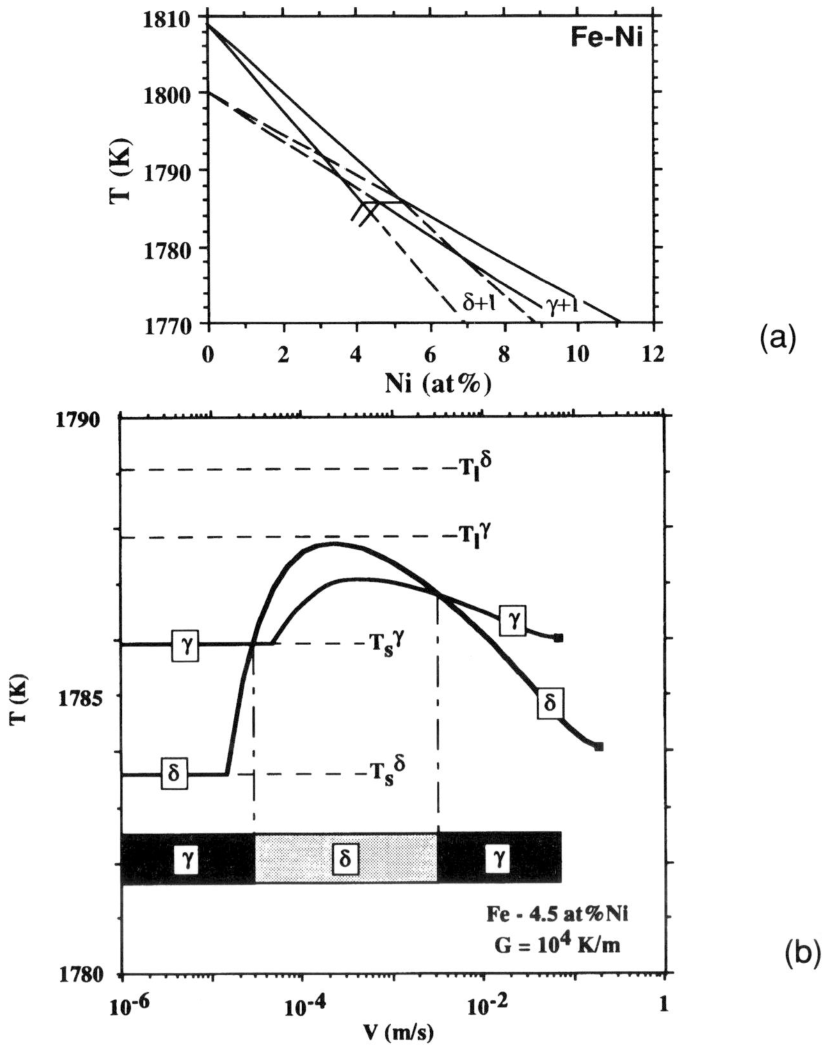

Figure 7. Phase selection in Fe–Ni alloys: (a) stable and metastable phase diagram for Fe–Ni alloys form,[19] modified for the correct metastable melting point of γ-Fe, and (b) interface response functions for δ and γ-Fe in Fe-4.5 at% Ni.

Another interesting point is obtained by the following prediction: Under steady state conditions a plane front grows at the solidus temperature. As discussed before at the lower weld pool side, such a plane front might develop due to the large *G/V* ratio. For an alloy with 4.5 at%Ni the solidus of γ is higher than the solidus of δ (Fig. 7b). Therefore the first phase to grow is γ if nucleation allows it. At the bottom of Fig. 7b the sequence of the phases forming in Fe-4.5 at%Ni is indicated (always under the assumption of a negligibly small nucleation barrier). The highest growth velocities calculated (end of the *T–V* curves for dendrites) represent the limit of absolute stability which might be reached in rapid welding processes.

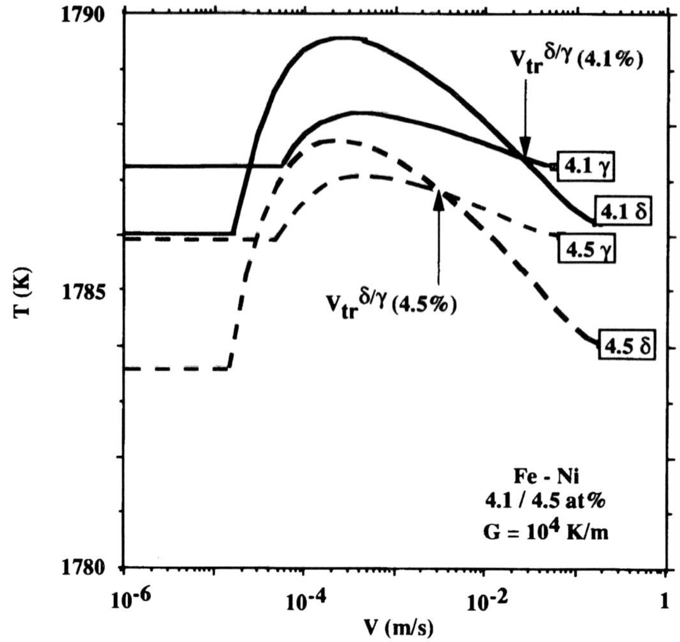

Figure 8. Interface response functions for δ and γ in two alloys of slightly different composition (Fe-4.1 at% Ni and Fe-4.5 at% Ni).

Figure 8 shows the calculated *T–V* curves for two compositions, 4.1 and 4.5 at%Ni. All the important effects known from austenitic steels can be found here. Interestingly there is a seven-fold increase in the predicted transition velocity from δ to γ when the composition decreases by only 0.4 at%Ni.

For a complementary discussion of the solidification behaviour of steels during welding the reader is referred to a review by David and Vitek.[20]

CONCLUSIONS

The discussion of phase and microstructure selection mechanisms shows that under rapid welding conditions some of the well established facts have to be revised. Metastable phases such as γ may form in a composition range where under normal welding conditions δ still forms. This becomes clear when one considers the fact that a phase with less solute rejection (larger *k*) is preferred at high growth velocities even if its liquidus temperature is lower than that of the equilibrium phase. This observation has important consequences for hot cracking: The Schäffler diagram has to be modified when the welding speed is increased, as recently proposed by David and Vitek.[20]

ACKNOWLEDGEMENTS

The author acknowledges the help of P. Gilgien in preparing this manuscript.

REFERENCES

1. W. KURZ and D.J. FISHER: *Fundamentals of solidification*, Trans Tech Switzerland, 3rd edn, 1992.
2. G. TAMMANN and A.A. BOTSCHWAR: *Z. Anorgan. Chemie*, 1926, **157**, 26.
3. W. KURZ and R. TRIVEDI: *Trans. ASME*, 1992, **114**, 450.
4. D. ROSENTHAL: in *'Congrés National des Sciences'*, Comptes Rendus, Bruxelles, 1935, **2**, 1277 and D. ROSENTHAL: *Trans. ASME*, 1946, **68**, 849.
5. M. RAPPAZ, M. GREMAUD, R. DEKUMBIS and W. KURZ: in *Proceedings European Conf. Laser Treatment of Materials*, ECLAT 1986, DGM Verlag, Oberursel, 1987.
6. G.P. IVANTSOV: *Dokl. Akad. Nauk SSSR*, 1947, **58**, 567, see also [1]
7. W. KURZ, B. GIOVANOLA and R. TRIVEDI: *Acta Metall.* 1986, **34**, 823.
8. M.J. AZIZ: *J. Appl. Phys*, 1982, **53**, 1158.
9. W.J. BOETTINGER and S.R. CORIELL: in .*Science and Technology of the Undercooled Melt* (P.R. SAHM, H. JONES and C.M. ADAM, eds.), Martinus Nijhoff, Dordrecht, 1986, p. 81.
10. R. TRIVEDI and W. KURZ: *Acta Metall.*, 1986, **34**, 1663.
11. M. BOBADILLA, J. LACAZE and G. LESOULT: *J. Cryst. Growth*, 1988, **89**, 531.
12. M. RAPPAZ, S.A. DAVID, J.M. VITEK and L.A. BOATNER: *Metall. Trans.*, 1990, **21A**, 1767.
13. W.W. MULLINS and R.F. SEKERKA: *J. Appl. Phys.*, 1964, **35**, 444.
14. H. NAKAGAWA, M. KATO, F. MATSUDA and T. SENDA: *Trans. Jap. Weld. Soc.*, 1970, **1**, 94.
15. M. RAPPAZ, S.A. DAVID, J.M. VITEK and L.A. BOATNER: *Metall. Trans.*, 1989, **20A**, 1125.
16. S.A. DAVID, J.M. VITEK, M. RAPPAZ and L.A. BOATNER: *Metall. Trans.*, 1990, **21A**, 1753.
17. V. KUJANPAA, N. SUUTALA, T. TAKALO and T. MOISIO: *Weld. Res. Int.*, 1979, **9**, 55.
18. M. WOLF and W. KURZ: *Metall. Trans.*, 1981, **12B**, 85.
19. O. KUBASCHEWSKI: *Iron – Binary Phase Diagrams*, Springer, Berlin, 1982.
20. S.A. DAVID and J. VITEK: *Intern. Mater. Rev.*, 1989, **34**, 213.

COMPUTER SIMULATION OF SOLIDIFICATION AND TRANSFORMATION IN AUSTENITIC STAINLESS STEEL WELD METALS

SHIGERU OHKITA, HIROSHI MORIMOTO and HIROSHIGE INOUE

Steel Research Laboratories, Nippon Steel Corporation, 20–1 Shintomi, Futtsu, Chiba, 299–12 Japan

ABSTRACT

Since weld metals are used mostly in the as-welded condition, their properties are governed directly by as-cast microstructure and microsegregation in weld metals. The solidification characteristics also affect the occurrence of hot cracking in weld metals. Many studies have been carried out concerning the microstructure and the microsegregation of solute elements in weld metals.[1–4] There have been many reports which estimate the extent of the microsegregation and the amount of retained delta ferrite in Fe–Cr–Ni austenitic stainless steels, but not many have investigated a 316 type (Fe–Cr–Ni–Mo) of stainless steel considering a thermodynamic equilibrium during solidification.

The toughness of a type 316 weld metal for the use of elevated temperatures[5,6] sometimes deteriorated by the decomposition of the retained delta ferrite. It was considered that the enrichment of Cr and Mo in the delta phase enhanced the precipitation of the sigma phase, Laves phase or carbide, resulting in the toughness deterioration.

The metallurgical investigation of multi-component stainless steels requires many experiments. This is time and money consuming. Thus, it is beneficial if a calculation model is used which enables us to compute the phase equilibria of a multi-component system through the various stages in the history of welding thermal. Koseki–Matsumiya's model which employs a finite differential method together with THERMOCALC (a computer software for phase diagram calculation),[9] is considered suitable for this kind of simulation.

The numerical analysis in this study estimates the solidification modes, the amount of retained delta ferrite and the concentration profiles of Cr, Ni and Mo in austenitic stainless steel weld metals of type 316.

MODEL OF SOLIDIFICATION AND TRANSFORMATION

A schematic illustration of the model is shown in Fig. 1.[7] A cross-section of a dendrite is divided into six sections, and one half length of an arm spacing is divided into 20 segments with an equal width. The flow of the calculation sequence is shown in Fig. 2.

The assumptions of this model are summarised as follows:

1. Unidirectional thermal conduction occurs parallel to the dendrite axis. Latent heat is considered during solidification, but all finite segments in a plane are considered to have a certain temperature at a certain time.
2. Local equilibrium is achieved at the liquid/solid and the delta/gamma interfaces.
3. Elements completely mix in the liquid phase during solidification.
4. The delta/gamma transformation kinetics and the distribution of solute elements at the delta/gamma interface are analysed locally within each FDM segment by using the Clyne–Kurz equation.[10] No mass transfer occurs between FDM segments during Clyne–Kurz analysis.
5. Equilibrium is calculated by Thermo-Calc with a thermodynamic database.
6. Mass transfer between segments is according to Fick's Law.

Physical constants [1,12] welding parameters and the chemical compositions of materials used in this work are listed in Tables 1, 2 and 3, respectively.

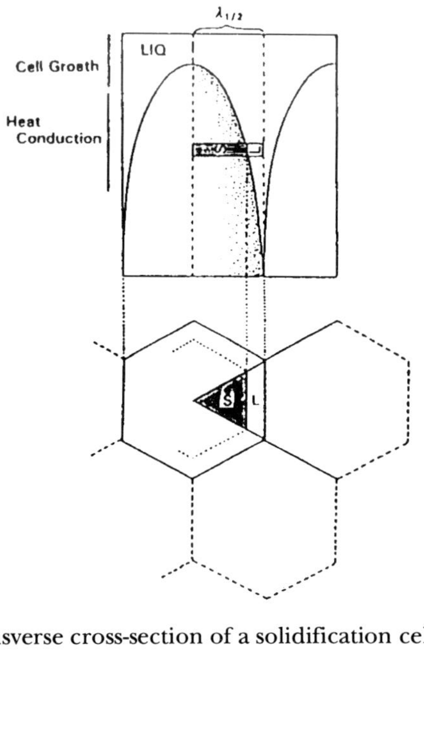

Figure 1. Model for transverse cross-section of a solidification cell.

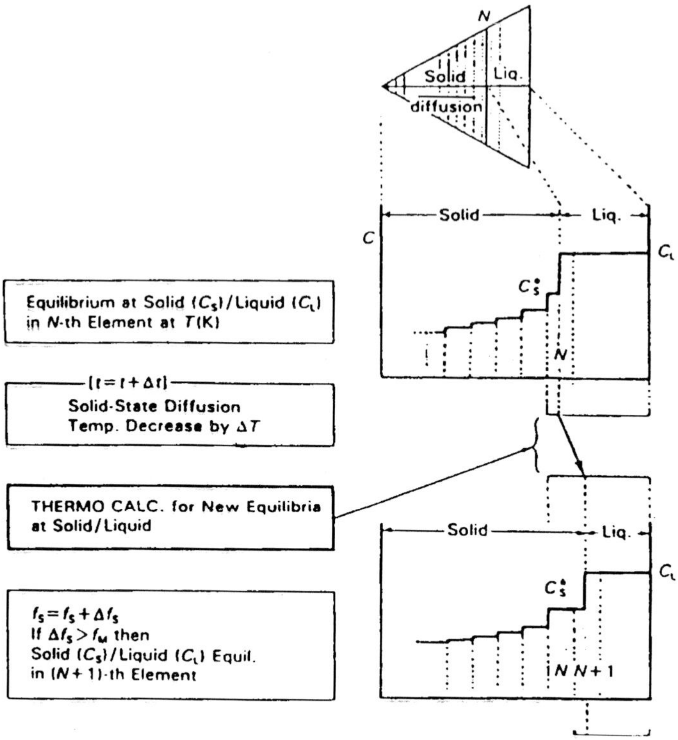

Figure 2. Schematic diagram of calculating method for solidification.

Table 1 Parameters used in calculation

Dendrite arm spacing		20 μm
Number of element		20
Heat input		$2000\,J\,s^{-1}$
Traveling speed		$0.15\,cm\,s^{-1}$
Density		$7.5\,g\,cm^{-3}$
Specific heat (liquid/ δ/ γ)		$0.82/\,0.75/\,0.67\,J\,K^{-1}\,g^{-1}$
Latent heat		$247\,J\,g^{-1}$
Thermal conductivity (liquid/solid)		$0.4/\,0.35\,J\,K^{-1}\,cm^{-1}\,s^{-1}$
Gas Constant R		$8.314\,J\,k^{-1}\,mol^{-1}$
Diffusion coefficient		$cm^2\,s^{-1}$
	in δ	in γ
D_{Cr}	2.30 exp(−239000/RT)	0.169 exp(−264000/RT)
D_{Ni}	2.40 exp(−242000/RT)	0.108 exp(−273000/RT)
D_{Mo}	0.66 exp(−22400/RT)	0.036 exp(−240000/RT)
D_N	0.0113 exp(−83000/RT)	0.360 exp(−157000/RT)

Table 2 Welding parameters employed

Welding type	Plate thickness (cm)	Arc current (A)	Arc voltage (V)	Travel speed (cm/s)
melt run*	1.0	200	10	0.15
deposit†	5.0	280	11	0.133

* Weld pool was quenched by melting Sn.
† With filler wire.

Table 3 Chemical composition of material considered (mass %)

Mark	C	Si	Mn	P	S	Ni	Cr	Mo	N
Weld-A	0.009	0.40	1.25	0.02	0.003	11.5	18.5	2.2	0.085
Weld-B	0.008	0.53	1.23	0.02	0.003	9.0	16.5	1.9	0.080
Weld-C	0.015	0.20	0.80	0.01	0.001	11.5	17.3	2.2	0.024
Weld-D	0.014	0.20	0.80	0.01	0.001	12.8	17.3	2.2	0.025
Weld-E	0.014	0.20	0.80	0.01	0.002	14.5	17.0	2.1	0.024

Calculations have been mainly carried out in the Fe–Cr–Ni–Mo and Fe–Cr–Ni–Mo–N system. The other elements were taken into consideration by using Cr equivalent (Cr_{eq} = % Cr + 1.5 x % Si) and Ni equivalents (Ni_{eq} = % Ni + 0.5 x % Mn + 30 x % C + % N + % Cu) instead of actual Cr and Ni contents (all mass %). A calculation has also been tried in a multi-component system to consider the possibility of applications for commercial stainless steels.

EXPERIMENTS

Two types of TIG welds were made for the assessment of the simulation results. One was multi-pass TIG welding with the filler metal deposited in a type 316 austenitic stainless steel with a double U-groove of 50 mm thick plate. The other was a TIG melt-run of type 316 steel of 10 mm thickness.

The former is mainly utilized for microsegregation; the latter for the identification of the solidification mode.

The concentration profiles of Cr, Ni and Mo in a transverse cross-section of dendrite were analysed by EPMA. The analysed voltage of 15 kV and a beam diameter of 0.3 μm were utilised for the measurement.

The amount of delta ferrite was measured by a ferrite scope.

RESULTS

The calculation of heat conduction was based on Rosenthal's equation[13] taking into account the latent heat of solidification. As shown in Fig. 3, the cooling rate was small during solidification, and became larger after solidification.

A solidification mode has been classified into four categories[14, 15] from the observation of as-welded microstructures. Those are austenitic single phase solidification mode (A-mode), austenitic–ferritic solidification mode (AF-mode), ferritic–austenitic solidification mode (FA-mode) and ferritic single phase solidification mode (F-mode). Solidification modes of the materials used were determined to be from the FA and AF modes from observation of the microstructure, as shown in Fig. 4. Calculated results in the Fe–Cr_{eq}–Ni_{eq}–Mo–N system showed good agreement with the optical microstructure. As shown in Fig. 5, weld-A is solidified at the ferrite phase at first, and then at both austenite and ferrite phase (FA-mode). In the case of

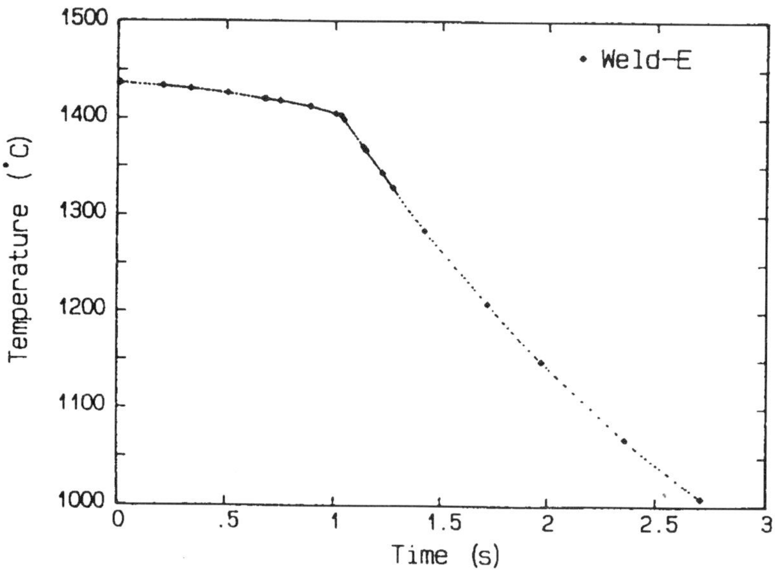

Figure 3. Calculated result of the cooling curve during solidification and transformation.

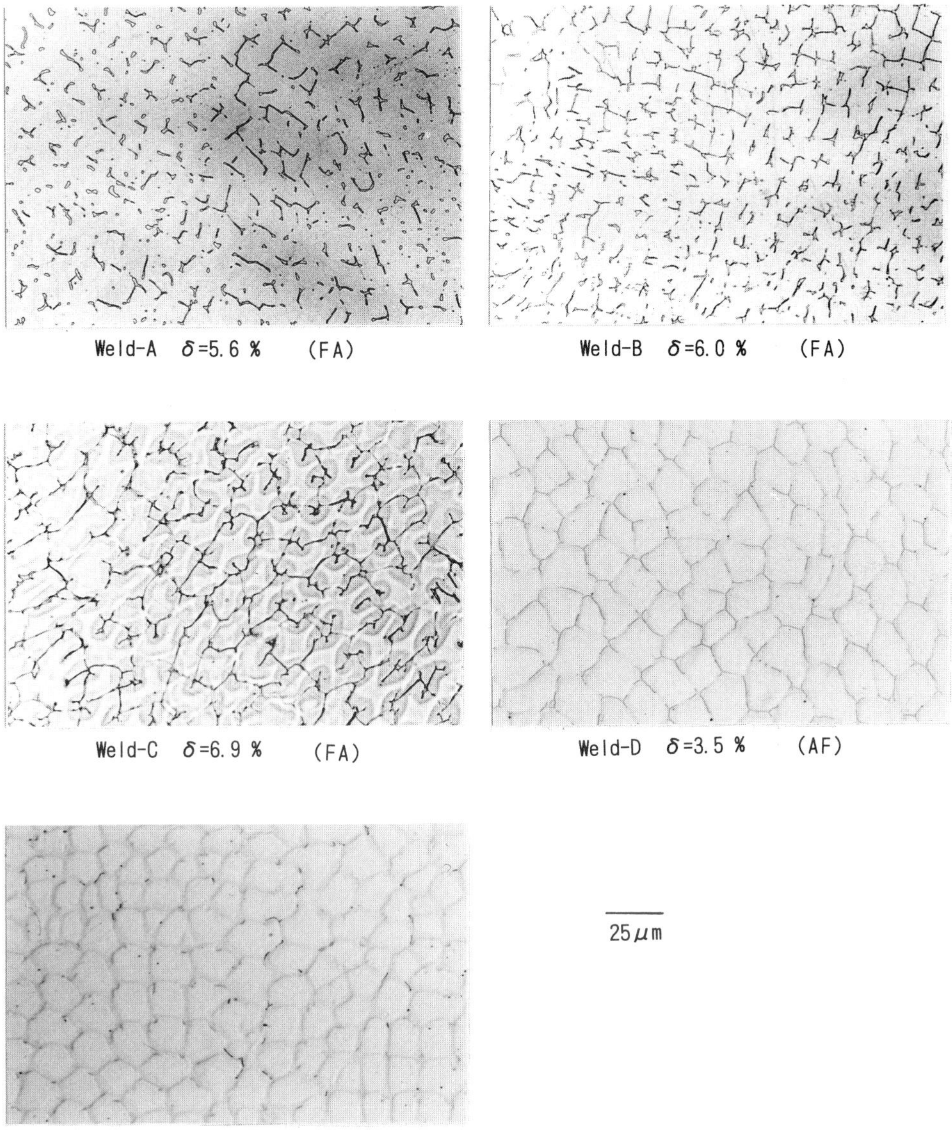

Figure 4. Microstructure of weld metals (calculated solidification mode).

weld-E, the primary phase of solidification became austenite, and ferrite phase appeared in the vicinity of the solidification terminal (AF-mode).

Figure 5 also indicates that the fraction of delta phase becomes maximum when the fraction of solid reaches 1.0 and then it decreases with temperature drop. It can be seen that the delta/gamma transformation almost finishes at 1273 K both in the AF and FA solidification mode (Figs 3 and 5). The amount

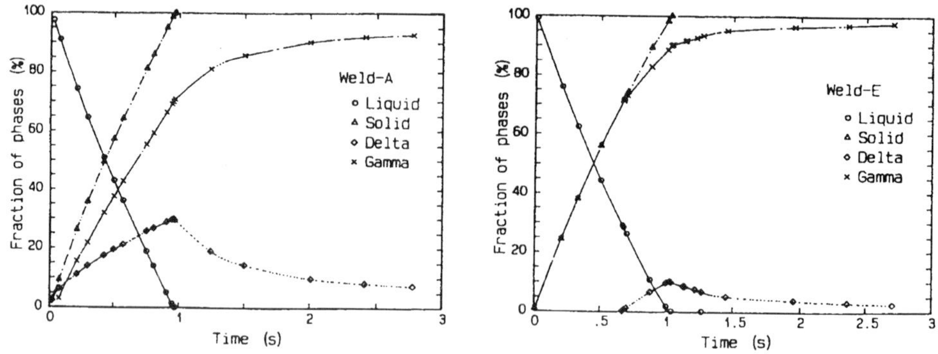

Figure 5. Changes in fraction of phases during solidification and transformation.

of retained delta ferrite decreases to 7.4 % and 2.4 %, in weld-A and weld-E, respectively. These calculated values are a little larger than those measured.

Figure 6 shows the profiles of delta fraction in weld-C and weld-D just after solidification and at 1273 K. In the case of weld-D, the amount of delta ferrite which solidified eutectically decreases monotonously during cooling down to 1273 K, but it does not disappear in the segments. In the case of the FA-mode of Weld-C, both primary and eutectic ferrite also transform to austenite, and the volume of the delta ferrite decreases greatly.

The calculated solute concentration profiles during solidification are shown in Fig. 7. The concentrations of Cr, Ni, and Mo in solidified segments hardly change during solidification because of the small diffusion coefficient of these elements. However, the profile of N becomes almost flat in all solidified segments during solidification. It is clearly shown that Ni and N are

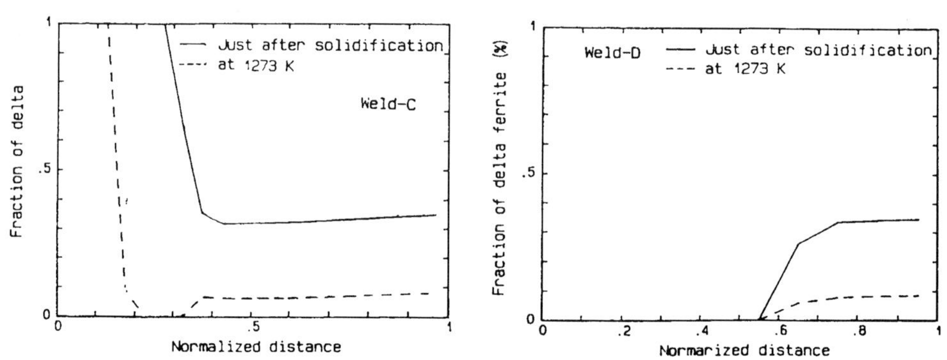

Figure 6. Profiles of fractions just after solidification and at 1273 K.

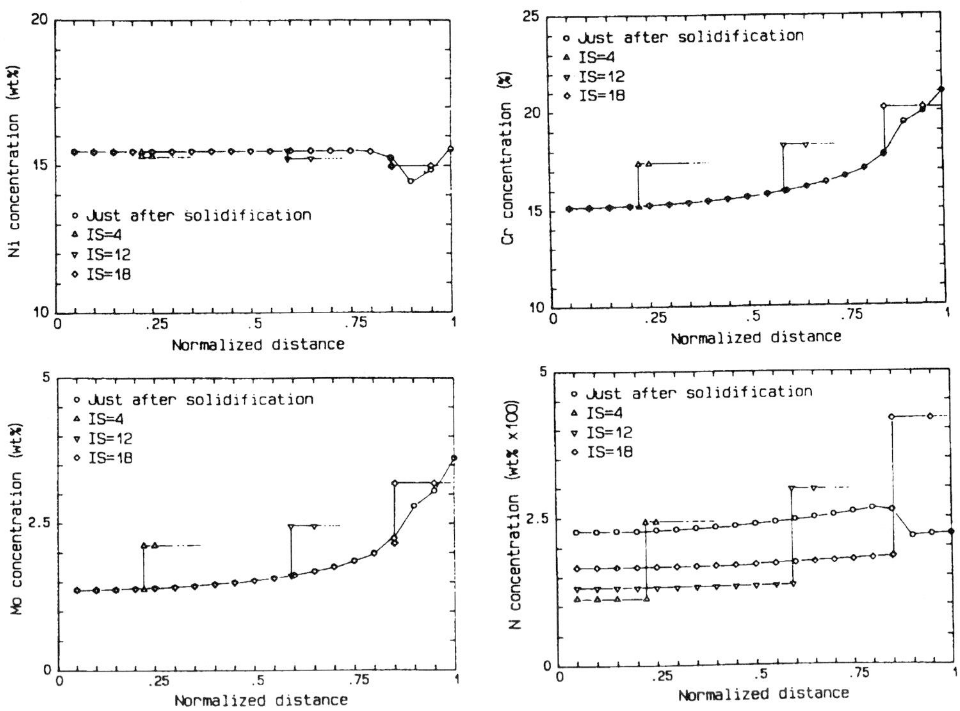

Figure 7. Profiles of elements during solidification (IS: segment no. where liquid/solid interface exists).

Table 4 Summary of the results

Mark	Solidification mode		retained δ-ferrite		Concentration of solute element (in δ/ in g)					
					Cr		Ni		Mo	
	calc'd	meas'd	calc'd	meas'd	calc'd	meas'd	calc'd	meas'd	calc'd	meas'd
Weld-A	FA	FA	7.4	5.6	27.55/ 16.64	24.36/ 17.90	4.12/ 15.38	5.58/ 12.12	3.80/ 1.77	4.08/ 1.78
Weld-B	FA	FA	11.7	6.0	24.57/ 14.89	22.13/ 16.42	2.90/ 10.75	3.61/ 9.98	2.75/ 1.43	3.48/ 1.32
Weld-C	FA	FA	8.6	6.9	25.76/ 15.56	–	3.56/ 13.32	–	3.36/ 1.67	–
Weld-D	AF	AF	5.0	3.5	29.01/ 15.15	–	4.36/ 12.69*	–	5.10/ 1.43*	–
Weld-E	AF	AF	2.4	0.02	30.30/ 15.01	–	4.79/ 14.63*	–	5.62/ 1.37*	–

* At dendrite core.

concentrated in the primary phase of austenite and the ferrite stabilizers of Cr and Mo are enriched in the ferrite phase just after solidification.

Figure 8 shows the result of EPMA measurements by a line scanning method in weld-A. The position of measured line is also shown in the figure. The result clearly indicates that the concentrations of Cr and Mo increase at the retained delta ferrite and that of Ni decreases in the delta phase.

The composition profiles in weld-A at 1273 K are shown in Fig. 9. Since simulations are carried out in the Fe–Cr_{eq}–Ni_{eq}–Mo system, concentrations of Cr and Ni have been revised to compare the calculated value with the measured one. The calculated values of the solute distribution have been in good agreement with the measured values.

The results of the simulation compared with the experimental results are summarised in Table 4. There is good agreement between calculated values of the solute distribution and experimental results. The simulated solidification modes in the Fe–Cr_{eq}–Ni_{eq}–Mo–N system also showed good agreement with the experimental data. The calculated amount of retained delta ferrite corresponded fairly well with the measured amount. The calculated values were larger than those measured in both AF and FA mode solidification, especially at large amounts of delta ferrite in FA mode solidification.

DISCUSSION

The accurate prediction of solidification mode, the estimation of element distribution during solidification and the subsequent delta/gamma transformation and calculation of the retained delta ferrite are possible using this model.[7,8] As has been shown, the model has sufficient functionality to simulate solidification and transformation of type 316 commercial steel welds. However, some improvements and discussion are necessary for practical application.

First, the most influential factor is the selection of the alloy system and the thermodynamic data. The solidification mode of weld A, for instance, which experimentally showed FA mode solidification, was located in the AF mode region in the Fe–Cr_{eq}–Ni_{eq}–Mo system, as shown in Fig. 10a. And the calculated value of the delta ferrite became larger than that measured. As for the simulations of alloy systems employed, it was pointed out that the phase diagram produced by Thermo-Calc did not completely correspond to the experimental results[7,16] even in the Fe–Cr–Ni ternary system. Thus, the reliability of thermodynamic data in the type 316 system would be decreased by the addition of the minor elements in the system. Provided that the thermodynamic data of the multi-element system are reliable, the precision of

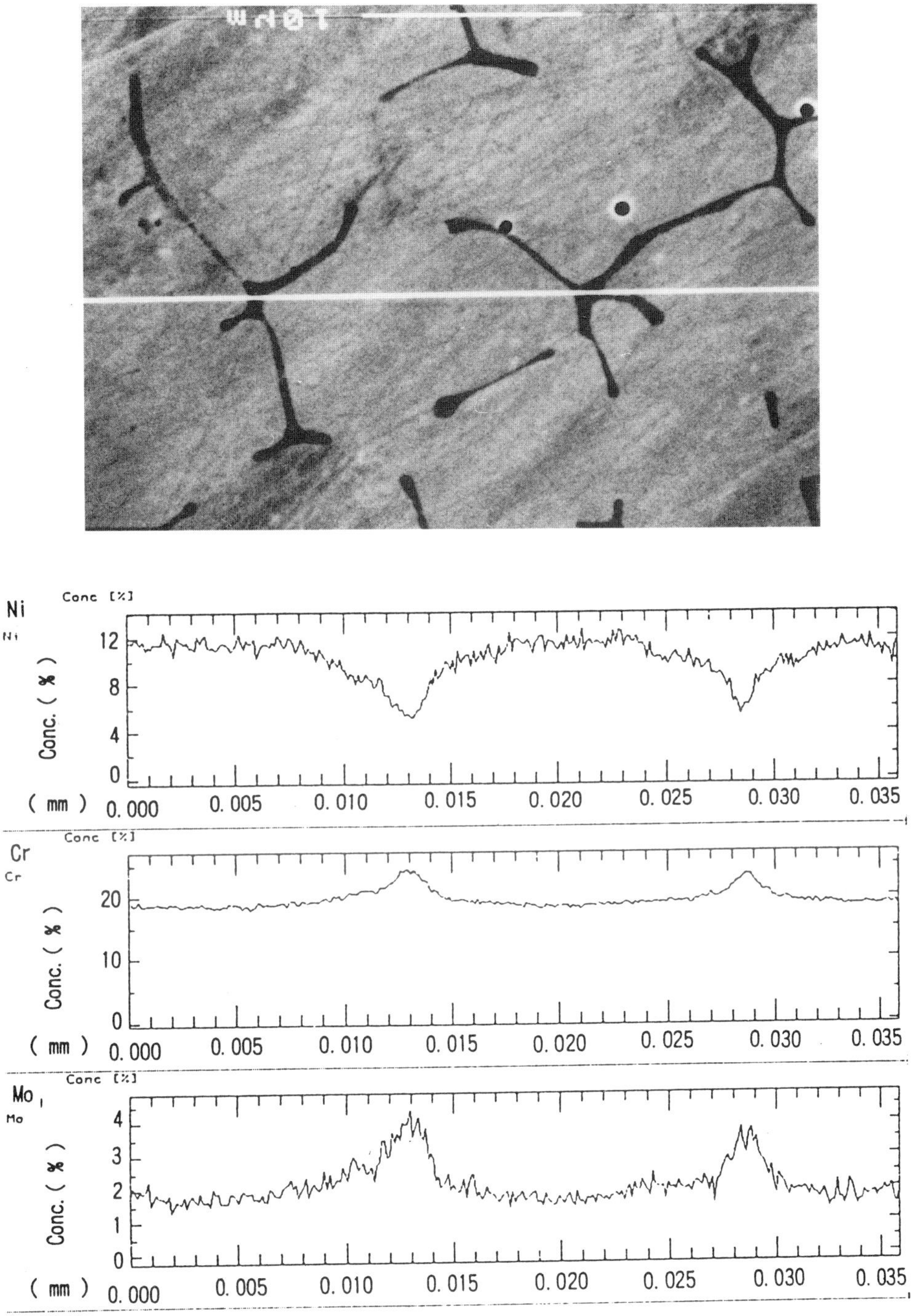

Figure 8. Result of EPMA measurement (weld-A).

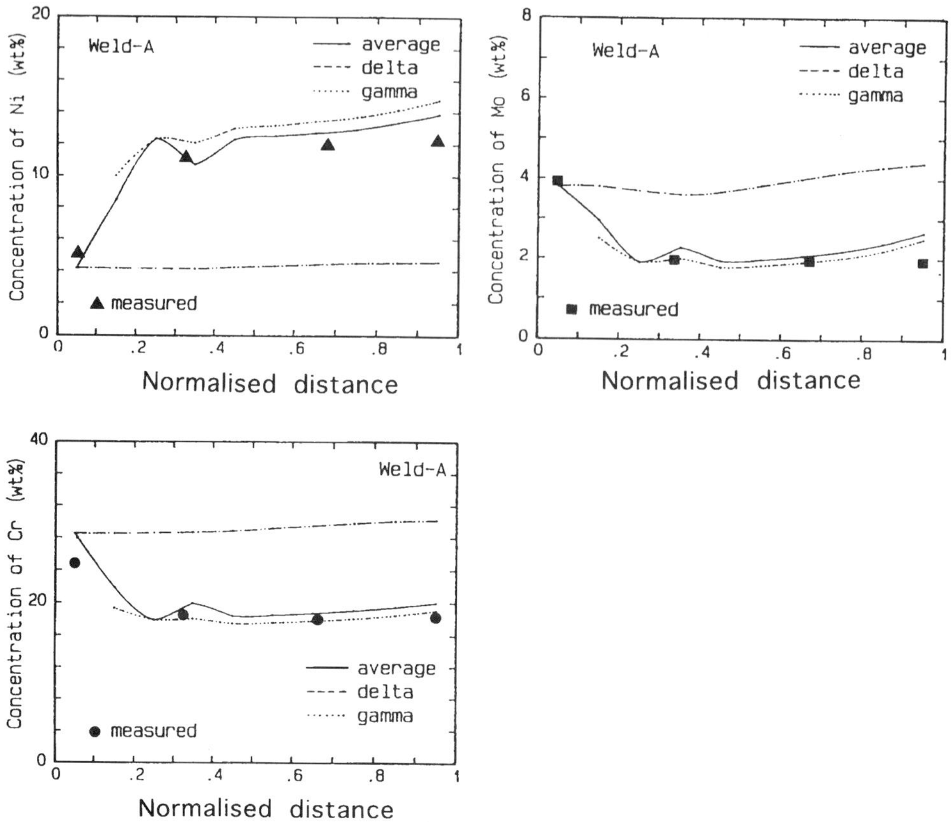

Figure 9. Comparison of the calculated and actual element profiles.

the simulation increases with the number of elements which are taken into consideration.

Secondly, however, the calculation time increases witn the number of elements included. This is especially so in the case of carbon and nitrogen which have large coefficients of diffusion which increase computing time because of the stability condition of FDM.

Thirdly, the appropriateness of the model to the weld solidification should be discussed. The solidification model used in this study assumed homogeneous liquid composition and multicomponent equilibria at the liquid/solid interface. But undercooling and kinetic effects are not taken into account in the calculation. The actual solidifying condition used in this study was conventional TIG welding and it was not subjected to rapid solidification. Therefore, the kinetic effect on phase transition between delta and gamma solidification can be ignored and the local equilibrium would be reasonable. The Clyne–Kurz equation was applied to each FDM segment, and the model was assumed to be homogeneous only in the delta phase. Usually every

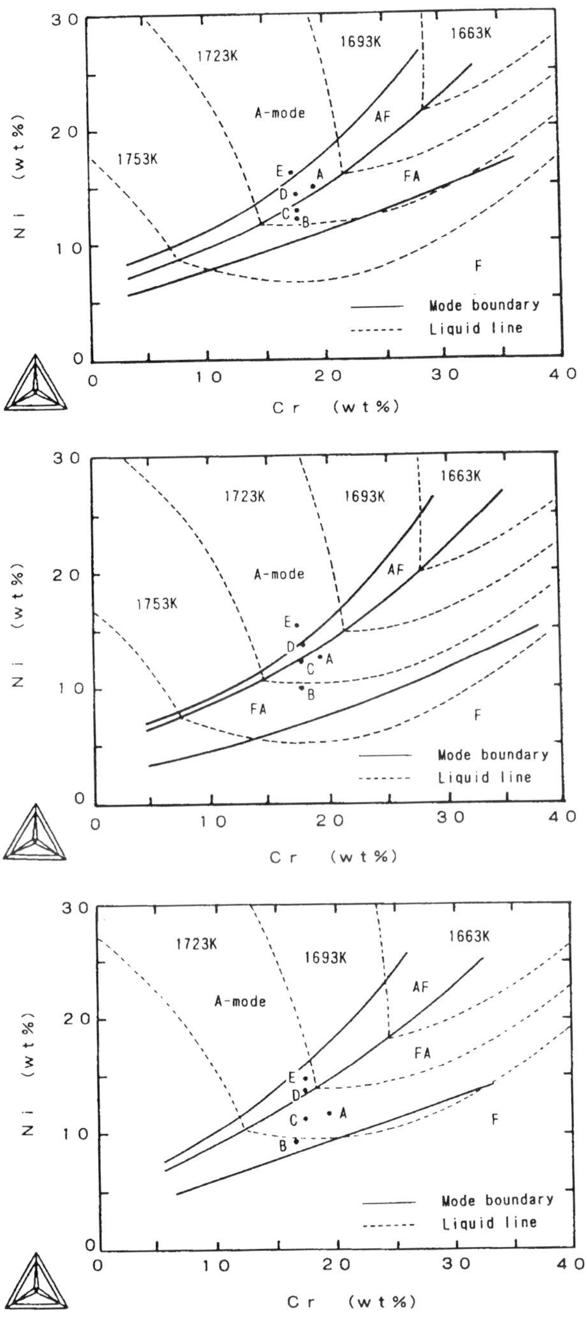

(a) Fe-Cr-Ni-2.2%Mo system

(b) Fe-Cr-Ni-2.2%Mo-0.085%N system

(c) Fe-Cr-Ni-2.2%Mo-0.085%N-0.009%C -0.45%Si-1.25%Mn-0%Cu system

Figure 10. Solidification modes of weld metals in different systems.

segment is assumed homogeneous everywhere. However, the large difference in diffusivity between gamma and delta during the weld thermal history and the limitation of the calculation size make it difficult to assume equilibrium even within the segment.

Several models which can predict the solidification phases and kinetic effect of dendritic growth have been proposed.[3, 18, 19] The solidification model used in this study is advanced in its use of thermodynamic data but needs modifying to improve its accuracy of prediction.

As for future work, together with the improvement of the current model, a kinetic analysis of decomposition of delta ferrite to the intermetallic compounds or precipitation is needed to complete this study. The delta ferrite is not preferable for toughness, but the sigma phase is more harmful. Modelling of delta ferrite decomposition seems to be possible by using this method of calculation. This kind of analysis seems to be effective for the prediction of weld properties.

CONCLUSIONS

A numerical model was applied to the solidification and delta-gamma transformation phenomena of type 316 austenitic stainless steel weld metals. The main results obtained are summarised as follows:

1. The prediction of the solidification mode is possible with good agreement by employing the principal elements in the calculation.
2. The simulated values of element distribution between the ferrite and the austenite after transformation showed good agreement with the experiments.
3. Some improvements are required for the precision of profiles of phases in the FA-mode solidification, especially in the case of a large amount of the retained delta ferrite.
4. The model is sufficiently robust to simulate the solidification and transformation of commercial steel welds which include such elements as C, Si, Mn, N, Cu and so on.

ACKNOWLEDGEMENTS

The authors thank Mr T. Koseki, Mr W. Yamada and Dr T. Matsumiya for advice in using the simulation program, and to Drs N. Yurioka and T. Tanaka for directing this work.

REFERENCES

1. D.L. OLSON: Prediction of austenitic weld metal microstructure and properties, *Welding Journal*, 1985, 281.
2. S.A. DAVID and J.M. VITEK: Correlation between solidification parameters and weld microstructures, *International Materials Review*, 1989, **34**, 5.
3. X.G. WANG, C. DUMORTIER and Y. RIQUIER: Simulation of solidification process and prediction of delta-ferrite distribution in continuously cast austenitic stainless slabs, *Proc. 6th International Iron and Steel Congress*, 1990, Nagoya, ISIJ.
4. Y. UESHIMA, S. MIZOGUCHI, T. MATSUMIYA and H. KAJIOKA: Analysis of solute distribution in dendrites of carbon steel with delta/gamma transformation during solidification, *Metall. Trans. B*, 1986, **17B**, 845.
5. T. NAKAZAWA *et al.*: Development of weld metals for low carbon medium nitrogen type 316 steel (in Japanese), *Report of the 123rd Committee on Heat-Resisting Metals and Alloys*, Japan Society for the Promotion of Science, 1991, **32**, 2.
6. T. NISHIDA *et al*: Development of 316 stainless steel with high creep rupture strength for FBR structural materials, *International Conference on Fast Reactors and Related Fuel Cycles*, Kyoto, Proceedings, 1991, **4**.
7. T. KOSEKI, T. MATSUMIYA, W. YAMADA and T. OGAWA: Numerical modeling for analysis of solidification and subsequent transformation of Fe–Cr–Ni alloys, *Metall. Trans. B*, to be published.
8. T. MATSUMIYA: Mathematical analysis of segregations and chemical compositional changes of nonmetallic inclusions during identification of steels, *Materials transactions*, JIM, 1992, **33**, 9.
9. B. SUNDMAN *et al.*: Proc. Conf. User application of Alloy Phase Diagram, L. KAUFMAN, ed., ASM, 1986, 185.
10. T.W. CLYNE and W. KURZ: Solute redistribution during solidification with rapid solid state diffusion, *Metall. Trans. A*, 1981, **12A**, 965.
11. Ed. by Jap. Inst. Metals; 'metal's data book, 2nd ed. (in Japanese)'
12. Tekko-no-Gyoko (Solidification of Steel), Supplement, Solidification Comm. of ISIJ, ISIJ, Tokyo, 1977, S32–S50.
13. F. MATSUDA: *Welding Metallurgy*, 1972, 179 (in Japanese).
14. T. TAKALO, N. SUUTALA and T. MOISIO: Austenitic solidification in austenitic stainless steel welds, *Metall. Trans. A*, 1979, **10A** 1173.
15. N. SUUTALA, T. TAKALO and T. MOISIO: Ferritic–austenitic solidification mode in austenitic stainless steel welds, *Metall. Trans. A*. 1979, **10A**, 512.
16. H. INOUE, T. KOSEKI and T. TANAKA: Thermodynamic analysis of solidification and transformation behavior in the weld metals of FA mode stainless steel (in Japanese), Preprints of The National Meeting of JWS, 1992, **52**, 278.
17. D.J. KOTECHI and T.A. SIEWERT: WRC-1992 constitution diagram for stainless steel weld metals: a modification of the WRC-1988 diagram, *Welding Journal*, 1992, 171.
18. H. FRIEDRIKSSON: *Solidification and Casting of Metals*, 1977, London, The Metals Society, 131.
19. M. BOBADILLA, J. LACAZE and G. LESOULT: 'Influence des conditions de solidification sur le deroulement de la solidification des aciers inoxydables austenitiques', *Journal of Crystal Growth*, (1988), **89**, 531–544.

SECTION II

MICROSTRUCTURAL MODELLING IN WELD METAL AND HEAT AFFECTED ZONE

POSSIBLE EFFECTS OF STRESS ON STEEL WELD MICROSTRUCTURES

H.K.D.H. BHADESHIA
University of Cambridge/JRDC
Department of Materials Science and Metallurgy
Pembroke Street, Cambridge CB2 3QZ, U.K.

ABSTRACT

Little is known about the effect of stress on the development of microstructure in steel welds. This paper contains an assessment of published data together with a description of the theory that is available for dealing with stress-affected transformations in steels. Attention is focused on those transformations which have the greatest potential for interaction with an externally applied stress. These include the solid-state transformation products of austenite, such as Widmanstätten ferrite, acicular ferrite, bainite and martensite.

INTRODUCTION

The ability of an external stress to influence the development of microstructure during solid-state transformation is well established in theory. There is also a great deal of research describing the contrary process, in which phase transformations induce the development of stresses; these are the so-called residual stresses, whose effects are particularly pronounced in constrained assemblies. In fact, residual stresses can arise even in the absence of any phase transformation. All that is required is that the temperature varies with position.[1] It is nevertheless true that experimental observations on steel weldments cannot be fully accounted for without a consideration of the phase changes which occur during the cooling of the weldment to ambient temperature.[2,3]

The purpose of the work presented here is to describe the relevance of these concepts to the welding of steels. There is a need for this subject to develop further before serious quantitative applications become viable, although very useful general principles already exist which can help in the assessment of weld phenomena. We therefore consider first the basic principles which are applicable to both the heat-affected zone and fusion zone of a weld, and which are useful in phase transformation theory *per se.*

PLASTIC DEFORMATION

The familiar mechanisms of plastic deformation are slip, mechanical twinning and diffusion-induced creep. For very small plastic strains, the first two of these deformation modes are conservative, i.e. they preserve an atomic correspondence between the deformed and undeformed parts of the crystal so that the crystal contains a memory of its original shape. All of these deformation modes are *lattice-invariant* because although they cause a change in the shape, the crystal structure remains as it was prior to deformation.

Phase transformations, by contrast, refer to changes in the crystal structure. However, like lattice-invariant deformations, they can also be associated with changes in the shape of the transformed regions.[4–7] The transformation can have a pronounced interaction with an externally applied stress only if it is accompanied by a shape change (a strain).

In steels, austenite transforms into many varieties of ferrite, which can be classified into two broad categories according to the mechanism of transformation (Fig. 1). Reconstructive transformations occur with the uncoordinated diffusion of all of the atoms, including iron.[8,9] They need not be accompanied by a change in chemical composition, but it is worth emphasizing that reconstructive transformations cannot be sustained without the diffusion of all atoms. For example, the freezing of water to ice is a reconstructive transformation where there is mass transport but no change in composition. Similarly, austenite in pure iron can transform into ferrite, either by a reconstructive mechanism involving diffusion, or by a martensitic mechanism. The latter case represents displacive transformations, in which the pattern of atomic arrangement is altered by deformation. There is no diffusion of iron or substitutional solutes during displacive transformations in iron alloys.

Both kinds of reactions lead to changes in shape (Fig. 2a). The simplest of these occurs during reconstructive transformations in which the densities of the parent and product phases are different. Hence, when austenite transforms into grain boundary allotriomorphic ferrite, there is a uniform expansion, and vice-versa.

Much more interesting deformations are associated with displacive transformations, where the shape change is in general described as an *invariant-pane strain* with a large shear component. Such a deformation leaves

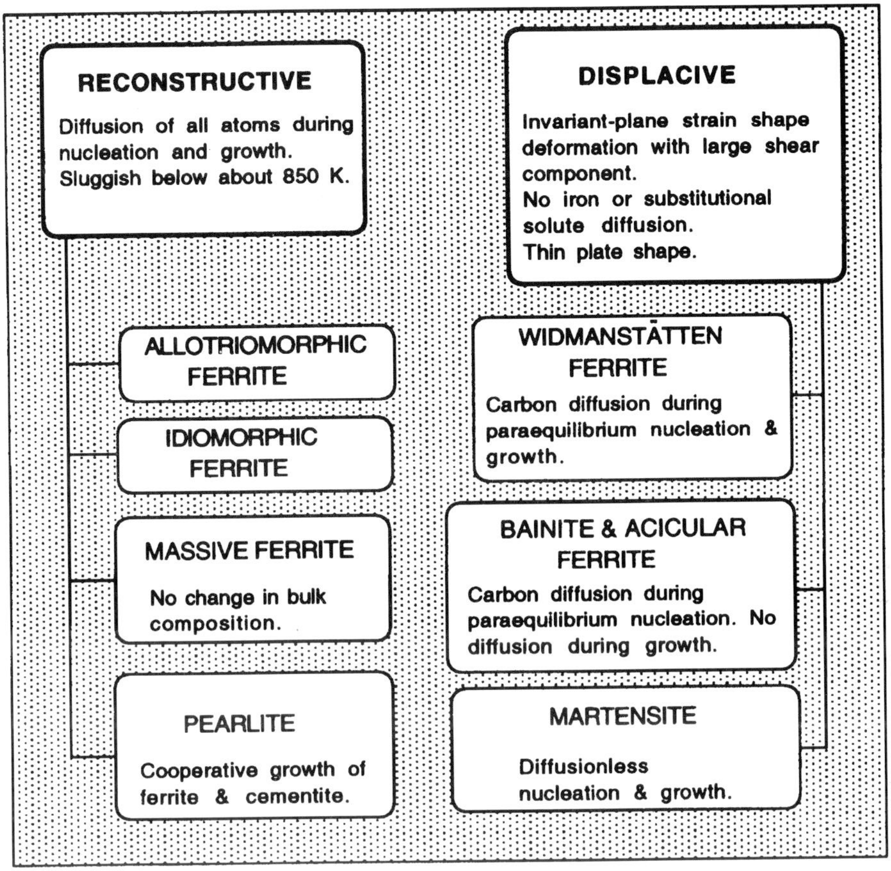

Figure 1. Classification of the transformation products of austenite.[9] All of the displacive transformation products cause a shape change on transformation. This shape change is an invariant-plane strain with a large shear component. The much smaller dilatational component of strain is directed normal to the habit plane (invariant-plane). The reconstructive transformations involve a lot of diffusion. Their formation is only accompanied by a volume change and unlike the displacive transformation products, they do not grow with a plate shape. Cementite and ε-carbides are not included in the above classification, but it is believed that they grow by a displacive transformation mechanism at low temperatures.

the plane of contact between the parent and product phases undistorted and unrotated. This plane is called the *habit plane*. The strain energy associated with a constrained invariant-plane strain (IPS) is minimized when the product phase has a thin-plate shape. This is why Widmanstätten ferrite, bainite, acicular ferrite and martensite in steels grow in the form of plates. The distinguishing features of a variety of deformation modes are compared in Table 1, and the detailed characteristics of the shape change due to transformation are emphasized in Table 2. The latter also includes some data

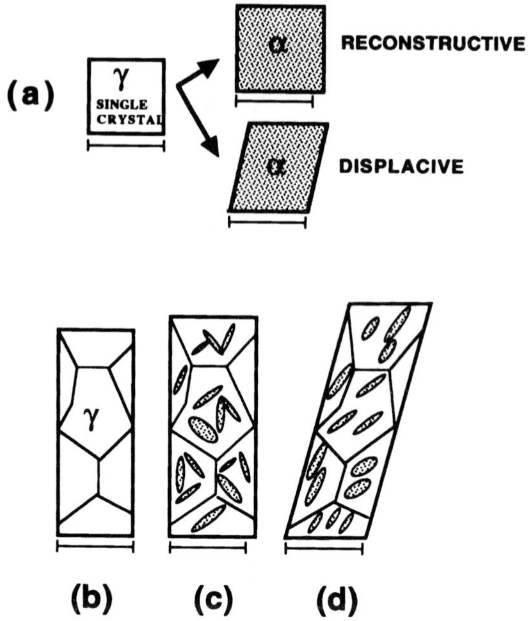

Figure 2. The shape changes accompanying unconstrained transformations. Note that the horizontal scale bars are all of the same length. (a) The two kinds of changes in shape that occur when a single crystal of austenite transforms to a single crystal of ferrite, as a function of the mechanism of transformation. (b) A polycrystalline sample of austenite. (c) A polycrystalline sample of austenite which has partially transformed by a displacive transformation mechanism into a random set of plates of ferrite. (d) A polycrystalline sample of austenite which has partially transformed by a displacive transformation mechanism into an organised set of plates of ferrite.

for mechanical and annealing twins in ferrite and austenite respectively. These are not, of course, transformation products, so that they do not cause any volume change; however, their growth mechanisms are analogous to those of displacive and reconstructive transformations respectively. The shear component of the IPS for the mechanical twin is seen to be substantially larger than all of the transformations listed. Indeed, it is only in iron–beryllium alloys (Fe_3Be) that martensitic transformation has a shear strain which is as large as that for mechanical twins in ferrite.[10–12] These empirical observations show that, in general, the shear associated with transformations in steels tends to be smaller than that caused by mechanical twinning.

The permanent strain caused by any transformation is called *transformation plasticity*. A phase change in a stress-free material is usually triggered by heat treatment, when the parent phase passes through an equilibrium transformation temperature. Alternatively, the application of a stress in isothermal conditions can trigger transformation in circumstances where it would not otherwise occur. Unusual effects can occur when stress and

Table 1 Characteristics of different modes of deformation

	Slip deformation	Mechanical twinning	Displacive transformation	Reconstructive transformation
Causes permanent change in shape	Yes	Yes	Yes	Yes
Invariant-plane strain shape change with a large shear component	Yes	Yes	Yes	No
Changes crystallographic orientation	No	Yes	Yes	Yes
Changes lattice type	No	No	Yes	Yes
Can lead to a density change	No	No	Yes	Yes

Table 2 Shape change due to transformation. An invariant-plane strain here implies a large shear component as well as a dilatational strain normal to the habit plane. s and δ refer to the shear and dilatational strains respectively. The values stated are approximate and will vary slightly as a function of lattice parameters and the details of crystallography

Transformation	Shape change	s	δ	Morphology
Allotriomorphic Ferrite	Volume change	0.00	0.02	Irregular
Idiomorphic Ferrite	Volume change	0.00	0.02	Equiaxed, facetted
Pearlite	Volume change	0.00	0.03	Spherical colonies
Widmanstätten Ferrite	Invariant-Plane Strain	0.36	0.03	Thin plates
Bainite	Invariant-Plane Strain	0.22	0.03	Thin plates
Acicular Ferrite	Invariant-Plane Strain	0.22	0.03	Thin plates
Martensite	Invariant-Plane Strain	0.24	0.03	Thin plates
Cementite Plates	Invariant-Plane Strain?	0.21?	0.16?	Thin plates
Mechanical Twins (α)	Invariant-Plane Strain	$1/\sqrt{2}$	0.00	Thin plates
Annealing Twins (γ)		0.00	0.00	Facetted

temperature work together. The transformation may occur at remarkably low stresses or at very small deviations from the equilibrium temperature. This is why even minute stresses can have a large influence on the development of microstructure. It is not surprising that transformation plasticity can be obtained at stresses which are much smaller than the conventional yield stress of the parent phase.

The plasticity associated with phase transformations has relevance in a wide variety of fields. Some of the important theory of transformation plasticity has its origins in the nuclear industry, where the thermal cycling of uranium through phase changes can lead to large deformations.[13, 14] The deformations occur under the influence of very small external stresses, leading to the popular interpretation that there is a mechanical weakening of materials as they undergo a change in crystal structure. Another example of transformation plasticity is when a suicidal virus infects a bacterium with its DNA information. It does this mechanically using a 'hypodermic needle' consisting of a cylindrical crystal. The crystal on transformation changes shape and hence penetrates the bacterium so that the virus can inject the bacterium with its DNA.[15] The virus thus ceases to exist!

A great deal of research followed the exploitation of transformation plasticity in TRIP steels,[16] which are designed to make a propagating crack do work in stimulating transformation.[17–25] This greatly increases the work of fracture. Recent applications include the design of automobile steels containing retained austenite, which transforms and plasticizes the steel during the forming process.[18, 19] The research on TRIP steels has led to one of the few promising mechanisms for the toughening of ceramics.[20–22] The subject of transformation plasticity is of fundamental interest[17, 23–27] and of use in the design of heat-treatments.[28–32]

As far as welds are concerned, residual stresses in engineering structures can determine the stability and safety of components. There is considerable activity in the experimental determination of these stresses and on computation methods which include a role of transformation plasticity in the development of residual stresses.[2, 3, 33, 34]

Since the classical papers by Greenwood and Johnson,[13, 14] the way in which transformation plasticity is included in the solution of engineering problems has focused on the role of the volume change. For a longitudinal stress σ, Greenwood and Johnson found that the longitudinal plastic strain is given by:

$$\varepsilon = \frac{5}{6}\frac{\Delta V}{V}\frac{\sigma}{\sigma_Y} \tag{1}$$

where σ_Y is the yield stress of the weaker phase and $\Delta V/V$ is the transformation volume strain. Clearly, it is the volume strain which is assumed to enable a transformation to interact with an externally applied or internally generated stress. This is a reasonable approach when considering reconstructive transformations. However, the most desirable microstructures in steels are those which are generated by a displacive transformation mechanism, with large shear strains (Table 2), which are not incorporated in the Greenwood and Johnson approach. This has rejuvenated some exciting areas of research, traces of which can be found in literature published as long ago as 1930,[35] but highlighted particularly by the work of Magee and Paxton,[36, 37] with more recent work by Fischer,[38] Leblond *et al.*,[39–43] Olson,[44] and Bhadeshia *et al.*[45] The work presented here is not therefore a review of the general area of residual stress development, a subject which has been covered in many excellent articles.[2,3,29,39–44] Instead, it deals with the micromechanics of phase transformations. How applicable is this to large assemblies? There is convincing evidence that particles which grow under the influence of stress form in an organized way, even when the parent phase is random polycrystalline. They consequently act in concert to greatly influence the dimensions and isotropy of large samples, to an extent which is far greater than might be expected from mere density changes. This is illustrated schematically in Fig. 2b–d.

THE DEFORMATION SYSTEM

It is common, during discussions of slip, to refer to the crystallographic indices of the slip plane and direction. We have seen that a reconstructive transformation occurs by diffusion. This diffusion involves the uncoordinated movement of atoms. It is then not possible to relate the combined macroscopic effect of all the atomic motions, to the details of the crystal structure in the manner just described for slip. Deformations caused by reconstructive transformations are more analogous to fluid flow than conservative slip.

The situation is quite different for transformations where there is a coordinated displacement of atoms to achieve the new crystal structure. Just as a combination of a plane and a direction constitutes a deformation system for slip or twinning, the habit plane and displacement vector of the invariant-plane strain accompanying transformation completely describe the deformation system responsible for transformation plasticity. The displacement vector describes the sense of the macroscopic displacements accompanying transformation, and along with the habit plane indices, also contains information about the magnitude of the shear component and dilatational component of the displacements. Typical data for the deformation systems associated with transformations are listed in Table 3.

Given the cubic crystal structure, and the fact that habit planes tend to be irrational, there will in general be 24 of these systems per austenite grain, and they may operate simultaneously to varying extents. Of course, unlike ordinary slip, the different deformation systems within an austenite grain cannot intersect, except in special circumstances where inter-variant transformations are possible, as is the case with some shape memory alloys. It follows that the

Table 3 The deformation systems associated with transformations. Typical habit plane and displacement directions for low-alloy steels. The indices all refer to the austenite phase. Note that the indices stated are approximate since the habit plane and displacement direction are usually irrational. The displacement vector does not quite lie in the habit plane because the dilatational strain is directed normal to the habit plane. The magnitude of the displacement, is given by *m*, which is the toal displacement including the shear and the dilatational components.

Phase	Habit plane indices	Displacement vector	*m*	
Martensite	(0.363 0.854 0.373)	[0.195 0.607 0.771]	0.185	Dunne and Wayman
Bainite	(0.325 0.778 0.537)	[0.159 0.510 0.845]	0.27	Ohmori
Widmanstätten ferrite	(0.506 0.452 0.735)	[0.867 0.414 0.277]	0.36	Watson and McDougall

ordinary notion of work hardening does not apply. Work hardening nevertheless manifests itself via a different mechanism, in which the stability of the austenite increases as it becomes ever more finely divided.

The Taylor/von Mises criterion[46,47] states that in any given crystal a minimum of five independent slip systems are necessary to produce an arbitrary shape change. A crystal in a polycrystalline aggregate has to accommodate the arbitrary deformations of neighbouring grains. Therefore, a polycrystalline material is brittle unless each grain contains at least five independent slip systems. Similar logic can be applied to the crystallographic variants of a phase generated by displacive transformation. The habit plane is predicted theoretically[48,39] and found experimentally[50] to have irrational indices. This means that there exist in principle 24 possible variants of the habit plane per grain of austenite (i.e. 24 independent deformation systems). Given this large number of transformation variants available per grain, the Taylor criterion leads to the conclusion that transformation plasticity can cause, or accommodate any externally imposed, arbitrary shape change assuming that a sufficient quantity of parent phase is available. It follows that polycrystalline samples can remain intact at grain boundaries when transformation plasticity is the sole mode of deformation.

THE MECHANICAL DRIVING FORCE

The interaction of an externally applied stress (which is below the yield strength of the parent phase) can manifest in two ways:

1. The stress can alter the driving force for the transformation.

2. It can change the appearance of the microstructure by favouring the formation of those variants which best comply with the applied stress.

Both of these factors are illustrated in Fig. 3, where a fine grained polycrystalline sample of austenite was stressed at a temperature *above* its normal martensite-start temperature.[51] The amount of martensite obtained is seen to vary directly with the magnitude of the applied stress. The stress manifests as a mechanical driving force whose contribution assists the chemical driving force, which on its own is inadequate to trigger transformation. Not only does the stress induce the formation of martensite, but only those variants which comply with the tensile stress grow in profusion (Fig. 3b). Thus, most of the plates are tempted to grow on those planes which are close to the plane of maximum shear stress (45° to the tensile axis). The microstructure would have been much more chaotic in the absence of the stress. The stress caused the alignment of plates, and a more ordered, organized microstructure results.

These considerations are dependent on the mechanism of transformation. For reconstructive transformations, it is only the hydrostatic component of stress that can interact with the shape change (which is due to just a density change). The response to an arbitrary stress is therefore not expected to be as large as that for displacive transformations in steels. In that case, the shear component of the IPS is relatively large and so is the corresponding mechanical driving force component.

We now focus on displacive transformations, using a method which is based on Patel and Cohen's work.[52] In the above discussion, the total driving force was partitioned implicitly into a 'mechanical driving force' and the more familiar chemical driving force.[26,52] This is based on the reasoning that the movement of a glissile interface is a combined deformation and transformation process. The work done by the external stress may be added to the chemical free energy change in order to obtain the net free energy difference.

The mechanical driving force is assumed to be given as the work done (ΔG_{mech}) by the external stress in producing the macroscopic shape deformation:

$$\Delta G_{mech} = \sigma_N \delta + \tau s \tag{2}$$

where σ_N is the normal stress on the habit plane and τ is the component of the shear stress on the habit plane which is parallel to the direction along which the shear displacements of the shape deformation occur (Fig. 4)†. The strains δ and s are the dilatational and shear components of the shape deformation.

† It was pointed out by Delaey and Warlimont,[53] that the Patel and Cohen treatment[52] assumes incorrectly that it is the maximum shear stress on the habit plane that interacts with the shape deformation, rather than the component of the shear stress resolved parallel to the shear displacements of the shape strain.

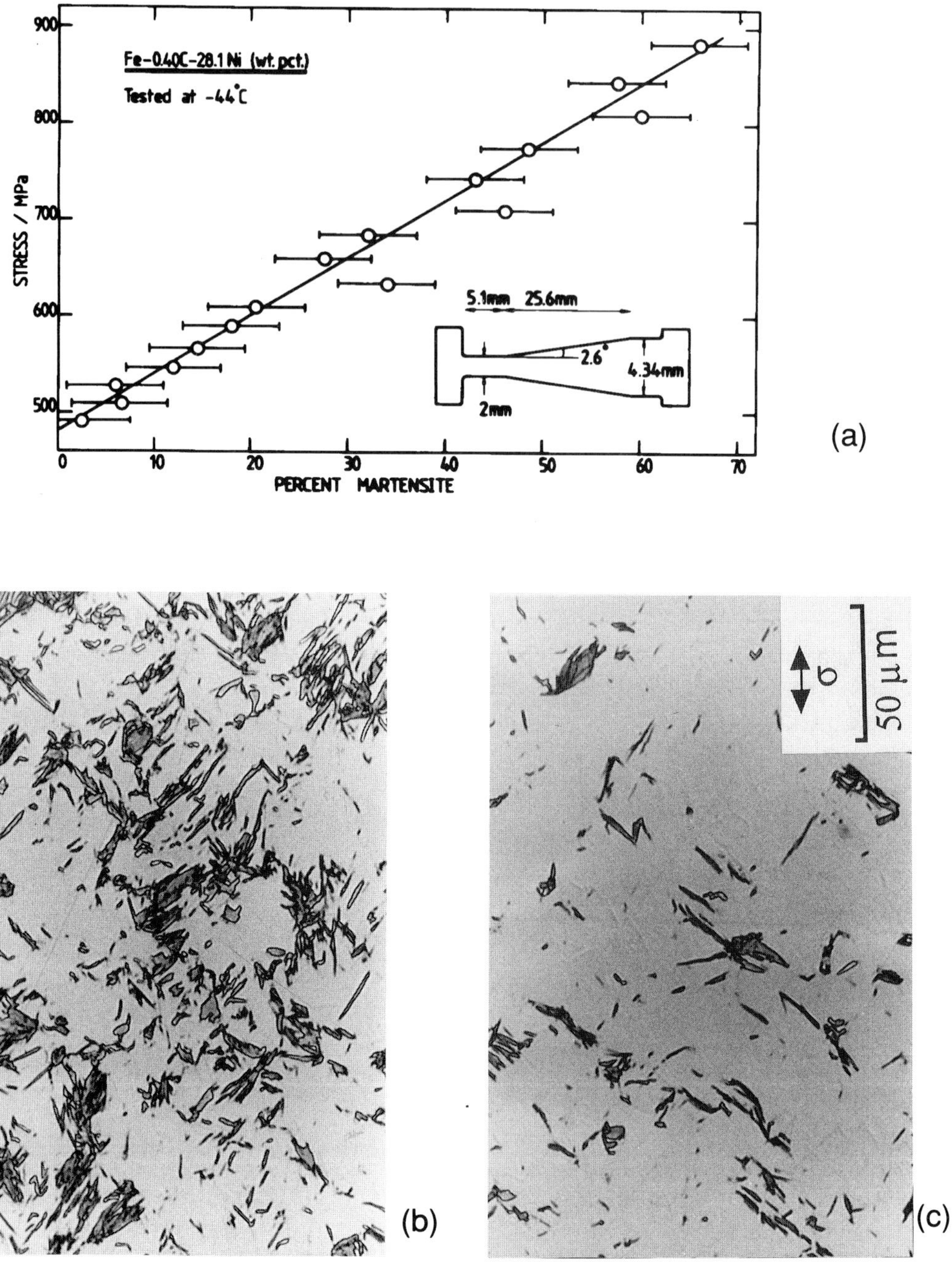

Figure 3. Stress affected martensitic transformation in a Fe–28Ni–0.4C wt.% alloy tested at a temperature above the martensite-start temperature. (a) Volume fraction of martensite as a function of stress. (b) Optical micrograph from a low stress region. (c) Optical micrograph from a high stress region. The arrow indicates the direction of the tensile stress.

Table 4 Typical values of the mechanical driving force coefficients.[44] The stress state for the crack tip is multiaxial, but the coefficient is calculated by expressing the stress in terms of the von Mises equivalent tensile stress.

Nature of stress	$\delta \Delta G / \delta \sigma$ J(mol MPa)$^{-1}$
Uniaxial tension	–0.86
Uniaxial compression	–0.58
Elastic crack tip	–1.42

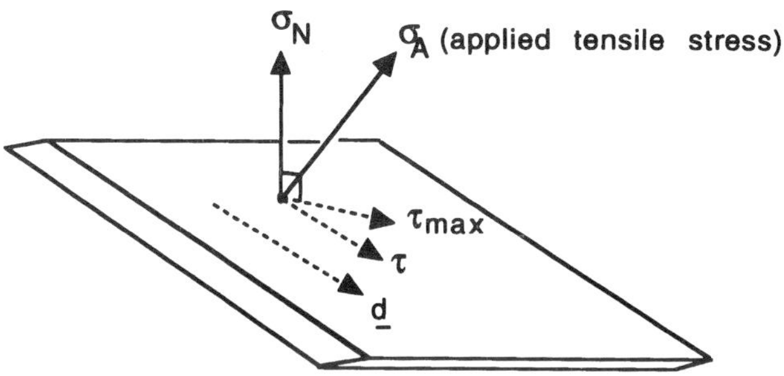

Figure 4. Resolution of the applied stress σ_A. The normal stress σ_N , and the shear stress τ, both act on the habit plane. The vector **d** is the direction along which the shear displacements of the shape deformation lie. τ_{max} is the maximum shear stress on the habit plane, but τ is given by resolving τ_{max} along **d**. Note that **d** differs slightly from the displacement vector of the invariant-plane strain, which includes a dilatational component in addition to the shear.

Some typical values of the mechanical driving force terms are given in Table 4. Given a free choice of some 12 to 24 crystallographic variants of the transformation product in each grain of austenite, the work done by the shear stress is always expected to be positive for uniaxial stress, whereas that due to the dilatational component depends on the sign of σ_N. For steels, this latter component is relatively small. Any observed consequences of stress must therefore reflect the dominant role of the shear component unless the stress is purely hydrostatic.

It follows from the Patel and Cohen analysis that since the shear stress remains positive irrespective of whether the sample is pulled in tension or uniaxially compressed, and since the shear component of the shape change is large, a uniaxial stress will always cause an increase in the temperature for displacive transformations in steels. Hydrostatic stress, on the other hand, has no deviatoric components and consequently only interacts with the dilatational component of the shape change. Thus, hydrostatic compression is expected and found to lead to a decrease in the transformation temperature (Fig. 5);

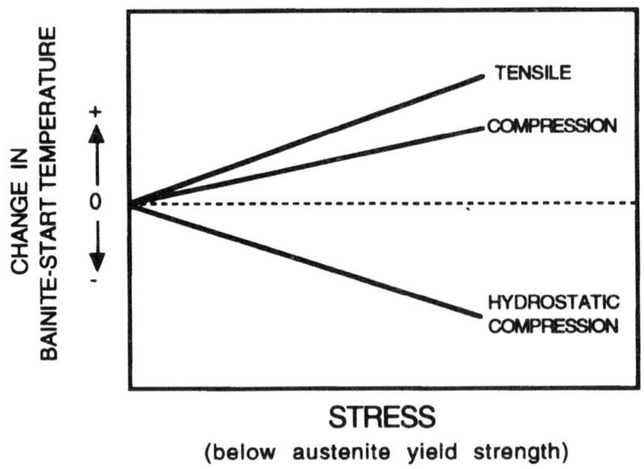

Figure 5. An indication of how the transformation-start temperature (for Widmanstätten ferrite, bainite, acicular ferrite or martensite) should vary as a function of the nature and magnitude of an applied stress whose magnitude is less than that of the yield stress.

some data[27] on the sensitivity of the transformation temperature to applied stress are presented in Table 5.

A reservation to the methodology described above is that shear stresses, unlike pressures, can not strictly be considered as state variables so that their use in thermodynamic equations can be uncertain.[26] This difficulty is, however, unimportant in the analysis as long as irreversible processes, such as diffusion or the multiplication of dislocations, do not act to relieve any of the shear stresses during the interval in which the experiment is conducted. What this means in practice is that, in the absence of transformation, the state of the system is not altered if the shear stress is changed and then restored to its original value.

A second complicating factor could arise if the stress influences the very nature of the transformation product, either by stimulating the formation of some metastable phase, or by decoupling groups of self-accommodating variants which would form in the absence of stress.[26] This could lead to a large modification of the chemical driving force term. As discussed later, there is some evidence to show that there are significant microstructural changes when

Table 5 Sensitivity of transformation-start temperatures in steels, to applied stress.[27]

Phase	Nature of stress	Sensitivity / K MPa^{-1}
Martensite	Presssure	–0.06
Bainite	Pressure	–0.09
Eutectoid	Pressure	–0.011
Martensite	Tensile	+0.06

bainite, acicular ferrite or martensite grow under the influence of an externally applied stress. Similar effects are expected for Widmanstätten ferrite (and carbides which occur at low temperatures) but they have not yet been confirmed experimentally.

Assuming that the interaction of the applied stress is with the macroscopic shape deformation, the stress must tend to favour the growth of those particular variants for which ΔG_{mech} is maximised. Hence, for a tensile stress, plates which have their habit planes inclined at approximately 45° to the tensile axis will tend to be favoured. The angle will not be exactly 45° because the displacement vector of the IPS is not quite parallel to the habit plane whenever there is a volume change accompanying transformation.

This assumes that the applied stress interacts only with the growth process. Its interaction with nucleation events could lead to a different criterion for variant selection.[26] Indeed, efforts at predicting the martensitic transformation texture from the crystallographic texture of the parent austenite are apparently more successful if it is assumed that variant selection depends on the Bain strain rather than on the macroscopic shape deformation.[54] The IPS deformation is unlikely to have developed at the nucleation stage, where the particle might be too small to sustain the lattice-invariant deformation that is necessary to convert the effects of the lattice deformation into what is macroscopically an invariant-plane strain. On the other hand, since the Bain strain is necessary to accomplish the lattice change, the texture prediction work indicates that variant selection may depend on the interaction of the applied stress with the nucleation process.

LIMITS TO STRESS-ASSISTED TRANSFORMATION

At temperatures close to that at which the equilibrium transformation occurs, an applied stress can assist reaction when the chemical driving force is insufficient to achieve the change on its own. There must exist a point, however, when the applied stress simply cannot provide enough mechanical driving force to complement the chemical term to give a driving force large enough to induce transformation. After all, the magnitude of the stress that can be applied is limited by the yield point of the parent phase. There are, therefore, limits to what can be achieved by the application of stress as a stimulus to transformation (Fig. 6). In fact, work hardening may also limit the influence of stress, since transformation can actually be retarded by plastically deforming the parent phase. This is known as mechanical stabilisation.

The highest temperature at which martensite forms during the cooling of austenite is the M_S temperature. This can be raised by the application of a suitable stress.[52] The maximum temperature at which martensite grows under

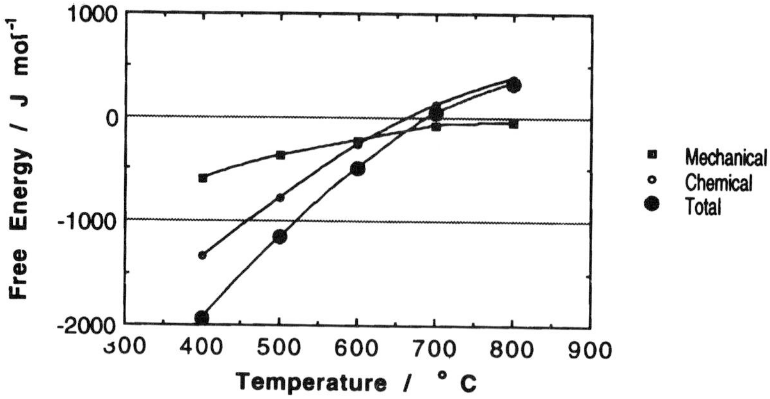

Figure 6 An illustration of the typical magnitudes of the chemical and mechanical driving forces for stress-affected transformation. The mechanical driving force is estimated for an applied stress which is equal to the yield stress of austenite. Since this yield stress decreases and becomes very small as the temperature rises, the contribution of the mechanical driving force becomes negligible at high temperatures. Therefore transformation becomes impossible as the temperature exceeds about 700°C.

the influence of stress is called the M_D temperature.[55] These phenomena are also revealed by studying the tensile strength of a sample containing austenite (reference 44, Fig. 7). At first the stress assists martensite formation so that the sample deforms readily. The formation of martensite at increasing temperatures above M_S becomes more difficult so that the strength rises towards that of austenite. Eventually, at M_σ the austenite begins to yield by slip. Finally, transformation becomes impossible above M_D.

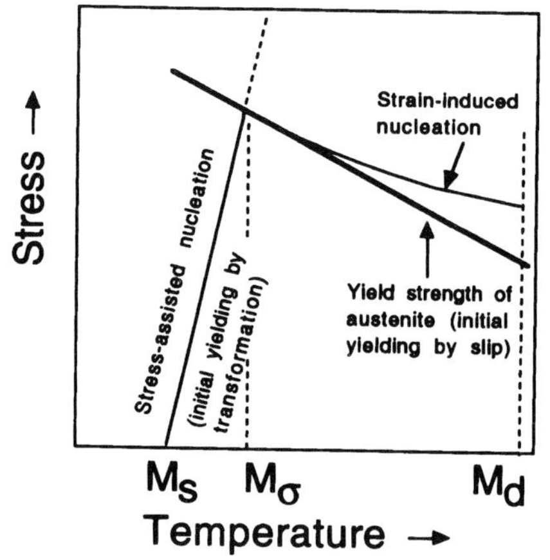

Figure 7 Stress-assisted and strain-induced martensitic transformation.[44]

There are no similar experiments for the other transformation products, but it is possible to piece together a lot of circumstantial evidence to show that the behaviour of bainite is probably similar to that of martensite.

Goodenow *et al.*[56] showed that the transformation stresses associated with the growth of lower bainite stimulated the growth of upper bainite at temperatures just above B_S, showing that stress can indeed raise the bainite-start temperature. It should in principle be possible to define a B_D temperature. Early work by Cottrell[57] gives some support to the B_D concept. He found that the maximum volume fraction of bainite obtainable by isothermal transformation of unstressed samples can be increased with the application of a tensile stress during transformation, but that the effect of the applied stress diminished as the bainite-start temperature was approached. Drozdov *et al.*[58] demonstrated that at a high enough transformation temperature above B_S (at which the austenite is still metastable), no amount of deformation causes the austenite to transform to bainite, a result which can be interpreted easily if it is assumed that the temperature concerned was higher than B_D.

These concepts are illustrated schematically in Fig. 8, which is for bainite but is based on similar ideas for martensitic transformations; the details have yet to be established systematically for bainite or other displacive

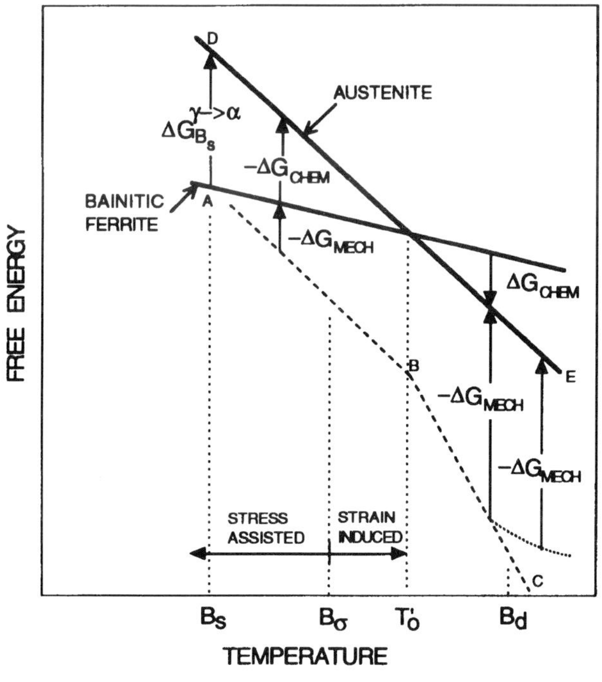

Figure 8 Illstration of the stress-modified transformation temperatures, B_S, B_σ and B_D

transformation products. The net driving force available for transformation, ΔG, is given by:

$$\Delta G = \Delta G_{\text{chem}} - \Delta G_{\text{mech}} \tag{3}$$

The model for the influence of stress assumes that the critical driving force needed to trigger bainitic transformation at zero stress (i.e. ΔG_{chem} at B_S) remains constant over the temperature range of interest. The application of a tensile or compressive stress assists the transformation by boosting the overall driving force ΔG with the term $-G_{\text{mech}}$, so that the B_S temperature rises continuously with the magnitude of the applied stress.

It is convenient to define a temperature B_σ, beyond which the austenite can no longer elastically support the applied stress prior to transformation. Clearly, beyond B_σ, the notion of using τ in the thermodynamic equations becomes highly dubious since the austenite inevitably undergoes plastic deformation prior to transformation. The deformation-induced defects might then influence the progress of transformation, perhaps by providing additional nucleation sites, although this kind of stimulation does not seem prominent in bainitic transformations where the predominant site for sheaf nucleation is almost always an austenite grain boundary. Following the terminology established for martensitic transformations, the region below B_σ is said to represent *stress-assisted* transformation, whereas *strain-induced* transformation describes the regime where the yield stress of the parent phase is exceeded.

As the stress is increased further, beyond the yield stress of the austenite, the bainite-start temperature continues to rise. When the T_0' temperature† is reached, the chemical driving force actually opposes transformation so that the mechanical component has to be larger than ΔG_{chem} at B_S. Of course, as the temperature rises, the yield stress of the austenite decreases, until it is no longer able to support a stress large enough to stimulate bainitic transformation; that temperature corresponds to B_D (Fig. 8).

GENERAL OBSERVATIONS

The effects of stress on martensitic transformations have been researched extensively, especially in the context of the shape memory effect.[55] Because the data on other higher temperature transformations are much more limited, but nevertheless important in welding, it is worth discussing how general observations on the next best researched transformation product (bainite) can

† At the T_0 temperature, γ and α of the same composition have identical free energies. The T_0' temperature is calculated in the same way, but allowing also for the stored energy of ferrite.[9]

give confidence in the interpretation of all the displacive transformations of interest.

EXTERNALLY APPLIED STRESSES

It is well known in the context of thermomechanical processing that deformation accelerates the rate of the bainite reaction.[57–63] There is also good evidence that the rate of reaction under stress increases with the rate of deformation.[58, 63] The degree of transformation to bainitic ferrite at any temperature is limited by the incomplete reaction phenomenon,[9] but the application of a tensile stress during transformation has been shown to stimulate further reaction beyond the limiting value.[57] The rate of reaction is also found to accelerate under the influence of a tensile stress as long as the stress exceeds a threshold value whose magnitude decreases with increasing temperature, although the significance of the threshold stress is not clear.[64] Umemoto *et al.*[65] have demonstrated an acceleration of transformation even when the applied tensile stress was rather small (Fig. 9).

INTERNALLY GENERATED STRESS

The stress-influencing transformation need not be applied externally; internal stresses generated due to other transformations also have an effect. Early studies indicated an acceleration in the rate at which upper bainite forms in specimens which are first transformed partially at a lower temperature.[66, 67] Martensite is the first phase to form on cooling a steel below the M_S

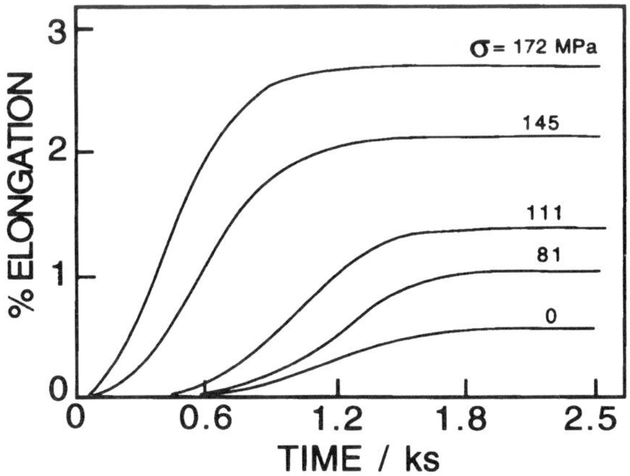

Figure 9. The overall kinetics of bainitic transformation as a function of an externally applied tensile stress. Assuming that the degree of transformation is related to the elongation, the data show an increase in the rate of reaction as a function of the magnitude of the applied stress.[65]

temperature, but after the initial burst of martensitic transformation and a suitable incubation period, the austenite undergoes accelerated decomposition to bainite.[68] This is because the austenite is deformed by the martensitic transformation. Supporting evidence is found in magnetometric studies, which have revealed that isothermal reaction below the M_S temperature leads first to the formation of the usual athermal martensite, followed by a small amount of isothermal martensite, and then some of the residual austenite begins to transform at an accelerated rate to bainite.[69] Similar results have been obtained by Radcliffe and Rollason[70] and it has also been demonstrated that the upper bainite reaction is accelerated by initial partial (displacive) transformation to lower bainite (Fig. 10). Their experiments also clearly illustrate the 'flat tops' of the bainite C-curves as highlighted first by Zener.[71] Another revealing observation is that both the nucleation and growth rates of bainite are accelerated by the proximity of a free surface.[72, 73] This observation emphasizes the role of the shape deformation of bainite, since any shape change can be accommodated better at a free surface, where the degree of physical constraint is smaller. In the absence of such mitigating circumstances, the matrix sometimes cannot tolerate the large displacements associated with the formation of bainite. It then relaxes by plastic deformation, which is driven by the shape change due to transformation.[74]

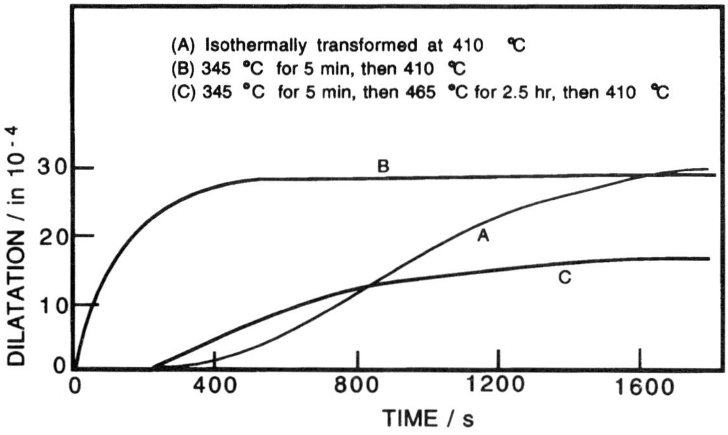

Figure 10. The influence of internal stresses on the rate of transformation at 410°C, in a Fe–0.31C–0.3Si–0.76Mn–3.07Ni–1.22Cr–0.49Mo wt.% alloy. Curve A represents isothermal transformation to upper bainite; curve B is for a sample which was first partially transformed to lower bainite and then to upper bainite, showing an acceleration of reaction rate at 410°C due to the internal stresses generated by the presence of lower bainite; curve C shows how annealing above the B_S temperature removes the stresses, and their accelerating influence on transformation kinetics.[56]

PLASTIC DEFORMATION AND MECHANICAL STABILIASTION

Deformation does not always accelerate transformation; it is well established for martensitic transformations that severe deformation causes the *mechanical stabiliastion* of the austenite. Mechanical stabilisation is due to the work hardening of the austenite as its defect density increases. The defects interfere with the motion of glissile transformation interfaces.

There are no systematic studies of mechanical stabiliastion for bainitic or other elevated temperature reactions. It has been reported that during the continuous cooling transformation of a low-alloy steel into a mixed microstructure, the decomposition of austenite to bainite was retarded when the austenite was deformed prior to transformation.[75] This could be interpreted as mechanical stabilisation, but there is unfortunately an alternative explanation for these particular experiments, an explanation which is consistent with the reported data. The deformation accelerates the formation of allotriomorphic ferrite, thereby causing an exaggerated enrichment of the residual austenite with carbon, and therefore a retardation of both the bainite and martensite reactions. Better evidence comes from experiments using hot-rolled samples, which have demonstrated that the austenite which is deformed to a larger extent tends to transform to bainite at a lower temperature during continuous cooling (Fig. 11). This effect can only be interpreted in terms of mechanical stabilisation because the alloy used had sufficient hardenability to avoid the growth of allotriomorphic ferrite or other interfering transformation products during cooling.

Perhaps the most convincing evidence for the mechanical stabilisation of bainite comes from studies of ausformed bainite. Samples of deformed austenite gave smaller maximum volume fractions of bainitic ferrite when compared with those transformed without any prior deformation. This is in spite of the fact that the deformed austenite transformed initially at a higher rate (Fig. 11). This effect did not appear during transformation at higher temperatures, presumably because the total volume fraction of bainite that can form is then reduced. Stabilisation therefore only manifests itself when the 'easy' regions of austenite are exhausted, i.e. those regions left unchanged by the inhomogeneous deformation. The non-uniformity of stabiliastion is reflected in the microstructure that develops. The sheaves of bainite that do form are found to be aligned along specific directions within individual austenite grains, in contrast with the usual bainitic transformation where sheaves grow in many different orientations (Fig. 12).

As is often the case with martensite plates in ausformed alloys, sheaves of bainite are sometimes found to be curved on a macroscopic scale. This is because of the deformation induced lattice curvature present in the parent austenite grains prior to transformation (Fig. 13).

Finally, it is worth mentioning two important experiments by Bhattacharyya and Kehl.[77] They first demonstrated that a tensile stress has little effect on

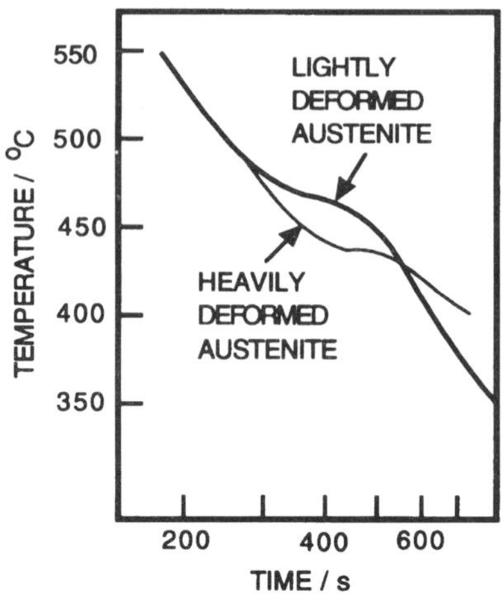

Figure 11. Plots of temperature versus time, for samples undergoing bainitic transformation during cooling. The deviations from linearity indicate the onset of transformation. The reaction is retarded in the austenite deformed to a greater degree before transformation, indicative of mechanical stabilisation. Data from Davenport.[76]

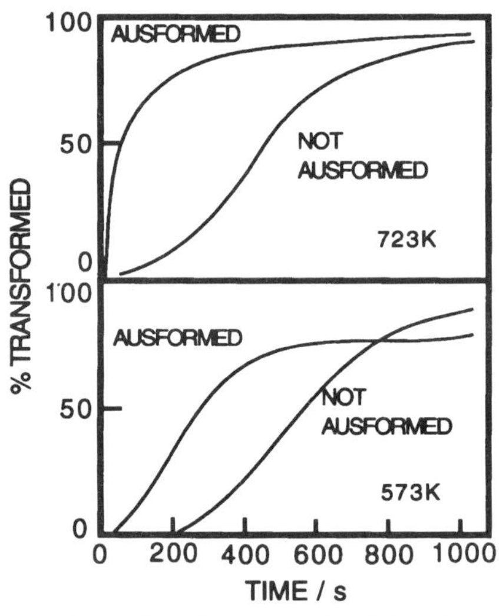

Figure 12. The effect of ausforming on the kinetics of the bainite reaction in a Fe–0.59C–2.01Si–1.02Mn wt.% alloy. Data from Tsuzaki *et al.*[59]

transformation if it is applied after more than about 30% of bainite has formed. The effect was attributed to the fact that the overall strength of the microstructure rises during transformation relative to the applied stress. An alternative explanation could be that it becomes difficult to stimulate variants favoured by the applied stress once many nuclei have been activated. Any favoured variants that do form will have a limited scope for development as the amount of austenite decreases.

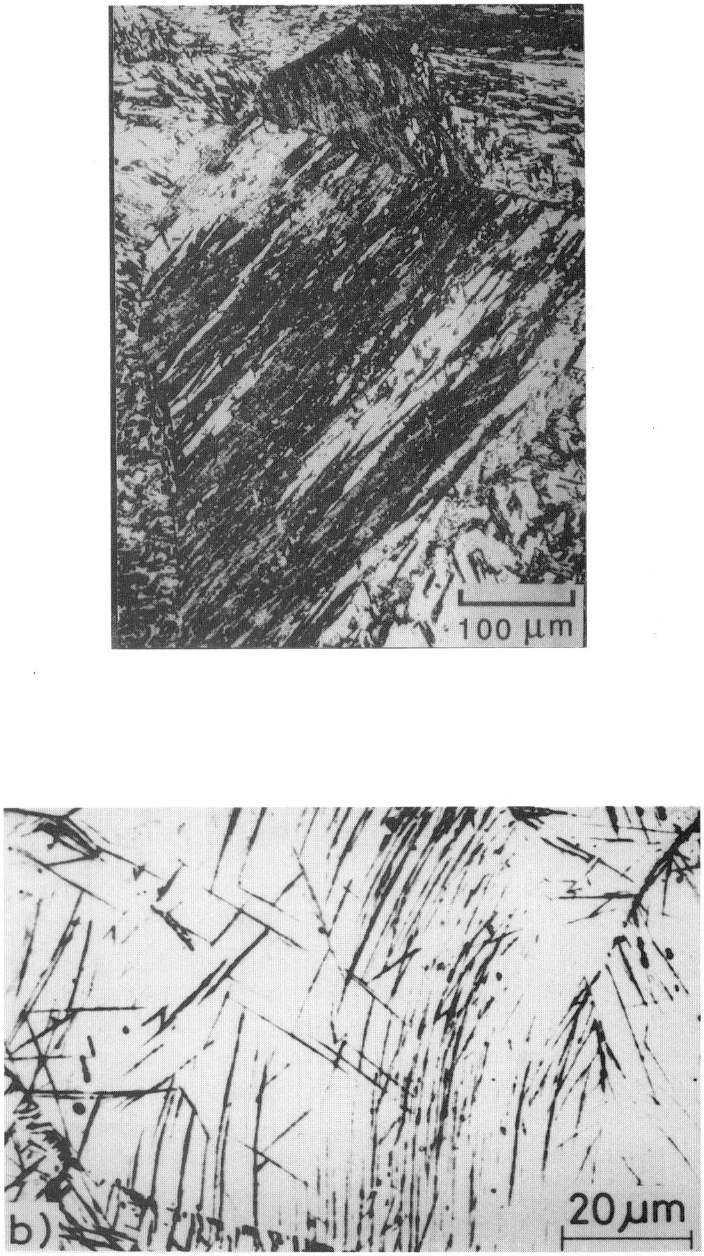

(a)

(b)

Figure 13. Optical micrographs illustrating the microstructure of an ausformed bainitic steel. (a) Alignment of sheaves of bainite in individual austenite grains. (b) Curved bainite sheaves reflecting the deformation-induced misorientations within the parent austenite grains.[59]

In the second experiment, Bhattacharya and Kehl showed that the removal of stress after some transformation has occurred does not decrease the rate of reaction. This might suggest that the primary influence of stress is in these experiments to stimulate nucleation rather than growth.

There is no doubt that a constrained weld undergoes plastic deformation in its austenitic state, as it cools. The strains involved are of the order of a few percent. Hence, the mechanical stabilisation effects are extremely unlikely. It is not surprising that macroscopically curved plates as discussed above (Fig. 13) are not found in welded samples. The intriguing question is: does the small amount of plastic deformation to be found in the austenite influence the kinetics of transformations as the weld cools? The probable answer is negative, since for martensitic transformations a small amount of plastic deformation enhances the nucleation rate. However, for the elevated temperature transformations it is unlikely that such dislocations are on their own terribly effective as heterogeneous nucleation sites. For example, allotriomorphic ferrite always nucleates at the austenite grain boundaries, as does bainite and primary Widmanstätten ferrite. Secondary Widmanstätten ferrite grows from allotriomorphic ferrite/austenite surfaces. Acicular ferrite nucleates at inclusions. It is much more likely that in constrained welds the major effects of the stress are via the mechanical driving force and through any changes in the degree of organisation in the microstructure.

THE EFFECT ON MICROSTRUCTURE

An applied stress will tend to favour the development of crystallographic variants which comply with that stress. This is analogous to the selective operation of one or more of the available slip systems in a crystal under stress; it is the systems with the largest Schmid factors which are favoured. Assuming that variant selection is similarly controlled by the interaction of the applied stress with the shape deformation, the stress should cause an alignment of the plates at roughly 45° to the tensile axis. This alignment has been observed in many experiments involving transformations in steels including weld deposits.[73]

MARTENSITE

This is illustrated in Fig. 13 where a fine grained polycrystalline sample of austenite was stressed at a temperature above its martensite-start temperature.[51] The amount of martensite obtained varies directly with the magnitude of the applied stress (i.e. the mechanical driving force). Not only did the stress induce the formation of martensite, but only those particular variants of martensite which complied with the applied stress grew in profusion. Thus, most of the plates attempted to grow on planes close to those of the maximum shear stress (approximately 45° to the tensile axis).

In the absence of an externally applied stress, each austenite grain could have transformed into one or more of the 24 possible crystallographic variants, leading to a more chaotic microstructure. The stress caused the alignment of plates, and a more ordered microstructure.

BAINITE

The observations are more difficult for bainite, partly because of the rapid rate of reaction under the influence of stress. The experiments have to be conducted at high temperatures. Further transformation may occur as the sample cools to ambient temperature, confusing the interpretation of the microstructure. Nonetheless, good evidence for microstructural alignment has been reported for bainite platelets especially at relatively large stresses.[65,77] All of these observations are based on polycrystalline samples, but that does not substantially alter the conclusions. There are so many variants available per austenite grain, that there is a high probability of a plate orientation lying close to the optimum with respect to the stress.

There are more subtle effects of stress on microstructure, even in the absence of any obvious plate alignment, at stress levels as small as 45 MPa. Variant selection leads to the development of a less chaotic microstructure.[60,62,45] Without stress, each grain of austenite transforms into many different orientations of bainite. Significantly fewer variants develop per austenite grain under the influence of stress, so that the selected orientations can grow unhindered and form thick packets of bainite plates. The sheaves are therefore found to be longer, and their number density per grain smaller when variant selection operates (Fig. 14).

A further effect on microstructure is when the austenite has been plastically deformed prior to transformation. Heterogeneous nucleation then occurs not only at the original austenite grain boundaries, but apparently also intragranularly on slip bands or other deformation heterogeneities.[62]

ACICULAR FERRITE

There is a lot of activity in the steel industry on the subject of 'acicular ferrite', for both welded and wrought alloys. Acicular ferrite is essentially intragranularly nucleated bainite.[78] The heterogeneous nucleation sites are known to be non-metallic inclusions, either added deliberately, or present as impurities. The microstructures of conventional bainite and acicular ferrite differ significantly: clusters of bainite plates (called sheaves) grow as a series of

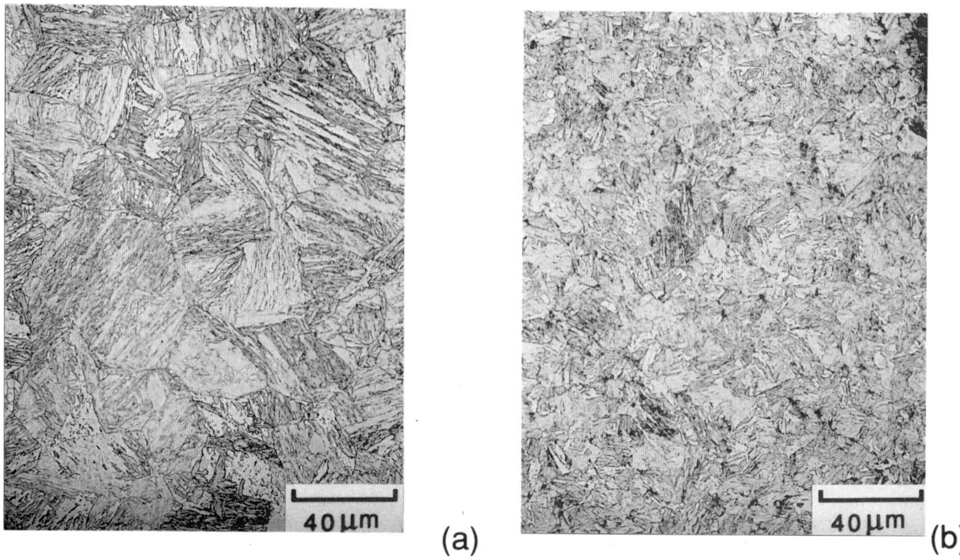

Figure 14. Light micrographs of bainitic microstructures generated in a Fe–0.12C–0.27Si–0.84Mn–0.14Ni–1.48Mo–2.86Cr wt.% alloy, by isothermal transformation at 400°C under the influence of stress. (a) Zero stress; (b) 95 MPa.[45]

parallel plates in identical crystallographic orientation emanating from austenite grain *surfaces.* By contrast, acicular ferrite plates nucleate intragranularly at *point* sites so that there is no room for parallel formations of plates to develop.

It is believed that the toughness is enhanced when the plates do not form packets, but lie on many differently oriented planes, like acicular ferrite. The feeling is that the latter microstructure causes the cleavage cracks to deflect frequently since a new crystallographic orientation is presented as each plate is encountered.

The dismantling of packets of parallel ferrite plates in bainitic or martensitic microstructures, into more chaotically orientated acicular ferrite plates, improves toughness. The increased chaos can only be achieved by introducing inclusions as intragranular nucleation sites. The greatest success in inducing the formation of acicular ferrite in this way has been achieved in weld deposits, which under normal circumstances contain relatively large concentrations of trace elements such as oxygen (about 300 ppm), aluminium, titanium, etc. introduced during fusion. The gain in toughness achieved in this manner seems to overcome any loss due to the presence of inclusions, in welds and low-strength steels.

It is important in the present context to note that the growth of acicular ferrite causes an invariant-plane strain shape deformation which has a large shear component.[79] And its response to an applied stress is as expected (Fig. 15). The number of variants which form in the presence of a uniaxial stress is reduced considerably, and the transformation strains become correspondingly anisotropic.[80]

An interesting observation is that the number of crystallographic variants of acicular ferrite, as found in the as-deposited microstructure of steel welds, is far less than the 24 that are theoretically possible in any given austenite grain[80] (Fig. 16). It is speculated that this is because of the stresses that develop as the weld cools, so that the growth of acicular ferrite occurs under the influence of some complex stress system. More work is needed in this area of welding metallurgy.

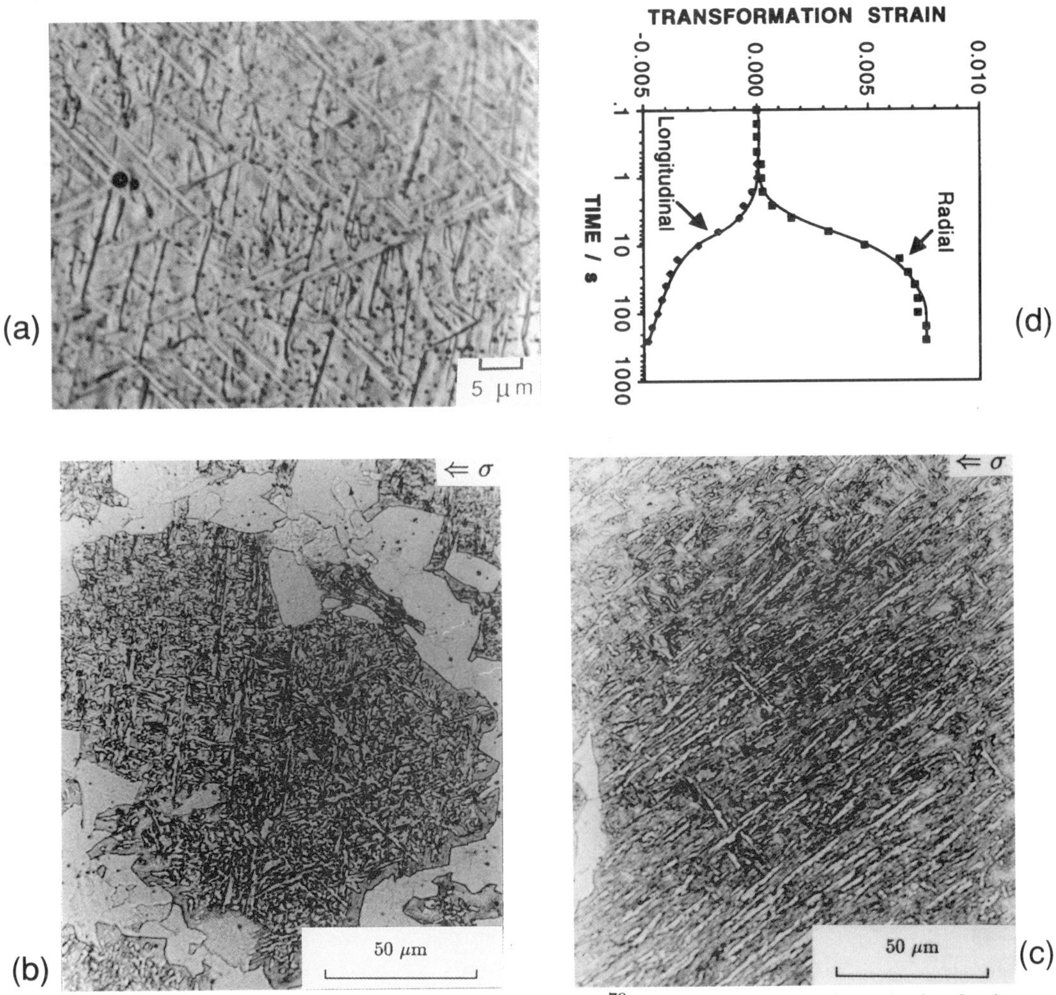

Figure 15. (a) Surface relief due to acicular ferrite.[79] An illustration of the acicular ferrite microstructure generated without stress (b), and under the influence of a small compressive stress (c). The micrographs represent longitudinal sections containing the stress axis which is identified as an arrow. (d) Anisotropic strains during the formation of acicular ferrite under the influence of a uniaxial compressive stress.[80]

CARBIDES

There is circumstantial evidence that the carbides which form during the tempering of martensite grow by a displacive mechanism. Such a mechanism must naturally involve the diffusion of carbon, but not of substitutional solutes

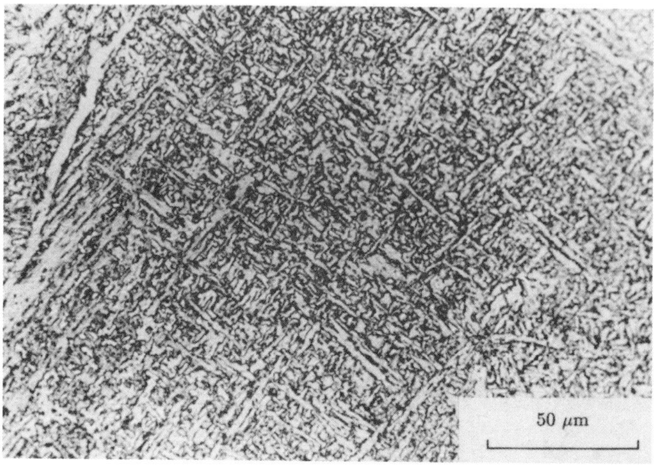

Figure 16. An example of the primary weld metal region of a Fe–Cr–C weld, showing non-random orientations of acicular ferrite plates. The total number of variants observed is far less than what is in theory achievable.[80]

or iron atoms. This is consistent with a large amount of data which show that the carbides associated with bainite and martensite do not partition substitutional solutes during transformation.[81–83] It is particularly interesting that the precipitation of cementite from martensite or lower bainite can occur under conditions where the diffusion rates of iron and substitutional atoms are incredibly small compared with the rate of precipitation (Fig. 17). The long-range diffusion of carbon atoms is of course necessary, but because of its interstitial character, substantial diffusion of carbon remains possible even at temperatures as low as –60°C. Thus, the formation of cementite in these circumstances must differ from the normal reconstructive decomposition reactions, which become sluggish at low temperatures. It has been believed for some time that the cementite lattice may be generated by the deformation of the ferrite lattice, combined with the necessary diffusion of carbon into the appropriate sites. The Fe/X ratio thus remains constant everywhere and, subject to that constraint, the carbon achieves equality of chemical potential; the cementite is then said to grow by *paraequilibrium* transformation.[84] The way in which the ferrite lattice could be deformed to produce the right arrangement of iron atoms needed to generate the cementite has been considered by Andrews[85] and Hume-Rothery *et al.*,[86] and the subject has been reviewed by Yakel.[87] Further high-resolution evidence supporting the idea that the carbide particles grow by displacive transformation (involving the diffusion

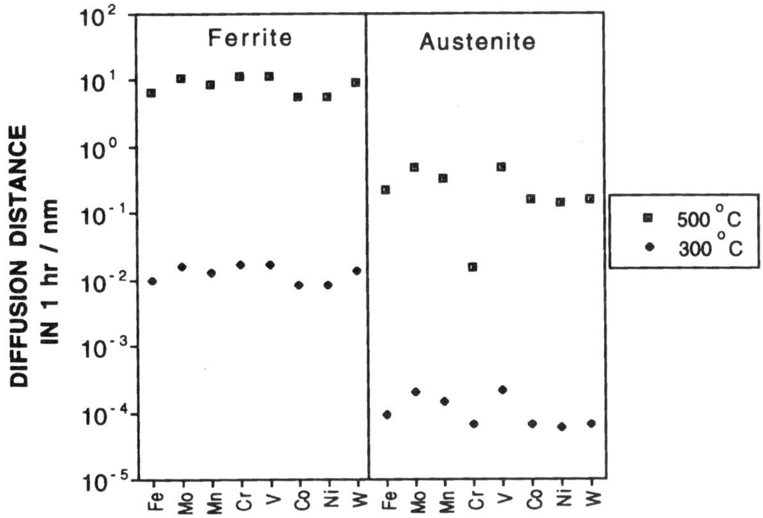

Figure 17. An illustration of the inconceivably small diffusion distances at the temperatures where some carbides precipitate.

of just carbon) has been published most recently by Sandvik,[88] Nakamura and Nagakura[89] and Taylor *et al.*[90, 91]

Figure 18(a) shows the microstructure obtained after tempering the quenched steel at 300°C for 30 min, without any externally imposed stress. Most of the martensite plates were found to contain more than one crystallographic variant of ε-carbide particles. These observations confirm the general feeling that a Widmanstätten array of carbides is obtained when carbon-supersaturated martensite is tempered. However, the number of carbide variants observed is surprisingly small, given that the orientation relationship between ε-carbide and the matrix is that of Jack.[92] Similarly, given that the ferrite/cementite orientation was always Bagaryatski, the number of possible variants should be larger than has been observed.

Recent work by Taylor *et al.*[91] has shown that the precipitation of transition carbides is influenced by a variety of precursor events. Thus, it is found that the number of carbide variants that form is less than that expected from an examination of just the crystallography, because precipitation is preceded by spinodal decomposition which causes carbon concentration waves which are exaggerated along the elastically soft directions of the matrix. The number of these soft directions is limited by the tetragonal symmetry of the martensite lattice. It has been suggested that the carbides then nucleate and grow by a displacive paraequilibrium mechanism along the carbon-enriched bands. They therefore observed only two carbide variants in any given martensite plate, even though the plane of precipitation $\{0\ 1\ 2\}_\alpha$ has a higher crystallographic multiplicity. A further factor which could limit the number of variants is of course the internal stresses arising due to the shape change of the martensite plates themselves.

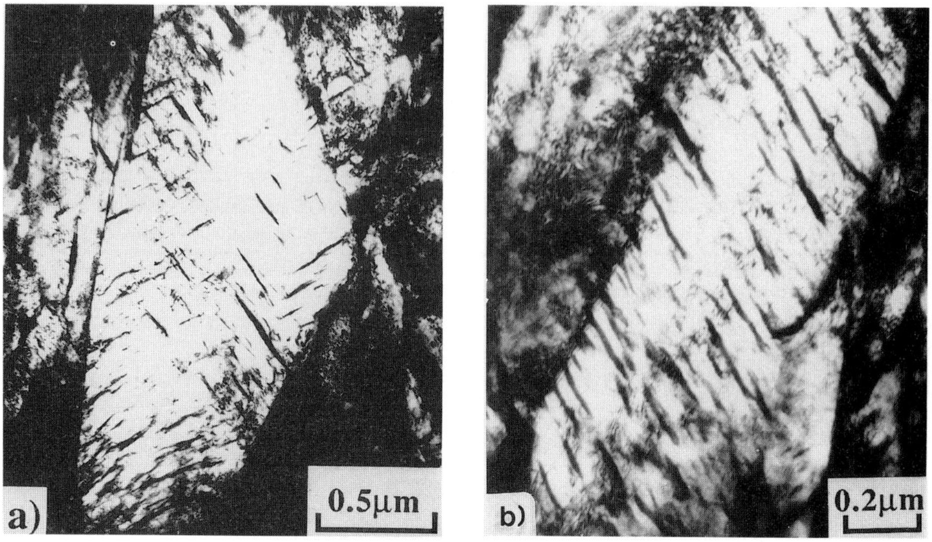

Figure 18. (a) Electron micrograph of a sample of martensite tempered at 300°C, showing the Widmanstätten pattern of carbide particles. (b) Electron micrograph of a sample of martensite tempered at 300°C, under the influence of a uniaxial compressive stress of 700 MPa, showing that the carbide particles precipitate in an organised array mostly consisting of a single crytallographic variant.[93]

Figure 18(b) shows the striking change in carbide precipitation behaviour, caused by tempering the martensite at 300°C under the influence of a compressive elastic stress of 700 MPa. The vast majority of martensite plates revealed only a single variant of ε-carbide.

It is evident that an externally applied stress can reduce the number of variants of carbide particles that precipitate in any given martensite plate. It follows that any internal stresses, due for example to the invariant-plane strain (IPS) shape change accompanying the growth of martensite, can have a similar effect. Furthermore, it is now generally accepted that the carbide particles formed during the tempering of martensite probably themselves grow by a displacive transformation mechanism which may involve the diffusion of carbon. This makes it even more probable that the precipitation of carbides should be influenced when it occurs in a stress field.

These results have significant implications on the interpretation of carbide precipitation in lower bainite. The lower bainite grows at relatively high temperatures where the driving force for carbide precipitation from supersaturated ferrite is likely to be smaller than that associated with martensite in the same steel. The chemical driving force will be further reduced since some of the excess carbon can escape into the residual austenite. It is not therefore surprising that there is a tendency to find just one crystallographic variant of carbide in platelets of lower bainitic ferrite.

To summarise, it appears that an externally applied stress can influence the precipitation of carbides in martensite. The number of carbide variants which grow in any given plate of martensite decreases when the virgin martensite is tempered under the influence of the stress. The effect becomes more prominent as the magnitude of the mechanical driving force increases relative to the chemical driving force for carbide precipitation. This explains why lower bainite is usually associated with a single variant of carbide within any individual plate of bainitic ferrite. It is speculated that the results may be of significance in the relief of residual stresses during stress relief heat treatments on welded assemblies.

HYDROSTATIC PRESSURE

There is general agreement that the application of hydrostatic pressure causes a retardation of the martensite[94] and bainite[95, 96] reactions in steels. The effect on the time–temperature–transformation diagram is illustrated in Fig. 19. The observed retardation is not in itself a feature unique to displacive reactions. All transformations which are accompanied by a reduction in density are expected to be retarded by hydrostatic pressure which opposes a volume expansion. The effect of hydrostatic pressure is two fold: it reduces the diffusion coefficients by decreasing the available free volume (although the details remain to be established), and it influences the free energy change for transformation. If ΔG_m is the molar Gibbs free energy change for a reaction, then since

$$\left(\frac{\delta G}{\delta P}\right)_T = V \tag{4}$$

it follows that

$$\Delta G_m\{P\} - \Delta G_m\{1\} = \int_1^P \Delta V_m \mathrm{d}P \tag{5}$$

where ΔV_m is the change in molar volume on transformation, V is the volume and P is the pressure. The way in which the free energy change for transformation is influenced by the pressure determines how the transformation temperature changes as a function of pressure. An alternative way of expressing this relationship is the Clausius–Clapeyron equation, whence the change in transformation temperature is given by

$$\mathrm{d}T^{\gamma\alpha} = T^{\gamma\alpha}\,\Delta V_m / \Delta H \tag{6}$$

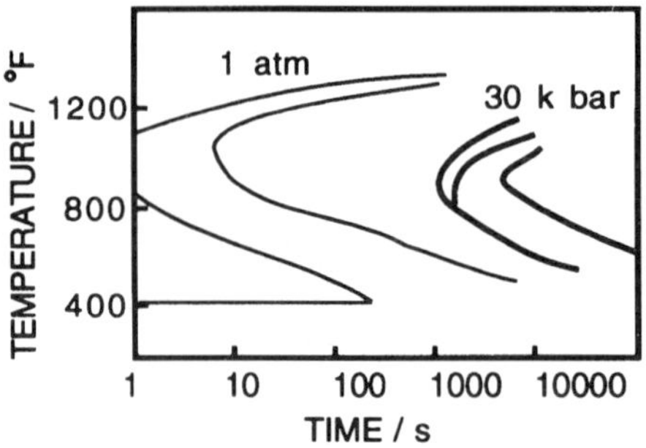

Figure 19. Isothermal transformation diagrams of Fe–0.82C wt.% at 1 atmosphere and at 30 kbar pressure.[96]

where $\Delta H^{\gamma\alpha}$ is the enthalpy change on transformation at the transition temperature $T^{\gamma\alpha}$. The equation is approximate in that its derivation depends on the assumption that the enthalpy change does not vary with temperature. With typical values of all the parameters, the variation in transition temperature with pressure should be approximately –0.01 K MPa^{-1}.[27]

MECHANICAL STABILITY OF RETAINED AUSTENITE

'Microphases' in welds and their heat-affected zones often consist of mixtures of phases such as martensite, carbides and retained austenite. The latter usually has a carbon concentration which is much larger than that of the alloy as a whole, since the carbon is partitioned into the residual austenite during elevated temperature transformations. If this austenite undergoes martensitic transformation induced by stress during welding or during service, then the resulting high carbon martensite is expected to be very brittle indeed. There is in fact a great deal of work in progress throughout the world, dealing with these local brittle zones.

The problem is likely to be most acute in steels where the precipitation of carbides is relatively sluggish, the residual austenite becomes enriched in carbon and a large proportion is retained on cooling to ambient temperature. The austenite, if it decomposes under the influence of stress, can be detrimental to the steel concerned since the resulting high-carbon, untempered martensite is expected to be very brittle. The larger blocky austenite is most likely to decompose to martensite even at relatively small stresses, Fig. 20.[97–102] The mechanical stability of any retained austenite is therefore important in obtaining good toughness in bainitic steels.

Miihkinen and Edmonds[103–105] have shown that the mechanical stability of the retained austenite decreases as the transformation temperature is increased. This is because the austenite can dissolve less carbon at higher temperatures. Similarly, substitutional solutes which reduce the solubility of carbon in austenite (such as manganese) is likely to make the austenite less stable to stress. The substitutional solute is never added in sufficient quantity

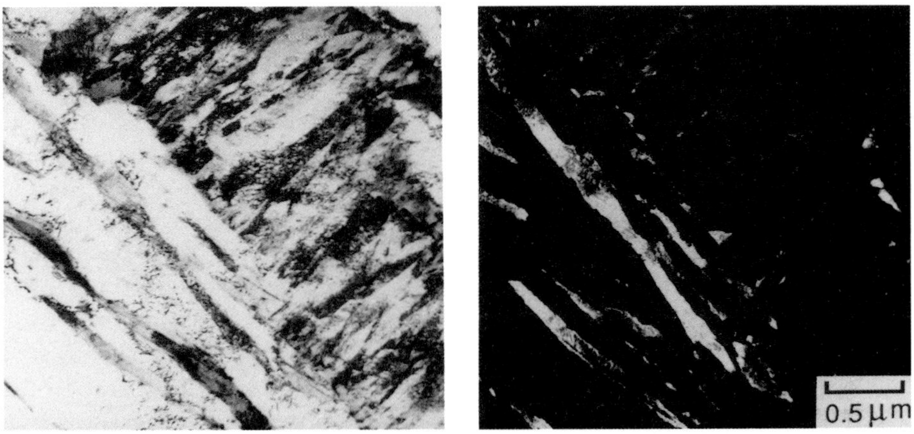

Figure 20. Electron micrographs illustrating the effect of applied stress (850 MPa) on a sample which initially had a microstructure of bainitic ferrite and retained austenite.[99] The larger regions of austenite transform to martensite but the films are preserved. (a) Bright field image showing a large region of stress-induced martensite; (b) corresponding austenite dark field image. Note that the sample was tempered prior to stressing in order to distinguish the martensite which forms during cooling from the bainite transformation temperature, from that which is induced by stress.

to compensate for the loss of austenite stability as its solubility for carbon decreases.

The way to improve local brittle zones consisting of mixtures of austenite and martensite, therefore, is to design alloys in which the solubility of carbon in the austenite is largest. This can easily be achieved using widely available phase diagram computation methods.

TRANSFORMATION UNDER CONSTRAINT: RESIDUAL STRESS

Residual stresses are mostly introduced unintentionally during fabrication. They are of particular importance in welded structures which as a consequence become susceptible to hydrogen embrittlement and other detrimental phenomena. Jones and Alberry[106, 107] conducted an elegant series of experiments to illustrate the role of transformations in steels, on the development of residual stress.

UNIV DUNELM

Using bainitic, martensitic and stable austenitic steels, they demonstrated that transformation plasticity during the cooling of a uniaxially constrained sample from the austenite phase field acts to relieve the build up of thermal stress as the sample cools. By contrast, the non-transforming austenitic steel exhibited a continuous increase in residual stress with decreasing temperature, as might be expected from the thermal contraction of a constrained sample.

When the steels transformed to bainite or martensite, the transformation strain compensated for any thermal contraction strains that arose during cooling. Significant residual stresses were therefore found to develop only after transformation was completed, and the specimens approached ambient temperature (Fig. 21).

The experiments contain other revealing features. The thermal expansion coefficient of austenite is much larger than that of ferrite, and yet the slope of the line prior to transformation is smaller when compared with that after transformation is completed (Fig. 21). This is because the austenite deforms plastically; its yield strength at high temperatures is reduced so much that the sample is unable to accommodate contraction strains elastically. On the other hand, when ferrite forms, its strength at low temperatures is higher, so that the slope of the stress/temperature curve (after transformation is complete) should be steeper and consistent with the magnitude of thermal contraction strains. All this has yet to be demonstrated quantitatively.

In the region of the stress/temperature curve where transformation happens, the interpretation of experimental data of the kind illustrated in Fig. 21 is very difficult. The predominant view amongst those involved in residual stress calculations, that the volume change during transformation gives the major contribution to transformation plasticity, is almost certainly incorrect for displacive transformations such as bainite or martensite. The shape change due to transformation has a shear component which is much larger than the dilatational term. Admittedly, this shear component should on average cancel out in a fine grained polycrystalline sample containing plates in many orientations. However, the very nature of the stress effect is to favour the formation of selected variants in which case, the shear component rapidly begins to dominate the transformation plasticity.

Finally, it is interesting to note that if transformation is completed at a higher temperature, then the ultimate level of stress at ambient temperature is larger, since the fully ferritic sample contracts over a larger temperature range. To reduce the residual stress level at ambient temperature requires the design of alloys with low transformation temperatures. The sort of high strength welding alloys used for making submarine hulls tend to have very low transformation temperatures (<250°C). This fact may be fortuitous, but such alloys should be less susceptible to cracking induced by the development of residual stresses.

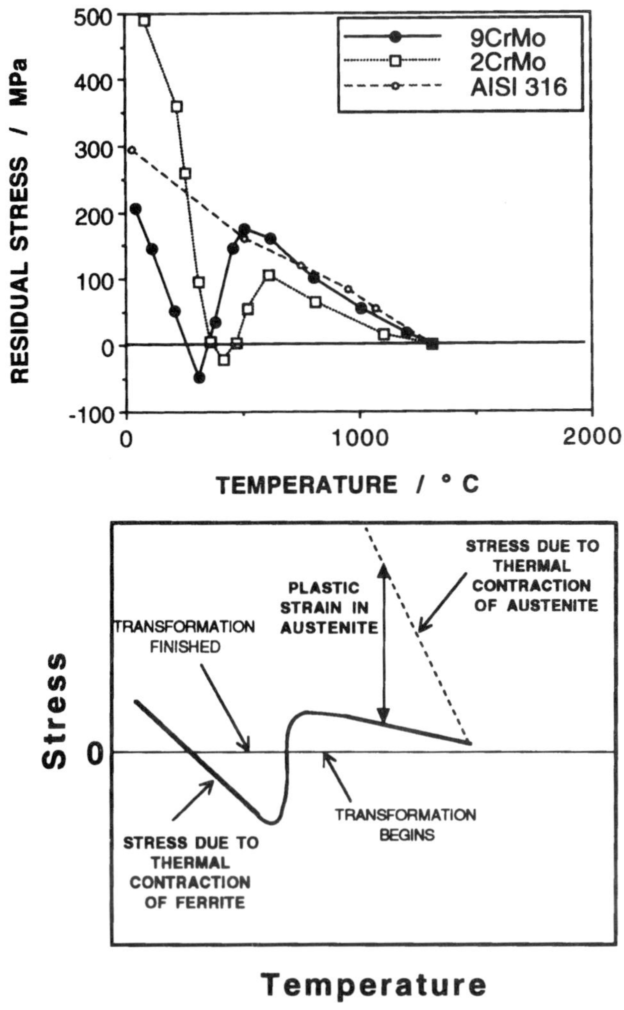

Figure 21. (a) Plot of residual stress versus temperature for a martensitic (9CrMo), bainitic (2CrMo) and austenitic steel (AISI 316). (After Alberry and Jones,[106,107]) (b) Interpretation of the Alberry and Jones experiments. The thermal expansion coefficient of austenite is much larger than that of ferrite.

ANISOTROPIC STRAIN AND TRANSFORMATION PLASTICITY

During their attempts to study the isothermal transformation of austenite using dilatometry, Davenport and Bain[35] had already noticed that 'the volume change (due to transformation) is not necessarily uniformly reflected in linear change in all dimensions'. They even found that the thickness of flat disc specimens actually decreased as the volume increased! These results were stated without interpretation; recent work has demonstrated that in polycrystalline samples which are crystallographically textured anisotropic transformation plasticity can be detected even in the absence of an applied stress.[45] When an unstressed polycrystalline sample of austenite is transformed, the shear components of the individual shape deformations of the large number of variants which form along any dimension should tend to cancel out on a macroscopic scale. Similarly, in the absence of stress, the dilatational component of the IPS shape deformation should tend to average out so that the volume expansion appears to be isotropic. Transformation plasticity (the major component of which comes from the large shear strain of the IPS) should therefore be minimised in a random polycrystalline sample, and due to volume change only. If on the other hand, the sample is not random, i.e. it is crystallographically textured, then the possibility of the individual shape deformations cancelling out over large distances is correspondingly reduced. Transformation will then lead to anisotropic strains even in the absence of any applied stress (Fig. 22).

To summarise, the observed plastic strain will be anisotropic when transformation occurs under the influence of stress. For reconstructive transformations, the maximum amount of plasticity cannot exceed the volume strain, as long as the stress does not exceed the yield strength of the weaker phase. For displacive transformations, the plasticity can be much larger and more anisotropic when the microstructure becomes non-random. For both cases, useful information is to be gained by measuring the transformation strains along more than one direction. Otherwise, for displacive transformations, it is not possible to separate the contributions from volume strain and shear strain. A deconvolution of the observed strain into these components is needed because it is only the volume strain which can be easily related to the extent of transformation in a dilatometric experiment. It is necessary to monitor the extent of transformation because it does not in many cases reach completion.

DECONVOLUTION OF STRAINS

It has been emphasised that when studying the evolution of microstructure under the influence of stress, it is useful to monitor the strains along more than one direction. The following analysis is for displacive transformations and assumes that the volume strain will be isotropic. Transformation plasticity is therefore attributed solely to the shear strains associated with displacive reactions. This does not follow convention, in the sense that the Greenwood and Johnson (and related methods) deal only with volume change effects. However, the purpose here is to focus on the effects of the shear component of strain due to displacive transformations.

For the case of radial symmetry, the radial and longitudinal strains (ε_R and ε_L respectively) can be related by:[45]

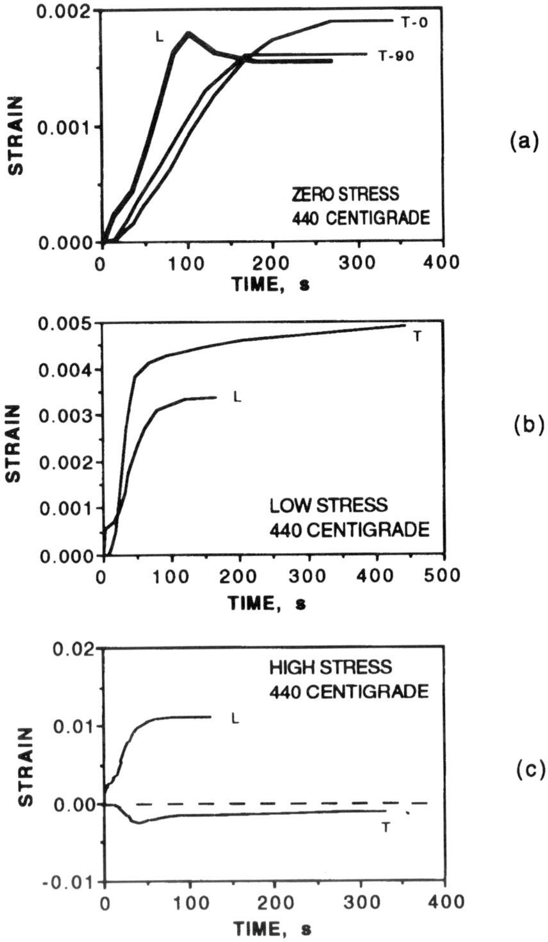

Figure 22. Dilatometric curves showing the dimensional changes during transformation to bainite in a cylindrical sample. T–0 and T–90 refer to the strains monitored along orothogonal transverse directions, and L to the strain along the longitudinal direction. (a) Transformation in the absence of an applied stress; (b) transformation under the influence of a tensile stress of about 45 MPa; (c) 90MPa.[45]

$$\frac{\Delta V}{V} = (1 + \varepsilon_L)(1 + \varepsilon_R)^2 - 1 \qquad (7)$$

$$\approx \varepsilon_L + 2\varepsilon_R$$

the approximation being valid only for small strains. Assuming that the volume change is isotropic, it follows that transformation plasticity in the longitudinal direction is given by:

$$\varepsilon_P = \varepsilon_L - \frac{\Delta V}{3V} \qquad (8)$$

Note that the sign of ε_P is negative for uniaxial compression, the state of stress relevant for the present experiments.

In fact, the volume change due to the growth of a bainite plate is not isotropic; the dilatational component of the shape deformation is directed normal to the habit plane. However, in a sample containing many randomly orientated plates of bainite, the overall macroscopic effect of stresses lower than the yield stress of austenite, may still be one in which the volume change *appears* isotropic. Furthermore, when the applied stress is larger than the yield stress, the magnitude of the transformation plasticity ε_P becomes much larger than the volume strain. Thus, the effect of any anisotropy in the volume strain becomes negligible by comparison.

MODELLING ANISOTROPIC TRANSFORMATION STRAINS

Consider a distribution of variants along all radial directions in a circle with the compression axis as its diameter.[109] The circle is divided into eighteen equal segments ($i = 1 \rightarrow 18$), each segment representing a particular orientation of bainite habit plane. The choice of eighteen segments is convenient and arbitrary. The compression axis of the sample is taken to be the z direction, the x and y directions being radially orientated; the unit vectors x, y, z define the orthonormal basis X of the sample, giving a corresponding reciprocal basis X^*. The shear and dilatational components of the invariant–plane strain accompanying the growth of bainite are approximately $s = 0.22$ and $\delta = 0.03$.[9] Thus, the 3×3 deformation matrix describing the shape deformation is given by:[48, 49, 108–110]

$$P = \begin{pmatrix} 1 + fse_1p_1 + f\delta p_1p_1 & fse_1p_2 + f\delta p_1p_2 & fse_1p_3 + f\delta p_1p_3 \\ fse_2p_1 + f\delta p_2p_1 & 1 + fse_2p_2 + f\delta p_2p_2 & fse_2p_3 + f\delta p_2p_3 \\ fse_3p_1 + f\delta p_3p_1 & fse_3p_2 + f\delta p_3p_2 & 1 + fe_3p_3 + f\delta p_3p_3 \end{pmatrix}$$

where p is the unit normal to the habit plane, and e is the unit direction along which the shear occurs. This can be written more succinctly as:

$$P = I + fs\begin{bmatrix} e_1 \\ e_2 \\ e_3 \end{bmatrix}(p_1 \;\; p_2 \;\; p_3) + f\delta\begin{bmatrix} p_1 \\ p_2 \\ p_3 \end{bmatrix}(p_1 \; p_2 \; p_3)$$

where I is a 3×3 identity matrix. A further reduction of notation is achieved using the Mackenzie and Bowles[48] notation:

$$(X\,P_i\,X) = I + f_i s[X;\, e_i](p_i\,;\, X^*) + f_i\delta[X;\, p_i](p_i\,;\, X^*) \qquad (9)$$

where the subscript i identifies a particular segment of interest and X and X^* represent the real and reciprocal bases of the coordinate system in which the deformation is described. The notation due to Mackenzie and Bowles[48] is discussed in detail in references 108, 109.

The components of the shear direction and the dilatation direction are given by:

$$[X;e_i] = f_i[-\cos(\theta_i)\ 0\ \sin(\theta_i)]$$

$$[X;p_i] = f_i[\sin(\theta_i)\ 0\ \cos(\theta_i)]$$

where θ_i represents the orientation of the habit plane of variant i. f_i is the volume fraction of bainite located in segment i.

A unit vector along the z direction changes to a new vector z' given by:

$$[X;z'] = \prod_{i=1}^{18} (X\,P_i\,X)\ [0\ 0\ 1] \tag{10}$$

whereas a unit vector along x changes to

$$[X;\,x'] = \prod_{i=1}^{18} (X\,P_i\,X)\ [1\ 0\ 0] \tag{11}$$

where $(x' - 1)$ and $(z' - 1)$ give the strains along the x and z directions. These are assumed to be equal to the strains ε_R and ε_L respectively.

It is expected that those segments which comply best with the applied stress transform most rapidly, whereas the others do so at a smaller rate, or not at all. This can to some extent be incorporated into the model by calculating the energy change U_i as the stress interacts with the shape deformation of a particular variant (i). Patel and Cohen's method[52] gives:

$$U_i = \frac{\sigma}{2}[s \sin 2\theta_i \cos \phi_i + \delta(1 + \cos 2\theta_i)] \tag{12}$$

where ϕ is the angle between the shear direction and the direction of the shear component of the applied stress as resolved on to the habit plane. To facilitate a 2-dimensional analysis, the value of ϕ_i is taken to be zero. A positive value of U_i adding to the chemical driving force ($\Delta G^{\alpha\gamma} = G^{\gamma} - G^{\alpha}$) for transformation; a negative value thus opposes transformation. Using these values of interaction energies, the model can be modified so that the segments transform in an order of decreasing U_i. There is, however, a further complication. The effect of stress should be largest when the interaction energy is large compared with the chemical driving force. To allow for this, the volume fraction f_i of each segment can be scaled according to the value of $U_i + \Delta G^{\alpha\gamma}$.

Note that for the model calculations, the transformation occurs with the most favoured variants growing first (Fig. 23). The model thus exaggerates the effect of stress, since in reality, for the sort of stress levels considered experimentally, no variant is likely to be entirely suppressed. In addition, the grains in a polycrystalline sample are 'randomly' orientated, so that perfect compliance with the applied stress is impossible. Nevertheless, the trends revealed by the model are expected to be correct.

The experimental data which need explaining, and their interpretation in terms of the model, are summarised in Fig. 24 and may be stated as follows:

1. Without any stress, in a random polycrystalline sample, the transformation strains are isotropic; this is easily understood since the shear components of randomly orientated plates tend to cancel out (Fig. 24a).
2. The application of a small stress at a high transformation temperature (i.e. small chemical driving force) causes the development of anisotropic

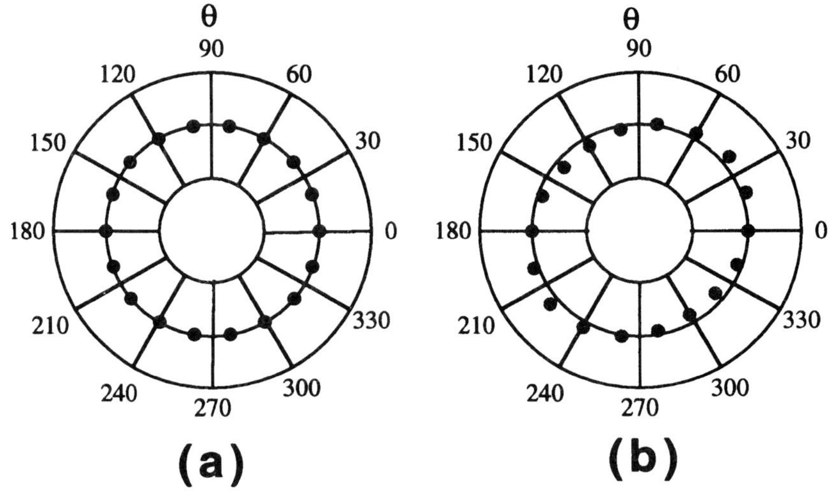

Figure 23. Illustration of the transformation behaviour inherent in the model.[110] The dots are intended to illustrate the area fraction of each segment, which in the calculations is scaled according to the value of $U_i + \Delta G^{\alpha\gamma}$. (a) Zero stress. All segments have equal area fraction, and the order in which they transform is irrelevant. (b) Applied stress of 40 MPa and $\Delta G^{\alpha\gamma}$ = 400 J mol^{-1}. The area fractions of the segments are no longer equal. The segments in which the distance of the dot from the origin is largest transform first.

strains, the transverse strain first being negative and then positive (Fig. 24b). The same effect is observed for a large stress and low temperature (i.e. a large driving force). The model explains this effect when it is assumed that the favoured variants form first, but that the stress is not large enough to suppress the eventual formation of other variants. The signs of ε_L and ε_R are always opposite for the favoured variants, but are identical for the rest of the variants. Therefore, the transverse strain is initially negative but then becomes positive as transformation progresses. The low stress/high temperature situation is equivalent to the high stress/low temperature case because in both of these circumstances variants which are not favoured cannot be suppressed. In the former case, the stress is too small for suppression whereas, in the latter case, the chemical driving force is too large to permit suppression.

3. When a large stress is applied at a high temperature, the favoured variants dominate. The strains are therefore always of opposite sign (Fig. 24c).

The model is therefore capable of qualitatively explaining all of the essential features of the formation of bainite under the influence of a small tensile stress. A uniaxial compressive stress (as used in the experiments described below) simply causes a reversal of the signs of the longitudinal and transverse stresses; there is also a minor effect from the unfavourable interaction between the compressive stress and the dilatational component of the IPS shape deformation.

The most interesting conclusion to emerge from the comparison of the model with experimental data is that transformation under the influence of a mild stress occurs sequentially. Variants which comply with the applied stress grow first, followed by those which do not. This also carries the implication that the interaction of the stress is with the growth process (i.e. the IPS shape deformation) rather than the strain field of the nucleus, which is likely to be different. it is worth noting that there are similar results for martensite, that most favoured variants grow first in the sequence of transformation under stress.[36, 45]

MAXIMUM STRAIN DUE TO SHEAR

An interesting observation illustrated in Fig. 25 is that the transformation plasticity increases disproportionately with the extent of reaction as the stress becomes larger. As the volume fraction of favoured variants increases with stress, the shear component of the IPS manifests itself more prominently in the measured transformation strains, since it is the randomness of the

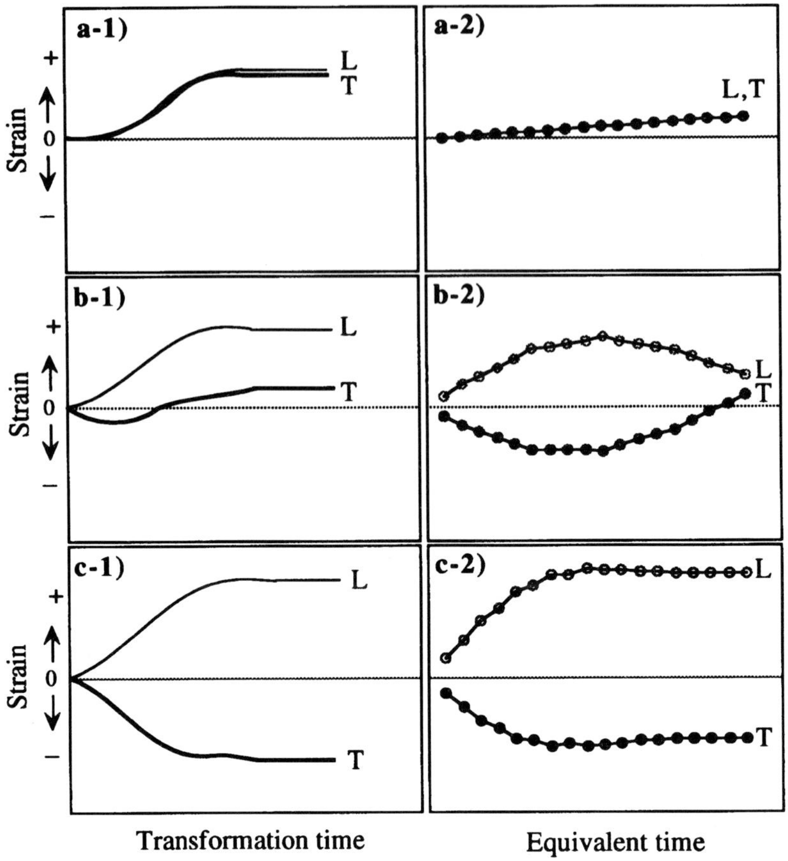

Figure 24. Schematic illustration of the reported[45] variations in longitudinal and radial strains during the isothermal formation of bainite under the influence of a tensile load, presented alongside predictions[110] from the crystallographic/thermodynamic model. The stresses are all intended to be below the austenite yield strength, and the data refer in this case to uniaxial tension. (a) Zero stress, any temperature; (b) small stress, low temperature; (c) small stress at high temperature or large stress at low temperature; (d) large stress, high temperature.

microstructure which leads to an overall cancellation of the shear strains. Thus, as increasing stress leads to more selective growth, transformation plasticity is expected to become more pronounced.

It is possible to estimate the maximum value of transformation plasticity that can theoretically be obtained if the entire specimen is induced to transform into the most favoured variant. Given that $s \approx 0.22$ and $\delta \approx 0.03$, the orientation θ of this variant is given by $\tan\{2\theta\} = s/\delta$.[37]. It follows that $\theta \approx 49°$ for uniaxial compression. For this orientation, the model described earlier yields $1 + \varepsilon_L = 0.920$ and $1 + \varepsilon_T = 1.125$, giving a transformation plasticity value of $\varepsilon_P = 2/3(\varepsilon_L$

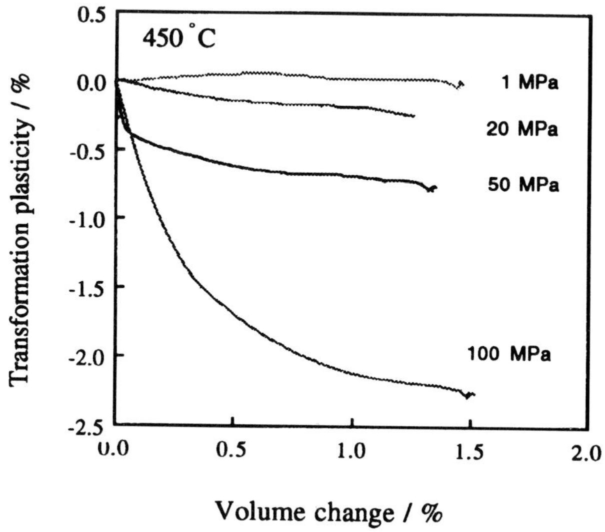

Figure 25. The relationship between the volume strain (which is a reflection of the amount of bainitic ferrite), the transformation plasticity and the applied compressive stress.

$-\ \varepsilon_T) \approx -0.14$. It follows that the transformation plasticity can in principle produce a strain as large as 14%, which compares with a value of less than 5.5% observed experimentally (Fig. 25). Thus, there exists a much greater potential for the microstructure to respond to an applied stress than has been possible to achieve in the present experiments. The difficulty is that with the relatively high temperatures associated with bainitic reaction, the applied stress is limited to a rather small value because of the low yield stress of austenite at high temperatures.

From the above discussion, and judging from previous studies, it appears that the microstructural changes induced by transformation under the influence of stress are often not expected to be very pronounced. The best measure of the microstructural response to stress is the anisotropy of the longitudinal and transverse transformation strains, rather than direct observations.

AUSTENITE YIELD STRENGTH

Relatively small stresses can at high temperatures be expected to cause some dislocation yielding of the austenite at the temperatures of interest. The measured proof stress of the austenite (0.2% plastic stain) in a low alloy steel[110] at 420 and 550°C is approximately 80 and 50 MPa respectively (Fig.

26). It follows that for many of the reported experiments, the austenite must undergo some plastic deformation before and during transformation. for comparison purposes, a compression curve for bainite at 420°C is also presented in Fig. 26.

SUMMARY

There is little doubt that transformations in steel are affected by both externally applied stresses, and those generated internally by prior transformation or heat-treatment. Alternatively, the occurrence of

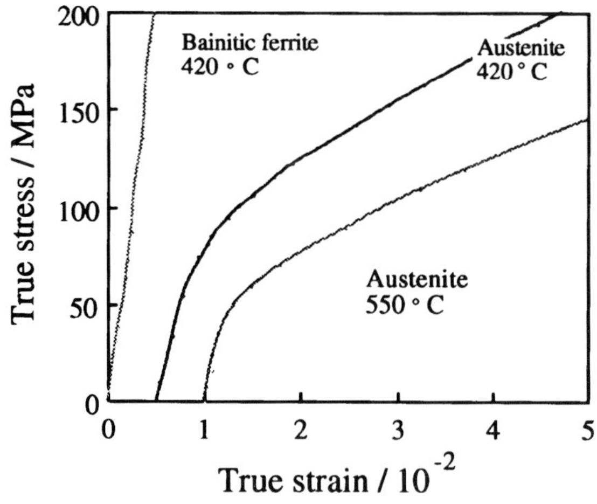

Figure 26. Experimentally determined stress–strain curves for austenite (prior to transformation) and for bainitic ferrite.[110]

transformations helps determine the state of stress in welded or heat-treated samples.

For reconstructive transformations, i.e. those in which there is mass transport at least over the scale of the transformation product, the interaction with stress depends largely on the difference in density between the parent and product phases. The maximum permanent strain due to transformation is then close to the volume strain given by the density change. The Greenwood and Johnson model of transformation plasticity is appropriate in these circumstances. The effects of stress on microstructural development during reconstructive transformation are rather small when the stresses are elastic. The stress may thermodynamically assist or oppose transformation, but because the austenite is weak at high temperatures where these transformations occur, the contribution to the driving force for transformation is always likely to be minor.

This interaction with stress for displacive transformations is rather different, and depends very much on the associated invariant-plane strain shape deformation with its large shear component. The shear component is much larger than the strain due to volume change. The transformation plasticity can therefore be much larger, if the shear components do not cancel out because of randomness in the microstructure. The fact that they do not do so in practice is manifested in large changes in the microstructure that develops under the influence of stress. For example, the growth of ferrite plates that best comply with the applied stress is favoured. The microstructure thus becomes non-random over the scale of the specimen, with considerable alignment of plates along planes of maximum shear stress.

The primary effect on microstructure during transformation under stress is that of variant selection, which at low stresses reduces the number of sheaves per austenite grain. Variant selection does not lead to an obviously aligned microstructure until larger stresses are applied, in which case the sheaves probably form on habit plane variants which are most parallel to the planes of maximum shear stress. Although deviations from the random microstructures that form in the absence of applied stress are often not easily detectable, especially at low stresses, they reveal themselves unambiguously in the form of anisotropic dimensional changes during transformation.

Although transformation plasticity always tends to be anisotropic, the degree of anisotropy is potentially much larger with displacive transformations.

SUGGESTIONS FOR FUTURE WORK

There are three key areas which need focused research if the micromechanics of phase transformations can be incorporated successfully into models for residual stress development. A possible route towards making such research useful in the problem of welding is illustrated in Fig. 27. Firstly, it is necessary to develop experimental techniques which reveal the role of the deviatoric components of the transformation strain. Secondly, the role of the applied stress in altering the degree of organisation in the microstructure needs to be investigated quantitatively. It is already established qualitatively that the microstructure becomes less chaotic when only favoured crystallographic variants grow under the influence of applied stress. Finally, it is necessary to further develop the model for the detailed interaction of the shape deformation due to transformation, with stress.

Any new research on the overall kinetics of stress-affected transformation should involve the measurement of at least *two* orthogonal transformation

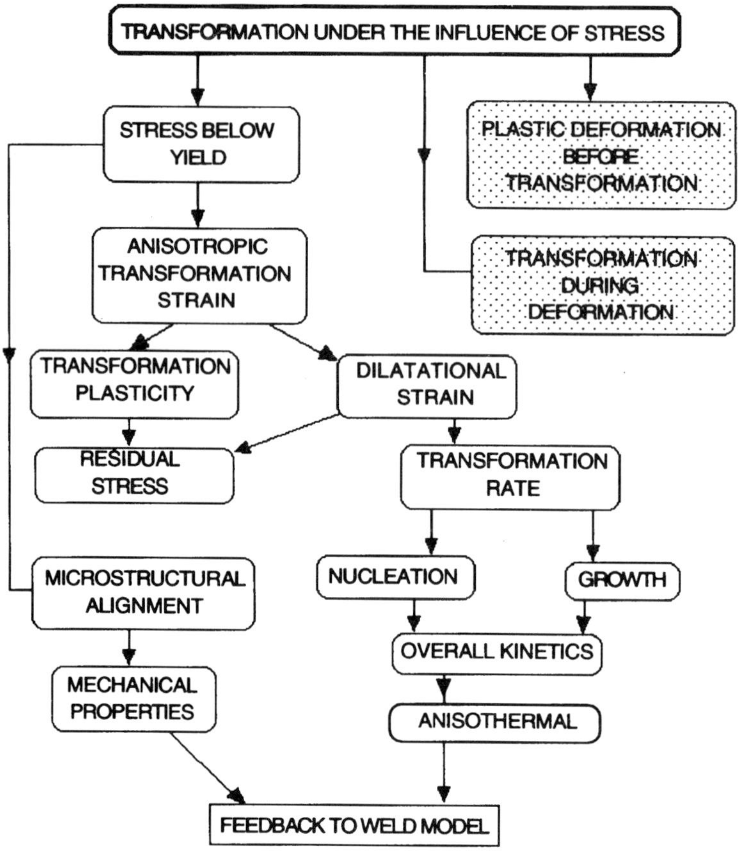

Figure 27. Flow chart illustrating a possible route towards the investigation and incorporation of stress effects into the modelling of steel weld deposits.

strains simultaneously during isothermal transformation. Conventional dilatometry measures just one. In addition, the volume strain could also be measured indirectly using electrical resistivity measurements. The resistance data can then be converted to the volume fraction of the product phase, and hence to a volume strain.

Measurements are necessary, using light microscopy, to monitor the extent of microstructural alignment as a function of stress and transformation temperature. The easiest method would involve the measurement of sheaf traces on longitudinal sections, but it is important to develop a computer model to better account for stereological factors. The model must allow different distributions of plates in three dimensions, and the effect of sectioning on the distribution of traces can then be examined numerically. This will enable the development of better metallographic techniques for the assessment of microstructural alignment, and the techniques should be confirmed using two-surface analysis.

Some samples should be metallographically polished prior to transformation under stress in an environmental chamber of a thermomechanical simulator, giving unique data on the stress/crystallography interaction. Such data are of importance in choosing variant selection criteria, and hence in calculating the mechanical driving force. Experimental data should be collected for a number of designed alloys, including those giving carbide-free transformations, and others in which mixed allotriomorphic/bainite microstructures are obtained. In this way, the effects of mixed reactions (which inevitably happen in 'real' situations) can also be characterised.

MECHANICAL PROPERTIES

There are some interesting experiments to be done. Bainite and acicular ferrite plates normally develop in many crystallographic orientations (a *chaotic* microstructure), but stress can make entire austenite grains transform into just one set of parallel plates in identical crystallographic orientation (an *organised* microstructure). It is widely believed that the toughness in welds is enhanced when the plates do not form packets, but point in many different directions. The feeling is that the latter microstructure causes the cleavage cracks to frequently deflect since new crystallographic orientations are presented each time a new plate is encountered.

By producing chaotic and organised microstructures by transforming without and with the application of stress, respectively, it should be possible to verify using simple mechanical tests whether the toughness is significantly higher for the chaotic structure.

ACKNOWLEDGEMENTS

The author is grateful to Professor C. J. Humphreys for the provision of laboratory facilities at the University of Cambridge, and to A. Matsuzaki, H. Harada and P. Shipway for many helpful discussions. Some of this work was carried out under the auspices of the 'Atomic Arrangements: Design and Control Project', which is a collaborative effort between the University of Cambridge and the Research and Development Corporation of Japan. The author is grateful to the Engineering and Physical Sciences Research Council for partly funding this work.

REFERENCES

1. P.W. FUERSCHBACK: *The Metal Science of Joining*, M.J. CIESLAK, J.H. PEREPEZKO, S. KANG and M.E. GLICKSMAN, eds. The Minerals Metals and Materials Society, Pennsylvania, USA, 1992, pp. 21–30.
2. J. GOLDAK: *Recent Trends in Welding Research*, DAVID, S.A. and VITEK, J.M., ASM International, USA, 1989, pp. 71–82.
3. L. TROIVE, L. KARLSSON, M. NÄSSTRÖM, P. WEBSTER and K.S. LOW: *Recent Trends in Welding Research*, S.A. DAVID and J.M. VITEK, eds.: ASM International, USA, 1989, pp. 107–112.
4. J.W. CHRISTIAN: *Decomposition of Austenite by Diffusional Processes*, V.F. ZACKAY and H.I. AARONSON, eds.: Interscience, New York, 1962, pp. 371–386.
5. H.M. CLARK and C.M. WAYMAN: *Phase Transformations*, ASM, Metals Park, Ohio, USA, 1970, pp. 59–114.
6. J.D. WATSON and P.G. MCDOUGALL: *Acta Metallurgica*, 1973, **21**, 961–973.
7. J.W. CHRISTIAN: *Physical Properties of Martensite and Bainite*, Special Report 93, Iron and Steel Institute, London, 1965, pp. 1–19.
8. H.K.D.H. BHADESHIA: *Prog. Mat. Sci.*, 1985, **29**, 321–386.
9. H.K.D.H. BHADESHIA: *Bainite in Steels*, Institute of Materials, London, 1992, 1–451.
10. G.F. BOLLING and R.H. RICHMAN: *Acta Metall., 1965,* **13**, 709–722.
11. G.F. BOLLING and R.H. RICHMAN: *Acta Metall.*, 1965, **13**, 723–743.
12. G.F. BOLLING and R.H. RICHMAN: *Acta Metall.*, 1965, **13**, 745–757.
13. G.W. GREENWOOD and R.H. JOHNSON: *Proc. R. Soc.*, 1965, **238**, 403–422.
14. R.H. JOHNSON and G.W. GREENWOOD: *Nature*, 1962, **195**, 138–139.
15. G.B. OLSON and H. HARTMAN: *Journal de Physique*, 1982, **C4–43**, 855–865.
16. W.W. BERBERICH, G. THOMAS, E.R. PARKER and V.F. ZACKAY: *Strength of Metals and Alloys*, ASM, USA, 1970, **3**, 894–899.
17. E. HORNBOGEN: *Acta Metall., 1978,* **26**, 147–152.
18. N. IMAI, N. KOMATSUBARA and K. KUNISHIGE: *Heat and Surface '92*, TAMURA, I., ed., Japan Technical Information Service, Tokyo, 1992, 25–28.
19. L. GIORDANO, P. MATTEAZZI, A. TIZIANI and A. ZAMBON: *Mat. Sci. Engineering A, 1991,* **A131**, 215–220.
20. P.M. KELLY: *Materials Forum*, 1990, **56–58**, 335–346.
21. W.M. KRIVEN: *Materials Forum*, 1990, **56–58**, 347–348.
22. T. SAKUMA: *Trans. Japan Institute for Metals*, 1988, **29**, 879–893.
23. O.A. NKARA and D.R.F. WEST: *Physical properties of martensite and bainite*, Iron and Steel Institute Special #Report 93, London, 1965, 183–192.
24. F.D. FISCHER: *Acta Metall.*, 1990, **38**, 1535–1546.
25. I. TAMURA: *Metal Sci.*, 1982, **16**, 245–254.
26. J.W. CHRISTIAN: *Metall. Trans. A*, 1982, **13A**, 509–538.
27. S. DENIS, E. GAUTIER, A. SIMON and G. BECK: *Mat. Sci. Technol.*, 1985, **1**, 805–814.
28. J-B. LEBLOND, G. BOTTET, J. DEVAUX and J-C. DEVAUX: *Mat. Sci. Technol*, 1985, **1**, 815–822.
29. S. SJÖSTRÖM: *Mat. Sci. Technol.*, 1985, **1**, 823–829.
30. F. ABBASI and A.J. FLETCHER: *Mat. Sci. Technol.*, 1985, **1**, 830–837.
31. F.M.B. FERNANDES, S. DENIS and A. SIMON: *Mat. Sci. Technol.*, 1985, **1**, 838–844.
32. T. INOUE and Z. WANG, *Mat. Sci. Technol.*, 1985, **1**, 845–850.
33. Z. WANG andT. INOUE: *Mat. Sci. Technol.*, 1985, **1**, 899–903.
34. J.B. LOSEFSSON: *Mat. Sci. Technol.*, 1985, **1**, 904–914.
35. E.S. DAVENPORT and E.C. BAIN: *Trans. AIMME*, 1930, **30**, 117–154.
36. C.L. MAGEE: Ph.D. Thesis Carnegie Mellon University, Pittsburgh, USA, 1966.
37. C.L. MAGEE and H.W. PAXTON: *Trans. Met. Soc. AIME*, 1968, **242**, 1741–1749.

38. FISCHER
39. J.B. LEBLOND, G. MOTTET and J.C. DEVAUX: *J. Mech. Phys. Solids*, 1986, **34**, 395–409; 411–432.
40. J.B. LEBLOND, J. DEVAUX and J.C. DEVAUX: *International J. Plasticity*, 1989, **5**, 551–572.
41. J.B. LEBLOND: *International J. Plasticity*, 1989, **5**, 573–591.
42. J.B. LEBLOND and J. DEBAUX: *Residual Stresses*, Elsevier, New York, 1989, pp. 1–6.
43. J.B. LEBLOND: *Internal Report CSS/L/NT/90/4022*, FRAMASOFT, Paris, 1990, pp. 1–12.
44. G.B. OLSON: *Deformation, Processing and Structure*, American Society of Metals, Metals Park, Ohio, 1982, pp. 391–424.
45. H.K.D.H. BHADESHIA, S.A. DAVID, J.M. VITEK and R.W. REED: *Mat. Sci. Technol.*, 1991, **7**, 686–698.
46. R. VON MISES: *Z. angew. Math. Mech.*, 1928, **8**, 161.
47. G.I. TAYLOR: *J. Inst. Metals*, 1928, **62**, 307.
48. J.K. MACKENZIE and J.S. BOWLES: *Acta Metallurgica*, 1954, **2**, 138–147.
49. M.S. WECHSLER, D.S. LIEBERMAN and T.A. READ: *Trans. Amer. Inst. Min. Metall. Engrs.*, 1953, **197**, 1503–1515.
50. A.B. GRENINGER and A.R. TROIANO: *Trans. Amer. Inst. Min. Metall. Engrs.*, 1940, **140**, 307–336.
51. H.K.D.H. BHADESHIA: *J. Materials Science*, 1982, **17**, 383–386.
52. J.R. PATEL and M. COHEN: *Acta Metallurgica*, 1953, **1**, 531–538.
53. L. DELAEY and H. WARLIMONT: *Shape Memory Effects in Alloys*, J. PERKINS, ed., TMS–AIME, Plenum Press, New York, 1975, 89–114.
54. R.K. RAY andJ.J. JONAS: *Int. Mater. Rev.*, 1990, **35**, 1–36.
55. I. TAMURA andC.M. WAYMAN: *Martensite, A Tribute to Morris Cohen*, G.B. OLSON and W.S. OWN, eds, ASM International, Ohio, 1992, 228–242.
56. R.H. GOODENOW, R.H. BARKALOW and R.F. HEHEMANN: *Physical Properties of Martensite and Bainite*, Iron and Steel Institute special report 93, London, 1969, 135–141.
57. A.H. COTTRELL: *J. Iron Steel Inst.*, 1945, **CLI**, 95P–104P.
58. Ya. B. DROZDOV, L.I. KOGAN and R.I. ENTIN: *Phys. Met. and Metall.*, 1962, **5**, 135–138.
59. K. TSUZAKI, T. UEDA, K. FUJIWARA and T. MAKI: *New Materials and Processes for the Future*, Proc. First Japan International SAMPE symposium and exhibition, Society for the Advancement of Machining and Processing, Chiba, Japan, 1989, pp. 799–804.
60. M.D. JEPSON and F.C. THOMPSON: *J. Iron Steel Inst.*, 1949, **162**, 49–56.
61. W.E. DUCKWORTH: *J. Met.*, 1966, **18**, 915.
62. V.A. DUBROV: *Phys. of Met. and Metall.*, 1969, **28, No. 2**, 126–131.
63. T.A. MUTIU, A.J. KINDERMAN and I.M. BERNSTEIN: *The Hot Deformation of Austenite*, BALLANCE, J.B., ed., TMS–AIME, New York, 1977, 410–427.
64. L.E. PORTER and P.C. ROSENTHAL: *Acta Metallurgica*, 1959, **7**, 504–514.
65. M. UMEMOTO, S. BANDO and I. TAMURA: *Proceedings of an International Conference on Martensitic Transformations*, Japan Institute of Metals, 1986, 595–600.
66. H. LANGE and K. MATHIEU: *Mitt K.W.I Eisenforchung*, 1938, **20**, 125.
67. W. JELLINGHAUS: *Arch. Eisenhuttenwesen*, 1952, **23**, 459.
68. R.T. HOWARD and M. COHEN: *Acta Metallurgica*, 1948, **176**, 384–400.
69. C.E. ERICSSON, M.S. BHAT, E.R. PARKER and V.F. ZACKAY: *Metall. Trans. A*, 1976, **7A**, 1800–1803.
70. S.V. RADCLIFFE and E.C. ROLLASON: *J. Iron Steel Inst.*, 1959, **191**, 56–65.
71. C. ZENER: *Trans. AIME*, 1946, **167**, 550–595.
72. T. KO: *J. Iron and Steel Inst.*, 1953, **175**, 16–18.
73. M.J. HAWKINS and J. BARFORD: *J. Iron Steel Inst.*, 1972, **210**, 97–105.
74. J.W. CHRISTIAN: *ICOMAT 79*, International Conference on martensitic Transformations, Cambridge, Massachusetts, 1979, 220–234.
75. C.M. VLAD: *Accelerated Cooling of Steel*, SOUTHWICK, P.D., ed., TMS–AIME, Warrendale, PA, 1985, 435–448.

76. A.T. DAVENPORT: *The Hot Deformation of Austenite*, J.B. BALLANCE, ed., TMS–AIME, New York, 1977, 517–536.
77. S. BHATTACHARYYA and G.L. KEHL: *Trans. ASM*, 1955, **47**, 351–379.
78. H..K.D.H. BHADESHIA: *International Trends in Welding Science and Technology*, S.A. DAVID and J.M. VITEK, ed., ASM International, OH, 1993, pp. 213–222.
79. M. STRANGWOOD and H.K.D.H. BHADESHIA: *Advances in Welding Science and Technology*, S.A. DAVID, ed., ASM International, OH, 1987, 209–213.
80. S.S. BABU and H.K.D.H. BHADESHIA: *Materials Sci. Engineering, 1992,* **A156**, 1–9.
81. BHADESHIA, H.K.D.H., *Mat. Sci. Technol., 1989,* **5**, 131–137.
82. S.V. TSIVINSKY, L.I. KOGAN and R.I. ENTIN: *Problems of Metallography and the Physics of Metals*, B. Ya LYBUBOV: ed., State Scientific Press, Moscow (Translation published by the Consultants Bureau Inc., New York, 1959), 1955, pp. 185–1991.
83. J. CHANCE and N. RIDLEY: *Metall. Trans. A*, 1981, **21A**, 1205–1213.
84. A. HULTGREN: *Jernkontorets Ann.*, 1951, **135**, 403.
85. K.W. ANDREWS: *Acta Metall.*, 1963, **11**, 939–946.
86. W. HUME-ROTHERY, G.V. RAYNOR and A.T. LITTLE: *Arch. Eisenhüttenwes.*, 1942, **145**, 143.
87. H.C. YAKEL: *Int. Met. Rev.*, 1985, **30**, 17–40.
88. P.J. SANDVIK: *Metall. Trans. A*, 1982, **13**, 789–800.
89. T. NAKAMURA and S. NAGAKURA: *Int. Conf. on Martensitic Transformations (ICOMAT-86)*, The Japan Institute of Metals, 1986, 1057–1065.
90. K.A. TAYLOR, G.B. OLSON, M. COHEN and J.B. VANDER SANDE, *Metall. Trans. A*, 1989, **20A**, 2749–2765.
91. K.A. TAYLOR, L. CHANG, G.B. OLSON, G.D.W. SMITH, M. COHEN and J.B. VANDER SANDE: *Metall. Trans. A*, 1989, **20A**, 2772–2737.
92. K.H. JACK: *Acta Crystallog.*, 1950, **3**, 392.
93. A. MATSUZAKI, H.K.D.H. BHADESHIA and H. HARADA: *G.R. Speich Symposium*, Iron and Steel Society, AIME, Warrendale, PA, 1992, pp. 47–52.
94. L. KAUFMANN: *Solids Under Pressure*, W. PAUL and D.M. WARSCHAUER, eds, McGraw-Hill, New York, 1963, 304–355.
95. W. JELLINGHAUS and H. FRIEDEWOLD: *Arch Eisenhuttenwesen*, 1960, **31**, 309.
96. T.G. NILAN: *Trans. Met. Soc AIME*, 1967, **239**, 898–909.
97. R.M. HORN and R.O. RITCHIE: *Metall. Trans. A*, 1978, **9A**, 1039–1053.
98. R.J. KAR, R.M. HORN and V.F. ZACKAY: *Metall. Trans. A*, 1979, **10A**, 1711–1717.
99. H.K.D.H. BHADESHIA and D.V. EDMONDS: *Metal Science*, 1983, **17**, 411–419.
100. H.K.D.H. BHADESHIA and D.V. EDMONDS: *Metal Science*, 1983, **17**, 420–425.
101. T. GEORGE, E.R. PARKER and R.O. RITCHIE: *Mat. Sci. Technol.*, 1985, **1**, 198–208.
102. I. TSUKATANI, S. HASHIMOTO and T. INOUE: *I.S.I.J. International*, 1991, **31**, 992–1000.
103. V.T.T. MIIHKINEN and D.V. EDMONDS: *Mat. Sci. Technol.*, 1987, **3**, 422–431.
104. V.T.T. MIIHKINEN and D.V. EDMONDS: *Mat. Sci. Technol.*, 1987, **3**, 432–440.
105. V.T.T. MIIHKINEN and D.V. EDMONDS: *Mat. Sci. Technol.*, 1987, **3**, 441–449.
106. W.K.C. JONES and P.J. ALBERRY: *Ferritic Steels for Fast Reactor Steam Generators*, British Nuclear Engineering Society, London, 1977, pp. 1–4.
107. W.K.C. JONES and P.J. ALBERRY: *Residual Stresses in Welded Constructions*, Welding Institute, Cambridge, 1977, Paper 2.
108. J.W. CHRISTIAN: *Theory of Transformations in Metals and Alloys*, Part I, 2nd ed., Pergamon Press, Oxford, 1975.
109. H.K.D.H. BHADESHIA: *Worked Examples in the Geometry of Crystals*, Institute of Metals, London, 1987.
110. A. MATSUZAKI, H.K.D.H. BHADESHIA and H. HARADA: *Acta Metallurgica et Materialia*, 1994, **42**, 1081–1090.

MODELLING OF WELDABILITY NEEDS AND LIMITS

B. BUCHMAYR
Institute for Materials Science and Welding Technology, Graz University of Technology, Kopernikusgasse 24, A-8010 Graz, Austria

ABSTRACT

A critical assessment is made concerning the role of mathematical modelling with respect to weldability problems. The capability of numerical models is analysed in order to define application fields where they are very powerful and others where they are not so predestined. The important role of models which deal with coupled phenomena is demonstrated, especially for the solution of complex problems in practice. It is also shown that the choice of an appropriate method and the effort spent for the model development is strongly dependent on the desired objectives. Furthermore, it is pointed out that the correct interpretation of the model results may be more important than its accuracy. Finally, the need for experimental support is emphasised in order to generate problem or model specific material data and to verify the model performance. From the obvious limits of numerical models, further demands to increase their capabilities are stated in the conclusions.

INTRODUCTION

There are several reasons which justify the use of numerical models for problems of weldability, because this subject is characterised by

— many interacting influencing parameters,
— multidisciplinarity of the subject and
— many of the underlying phenomena can be numerically described.

It is therefore not surprising that more and more welding experts and scientists try to apply mathematical models to solve particular aspects of weldability.[1–5] Although many interesting developments were published and additional knowledge could be gathered there is still a long way to achieve full acceptance by welding engineers. In order to illustrate the pursuit of knowledge by welding engineers with regard to weldability aspects, which may be a subject of modelling, some questions and practical demands should be pointed out. Experts for numerical analysis of weldability are confronted by practical engineers with the following questions and demands:

— Can we leave a particular weld defect in a component? (Is our current knowledge about concepts of fitness for purpose applicable in practice?)
— Can you predict the remaining loading capacity in weldments with high residual stresses?
— Specify how much toughness is required for a given case.
— Should we under(over)match the filler metal?
— We cannot preheat; is there a danger for cold cracking?
— Determine an optimum chemical composition for excellent HAZ toughness.
— Determine how the width of the intercritical zone of a T91 weldment affects the overall creep strength.

These are only some examples of important questions, which are however very challenging for materials experts, who develop or use computer models.

In this paper, a critical analysis will be given concerning typical application fields for numerical modelling including some examples as well as less predestined problem fields. It will be shown that model development work in the field of weldability needs more and more coupling of phenomena and interdisciplinary team work. Finally, the lack of relevant and accurate material data will be raised.

PHENOMENA PREDESTINED FOR COMPUTER MODELLING

In principle, there are particular phenomena involved during welding which can be described easily by numerical procedures. Into this category fall all phenomena which are based on physical or chemical fundamentals, like those described in Table 1. We know the underlying mathematics to solve a particular problem and we try to apply correct boundary conditions. Additionally, we make improvements regarding physical correctness and incorporate more and more coupled effects.

Based on these well-established methods and the availability of the numerical tools (finite element codes, finite difference schemes etc.) typical and successful application fields for computer simulations in welding technology become obvious, like the

- — calculation of temperature fields (FEM/FDM, 2-dim./3-dim.);
- — flow phenomena in the weld pool (including free surface phenomena);
- — simulation of solidification behaviour of welds (including micro-segregation);
- — calculation of residual stresses and distortions (including multilayer weldments and taking into account the different properties in the HAZ subzones);
- — prediction of microstructural changes in the HAZ (phase transformation, precipitation kinetics, grain growth);
- — ternary diffusion in dissimilar welds (carbon depletion and carbide seam formation in CrMoV welds);
- — hydrogen diffusion, taking also into account H-traps;
- — material transfer from the electrode to the pool;
- — simulation of thermal spraying and many more.

To illustrate the capability of numerical models in these application fields some examples will be shown in the following sections.

TRANSFORMATION IN THE HAZ

A chronological sequence of microstructural changes in the HAZ is shown in Fig. 1. The property-determining processes involved are modelled as a function of the main-influencing parameters, like chemical composition, heat input and post-weld heat treatment.[2] The microstructural features are then correlated with the mechanical properties. The most important result of physical based modelling is that the microstructural changes can be predicted and experimental investigations can be reduced. A causal connection can be established which reflects the nature of the process. Although the interim

Table 1

Phenomena, predestinated for computer modelling

all phenomena, which are based on physical/chemical fundamentals, like

phenomenon	math. component	remarks
fluid flow	Navier–Stokes eqn	complex vector-tensor manipulation
heat conduction	Fourier's Law	simple, complex if phase change is involved
diffusion	Fick's Laws	simple for one, complex for multiple comp.
heat & mass transfer	Convective transport	combines fluid flow, heat and/or mass transfer
eletrodynamics MHD	Maxwell's eqn	very complex problems, i.e. plasmas etc.
phase diagrams	Thermodynamics	interaction data are needed
rate prediction	Kinetics	needs experimental input
stresses / deform	Constitutive eqns	very complex, needs realistic input

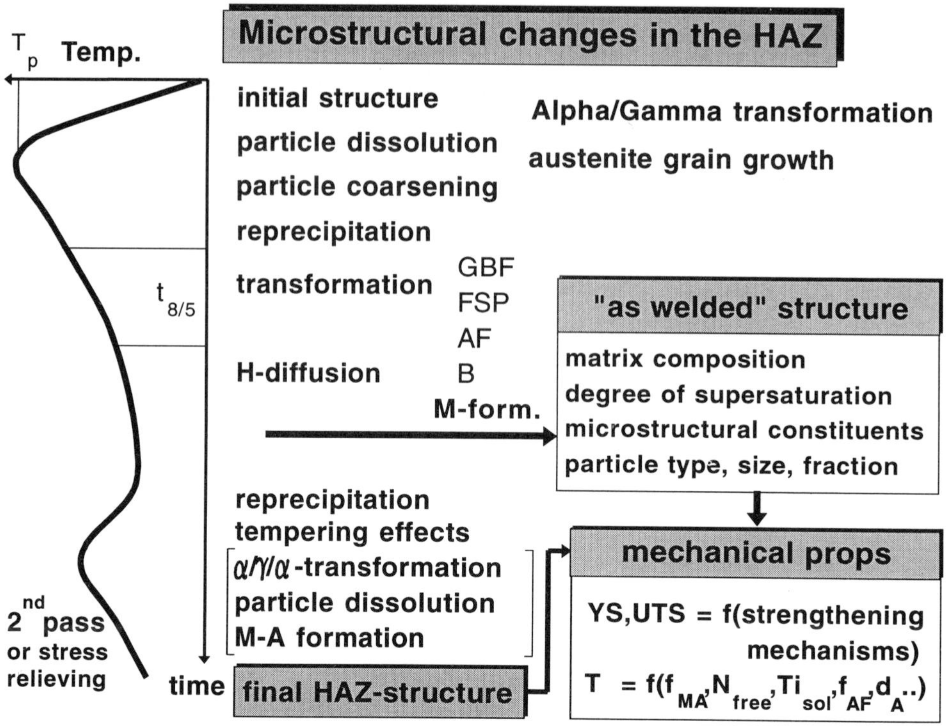

Figure 1. Mathematical modelling of the HAZ microstructure and properties taking into account the most significant processes and parameters.[2]

results are key data for the assessment and hints for further process development work, the final correlation with the mechanical properties still need multivariate regression techniques to obtain the desired accuracy.

AMOUNT OF SOLUTED NITROGEN AND ITS INFLUENCE ON TOUGHNESS

Using a regular solution model for the description of the thermodynamical equilibrium (as assumed for high temperatures), the amount and composition of precipitates can be predicted.[6] Figure 2 shows the result of such a calculation for different peak temperatures,[7] which correspond to different distances from the fusion line. Due to the rapid cooling to room temperature it can be assumed that most of the nitrogen stays in solid solution. During reheating nitrides are precipitated, which affects the toughness properties. The strong effect of nitrogen on the transition temperature, measured by Charpy-impact tests after PWHT is shown in Fig. 3. From this correlation, we can conclude that the prediction of property-determining parameters can be very important although a model-based prediction of the effects on the toughness behaviour is not yet feasible.

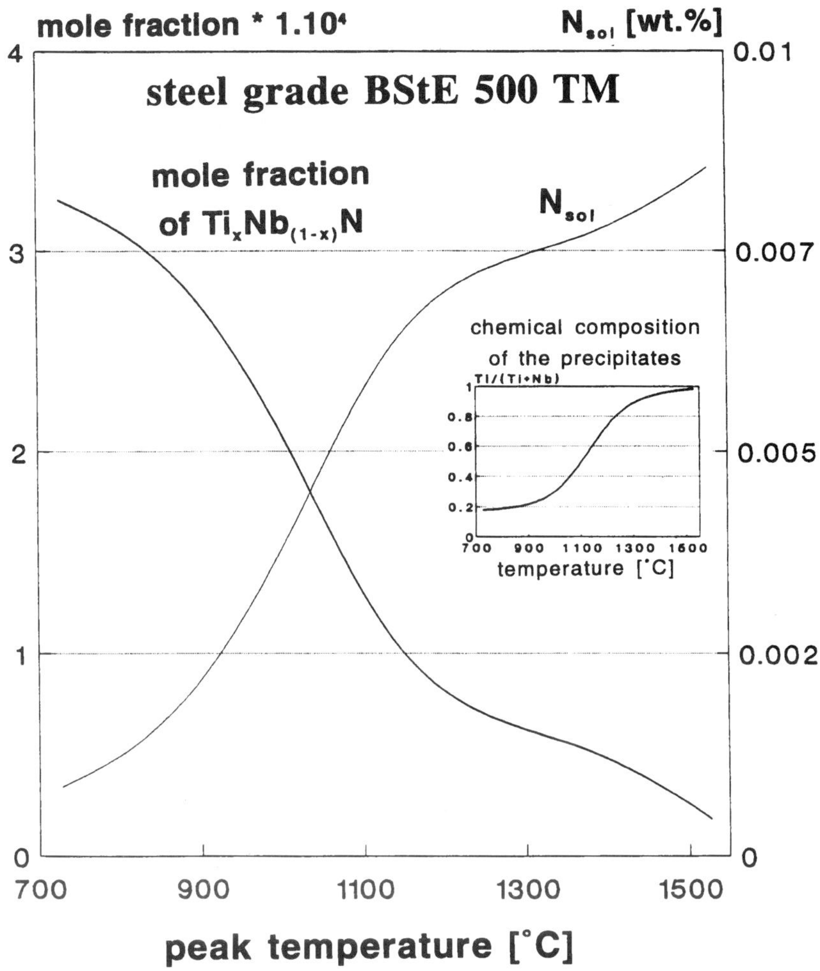

Figure 2. Predicted amount of nitrogen in solid solution and composition of carbonitrides as a function of peak temperature.[7]

MODELLING OF DISSIMILAR WELDS AND GRADED JOINTS

Based on the fundamentals of ternary diffusion coupled with the prediction of carbide dissolution as well as precipitation, the microstructural changes in dissimilar weldments of heat-resistant steels can be described.[8,9] Figure 4 shows one result in the form of the carbon profile across the fusion line. Due to the improved understanding gathered by this modelling work, a new design for such critical weldments in thermal power plants could be derived. A graded joint (2.25%CrMo/AISI316) was developed with stepped composition from the low alloy to the high alloyed side.[10] The carbon distribution is calculated as a function of powder particle size and volume fraction of either ferritic or austenitic particles. The microstructure in the region with 11vol.% austenitic particles is shown in Fig. 5. The corresponding numerical results for the

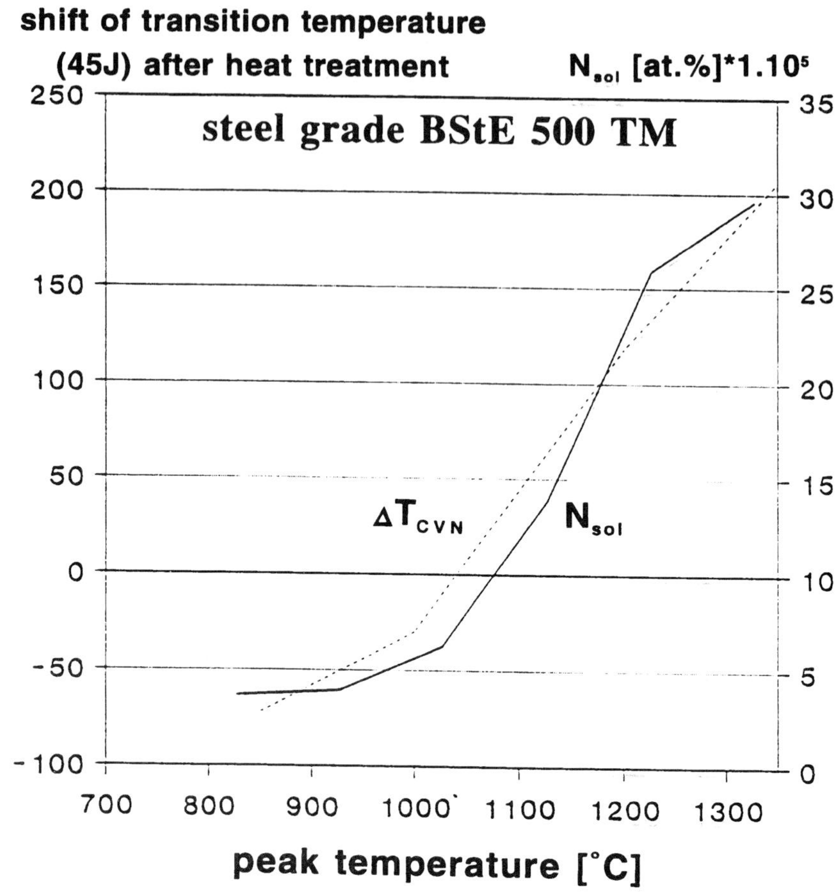

Figure 3. Direct correlationship between transition temperature and amount of free nitrogen in a weldment.[7]

amount of carbon, chromium and $M_{23}C_6$-carbides across the particles is given in Fig. 6.

SIMULATION OF THE MAG-PROCESS

A very impressive PC-based model for the prediction of the weld pool shape was developed by Dilthey *et al.*[11] The 'MAGSIM'-program calculates the thermal history, the solidification behaviour as well as overfill and undercut as a function of the welding parameters. Using the model results, a classification of the weldment quality is made with respect to the German DIN specification. Furthermore, the effect of scatter in the welding parameters on the weldment

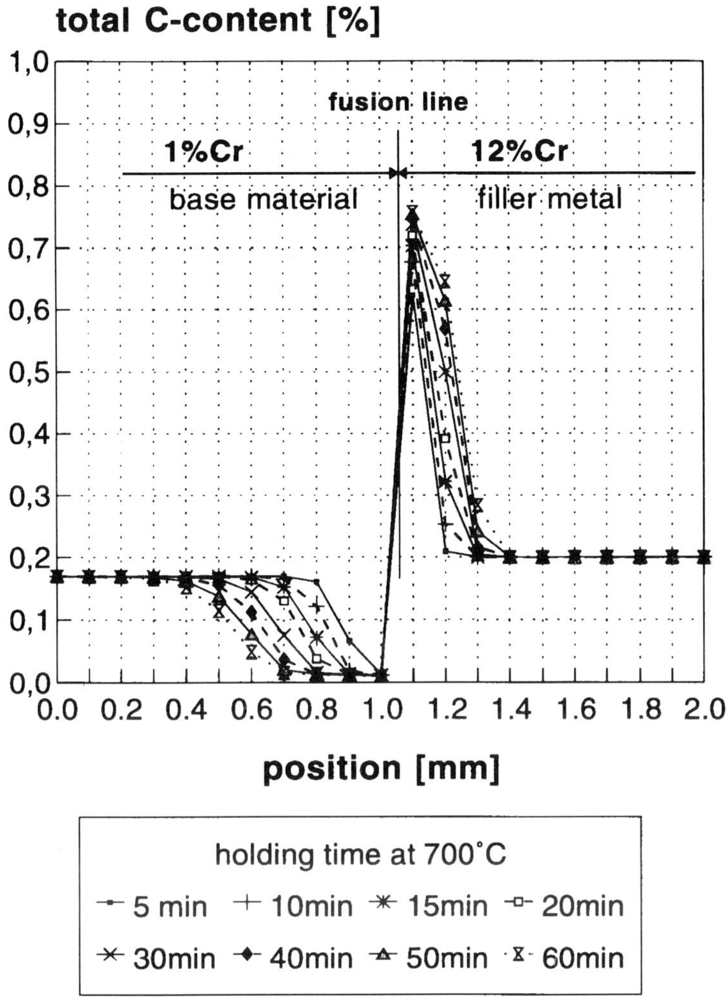

Figure 4. Predicted carbon profile at the fusion line of a dissmilar weld between a 1%CrMoV base metal and a 12%CrMoV filler metal.[9]

quality is predicted using an additional statistical approach. An example of the user interface is shown in Fig. 7.

PHENOMENA, LESS PREDESTINED FOR MODELLING APPLICATIONS

It is obvious that those problems, which are not yet fully understood in the physical sense, cannot be modelled in a sufficient way. Oversimplified

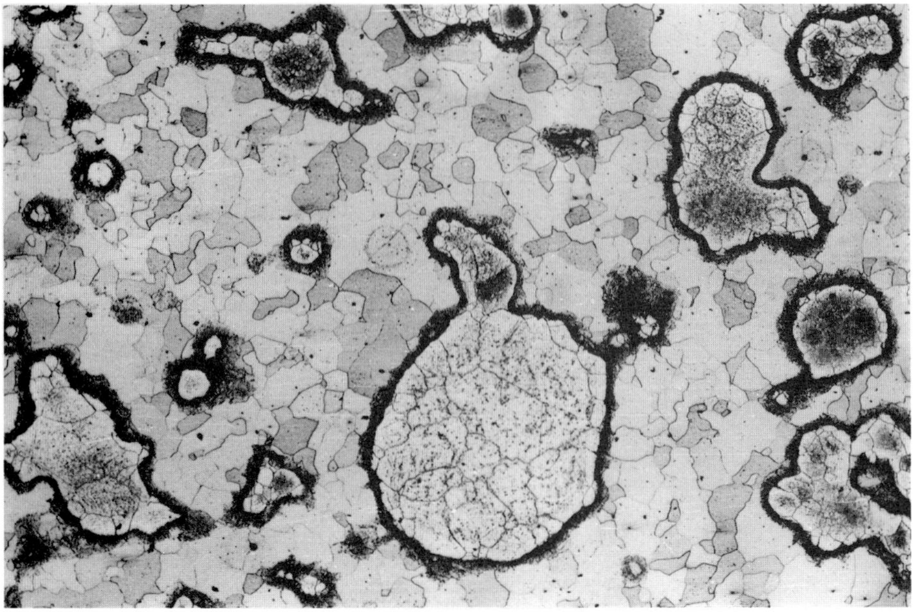

Figure 5. Microstructure of a graded joint (2.25%Cr Mo/AISI316) after hot isostatic pressing and heat treatment[10], magnif. 100x.

approaches may be worthless and cannot contribute to any solution of the problem.

Many phenomena in welding metallurgy are very complex, i.e. there are many synergistic effects, which cannot be described by a simple set of equations. However, certain effects cannot be neglected and therefore semi-empirical approaches have to be used, which need adjustment based on experimental investigations.

Into this category, the following phenomena can be assigned:

— structure–property relationships (strength = *f*(temperature, strain, strain rate, chemical composition, microstructure);
— toughness behaviour (even multiple regression approaches provide only moderate predictions and show a rather limited applicability);
— mechanistic approaches for hot, cold and stress relief cracking;
— real microstructures (e.g. different ferrite morphologies);
— kinetic processes always need adjustment to measured values (transformation, precipitation, grain growth, recrystallization kinetics, etc.);
— damage mechanisms / physical meaningful failure criteria;
— complex coupled problems and multicomponent problems;
— behaviour and properties of interfaces (e.g. grain boundaries);
— in general, all problems where there is no physical background.

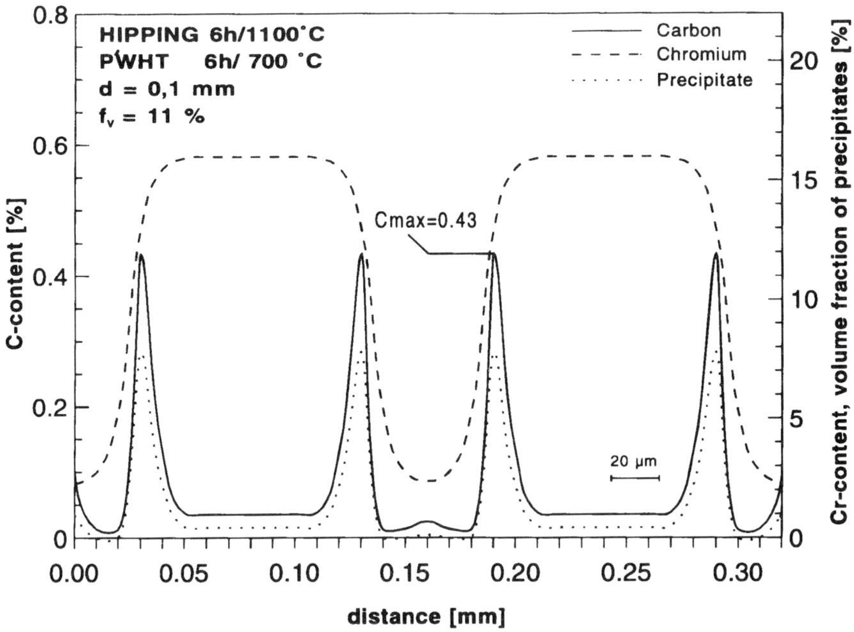

Figure 6. Predicted carbon- and Cr-concentration as well as volume fraction of $M_{23}C_6$-carbides in the graded joint 2,25%Cr Mo/AISI316.[10]

To illustrate these examples, various microstructural constituents in weld microstructures are used as shown in the IIW-Doc. No. IX-1533-88. The definition comprises seven different morphologies for ferrite (PF, PF(G), PF(I), AF, FS(A), FS(NA), FC). Except for the derivative concept and some improved understanding of the ferrite formation, mainly worked out by Bhadeshia,[12] we have no comprehensive model to predict the ferrite shape (e.g. acicular ferrite), although we know that it is very important with respect to toughness behaviour.

Furthermore, it has to be pointed out that the predictability of the toughness behaviour is very poor. Especially if we consider microalloyed structural steels, the synergistic influences and optimum contents of Ti, Nb, V, B and N are very difficult to understand and to model. It will still need some time in order to predict optimum combinations of micro-alloying elements, like those derived from experimental investigations[13,14,15] as shown in Fig. 8. From the metallurgical point of view, we can define the sequence of relevant mechanisms and their influencing factors and try to model these modules, at first individually and then in combination. The global structure of such a model[16], which should be able to predict the HAZ toughness, is shown in Fig. 9.

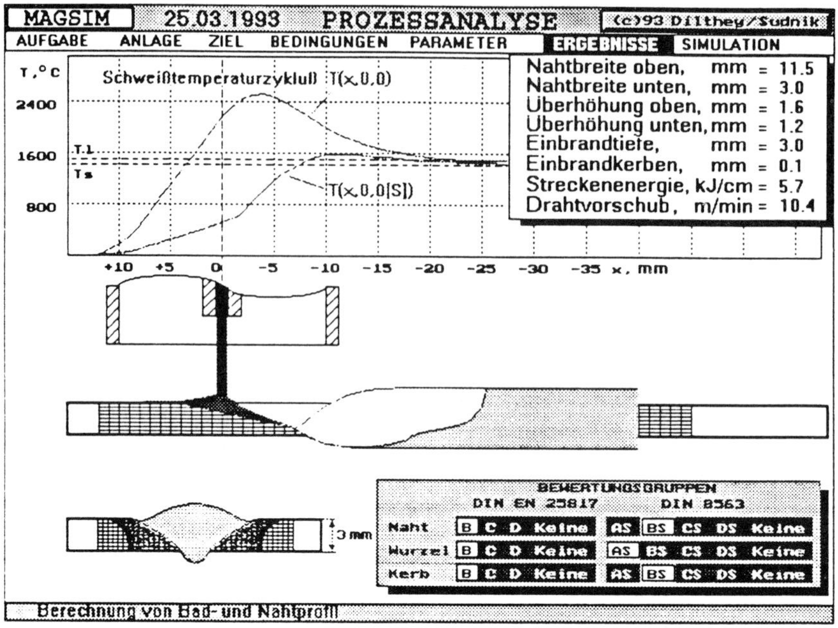

Schweißparameter: Schutzgas M21; Blechdicke s = 3 mm; Spaltbreite b = 0 mm;
Stromstärke I_S = 295 A; Schweißspannung U_S = 27 V;
Schweißgeschw. v_S = 89 cm/min; Draht-∅ d_{Dr} = 1,2 mm;
Kontaktrohrabstand Z_k = 14 mm

Figure 7. User interface of the PC-simulation program MAGSIM to predict the weld form and weldment quality.[11]

COUPLED PHENOMENA

Due to the nature of welding metallurgy, there are manifold processes and influencing factors involved, which need an effective, interdisciplinary cooperation between different experts. It is evident that any model development will also need a lot of experimental support. Into this category of coupled problems belong:

— Residual stresses and distortion, with the subfields; transient temperature field calculation; prediction of microstructural evolution; FEM calculation of the stress–strain behaviour.
— Cold cracking, with modules for microstructure after welding; residual stress calculation; hydrogen diffusion criteria for crack formation.
— Hot cracking, with various submodules.
— Stress relief or reheat cracking, with coupling of stress relieving and changes in the microstructure.
— Creep rupture in the HAZ, with submodules for microstructure, tempering effects (particle kinetics), mechanical behaviour and damage mechanism.

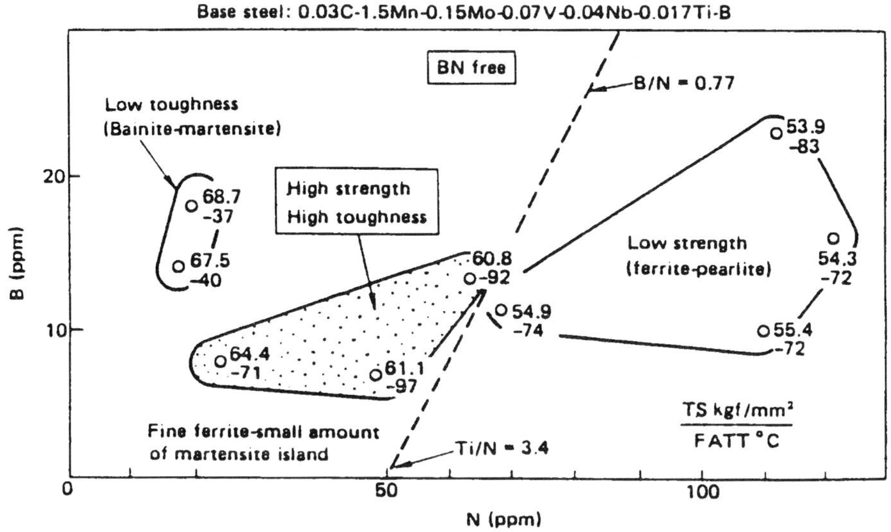

Figure 8. Optimum content of nitrogen and boron for excellent toughness behaviour of microalloyed HSLA steels.[14]

In order to illustrate the coupling of modules for the prediction of residual stresses in weldments, Fig. 10 shows the main modules and their interactions.[2] Usually, one can provide a two-step approach, i.e. at first the coupled prediction of temperature and microstructure as a function of time and space and, second, the prediction of the mechanical response taking into account the results from the first step. A schematic flow chart of the essential calculation steps within this two-step approach is shown in Figs 11 and 12.

Even the residual stress response of multilayer weldments can be considered, as demonstrated in 17 using the FEM-package SYSWELD.

Although many problems can be solved by numerical models, we are still not able to solve the problem of cold cracking in a physical correct way. If we consider that the main influencing parameters for cold cracking, namely microstructure, residual stresses and hydrogen content, can be calculated using a FEM-approach, we are however not yet able to predict the risk of cold crack formation, the time of this event or the locus of crack initiation. But this is not a draw back from the modelling side, it is caused rather by the lack of physical models about cracking criteria.

For modelling experts, it is therefore frustrating to admit that the step back to empirical equations provides more useful information than may be used. It means, empirical equations for the prediction of preheating temperature or critical HAZ hardness to avoid cold cracking are better tools.

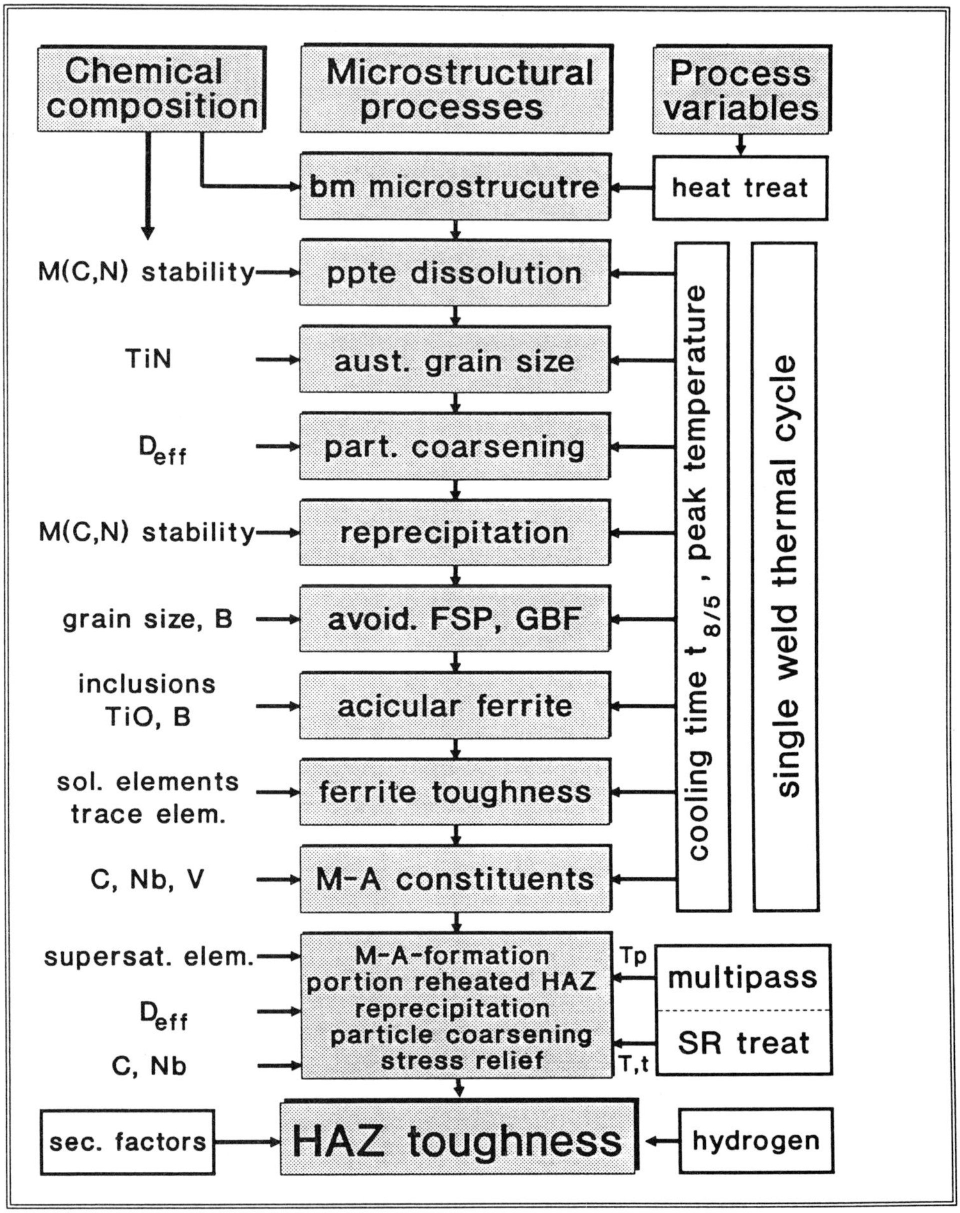

Figure 9. Global model structure to predict the HAZ-toughness as a function of the underlying processes and influencing factors.[16]

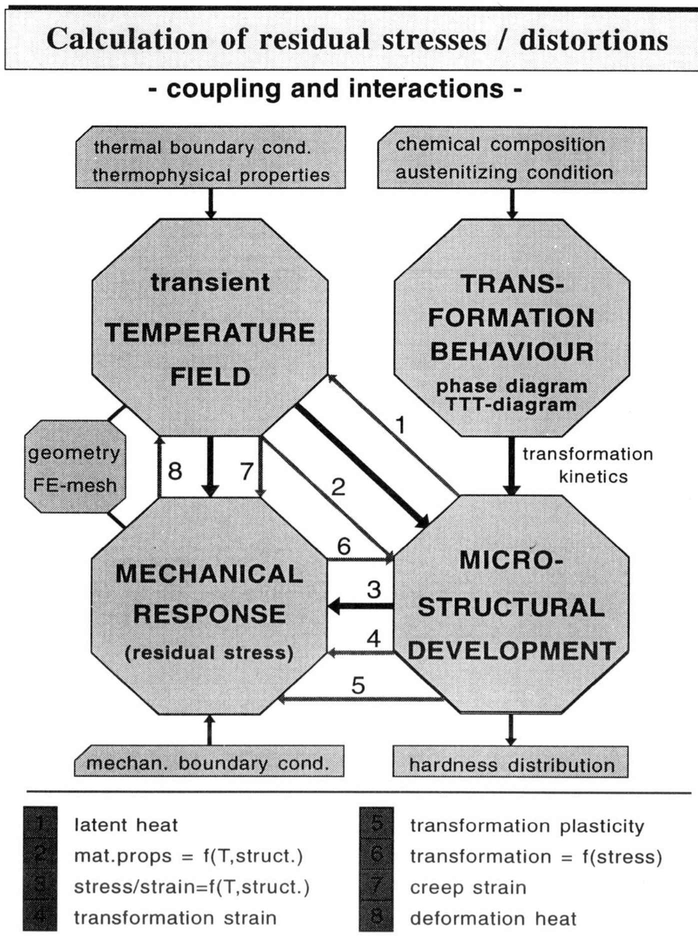

Figure 10. Coupling of the modules for temperature, microstructure and mechanical response in order to predict residual stresses.[2]

THE ROLE OF WELL DEFINED OBJECTIVES FOR THE MODEL DESIGN

The aforementioned situation regarding model capability in the case of cold cracking clearly demonstrates the limits of numerical models. Very sophisticated models are considered to be worthless if they do not contribute to a particular problem. If we know the distribution of stresses, microstructural constituents and hydrogen in a weldment, we still cannot say whether the weldment will reveal cold cracks. On the other hand, we are able to say that when we apply a particular preheating then we will have no cracking problem. What we can conclude from this is that at the beginning of any model development, we have to consider the presumed results and benefits from such a work. It depends on the objective which method will provide most

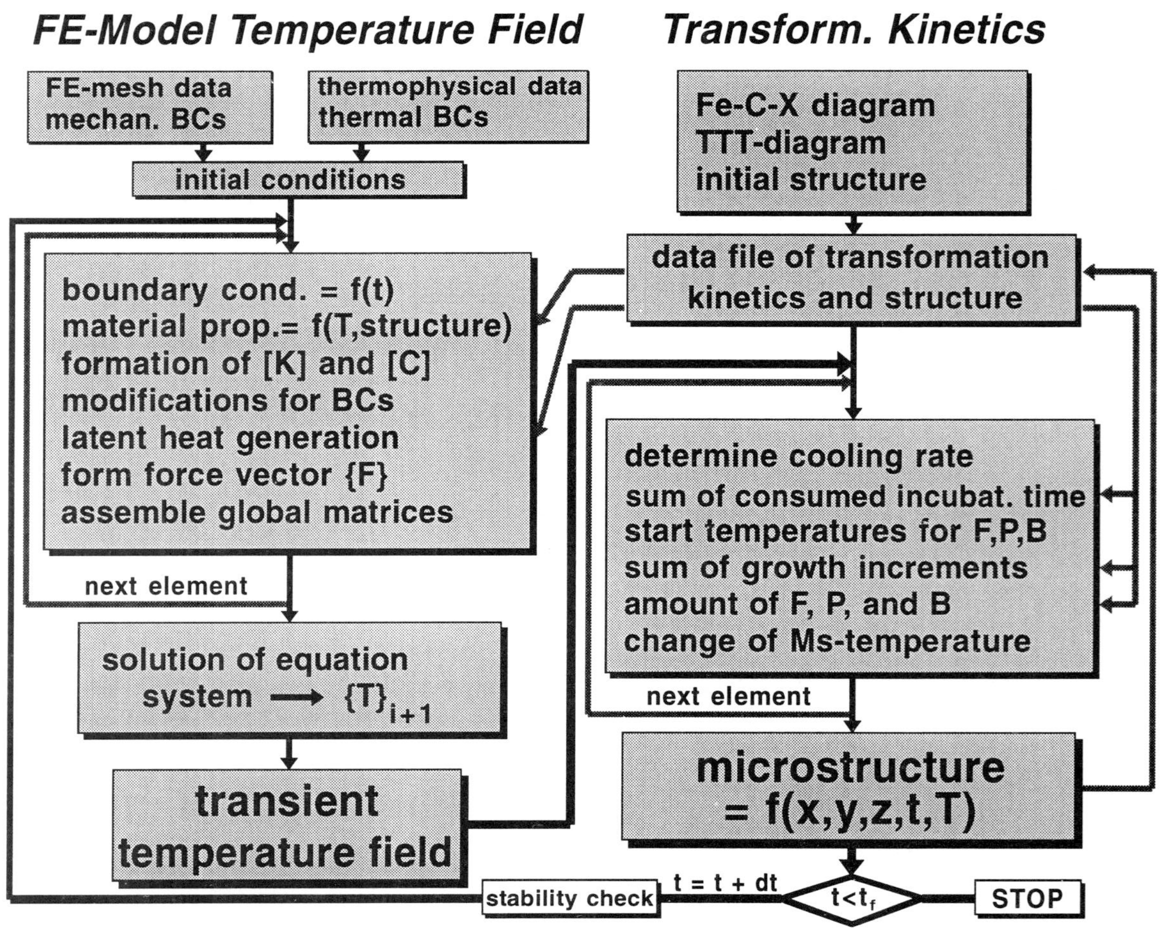

Figure 11. Coupled prediction of temperature and microstructural evolution as a prestep for the residual stress calculation.[2]

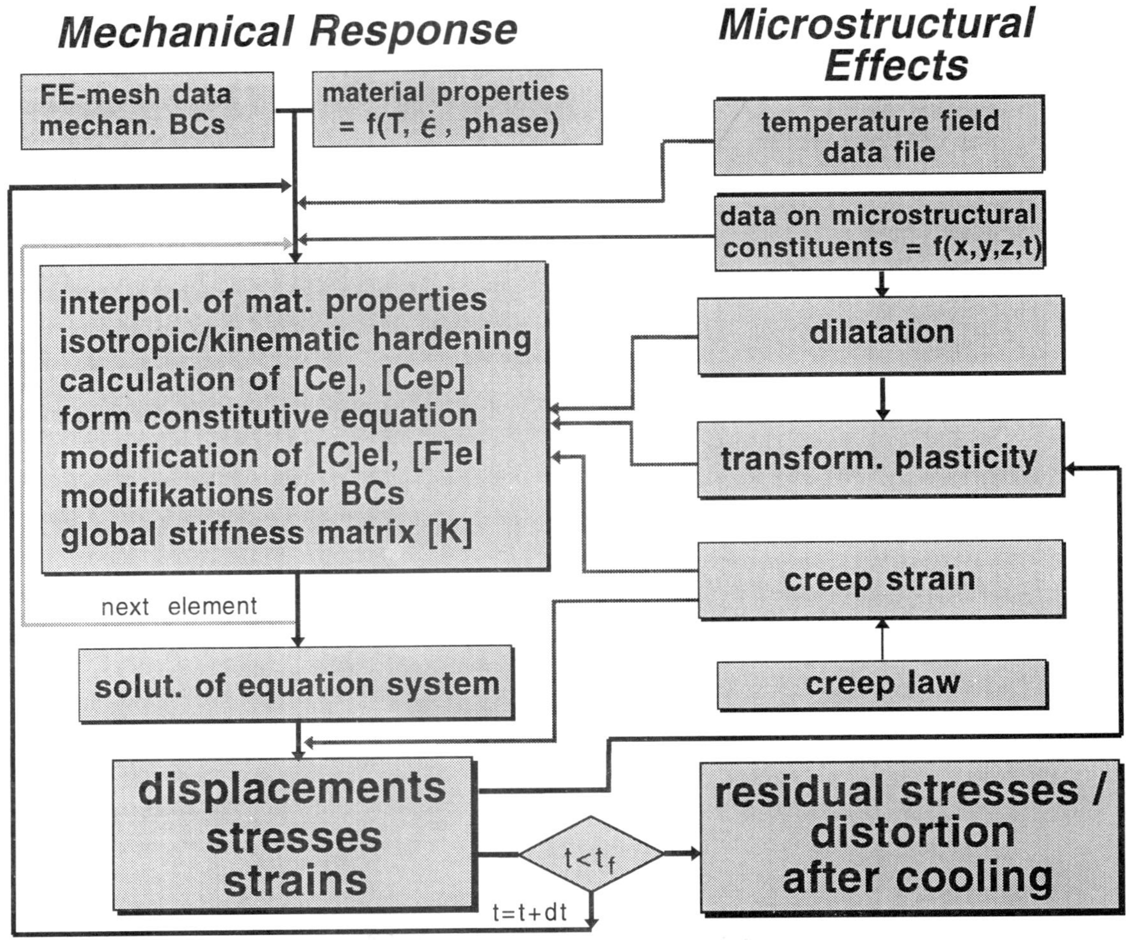

Figure 12. FE-prediction of residual stresses taking into account the microstructural effects.[2]

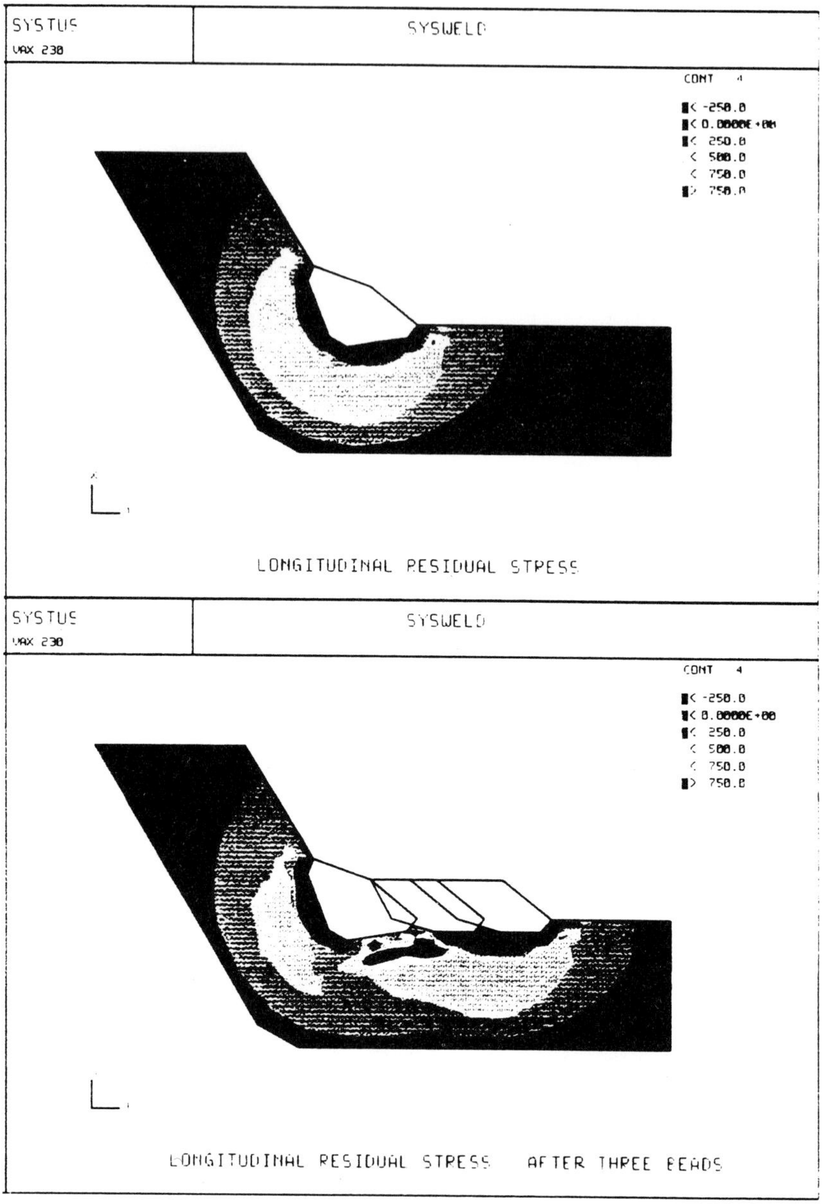

Figure 13. Predicted residual stress distribution in a multilayer weldment using the FE-software SYSWELD.[17]

information and support for a given problem. In some cases it is worth using more sophisticated ones, in others it is better to apply statistical multivariate techniques.

For R&D departments, it is becoming now more a standard procedure to use the quality function deployment (QFD) method as a project design tool for goal (customer)-oriented development steps. The main benefit from using this approach is the information from the manifold effects and parameters as to which are most sensitive or which are most important to achieve an improvement (Pareto principle). By applying this planning tool, most practical tasks can be solved in an effective manner.

Another aspect which is worth mentioning concerns the role of the accuracy of model predictions. When we develop a particular model, it is evident that we make assumptions and short cuts by neglecting weak interactions and so on. In so doing, we already take into consideration the balance between effort and benefits. The question of accuracy is therefore of secondary importance in modelling work at the beginning and can be improved subsequently by adjusting proper parameters. But we have to keep in mind that the main contribution to a problem comes from a consequent and logical interpretation of the model results. Only if we are able to derive and apply appropriate measures to solve the problem, have we done our job sufficiently.

Therefore it is not a question of using sophisticated models in every case; rather a question of the correct use of an appropriate method. In some cases, some simple experimental tests or a simple empirical equation can provide the correct answer. This analytical capability should be a prerequirement for any modeller.

LACK OF RELEVANT AND ACCURATE MATERIAL DATA

Using the approach of metallurgical modelling of welding phenomena, we need a lot more physical data which can be assigned to very specific problem areas. Some of the data can be taken from other domains with similar mechanisms, like the many approaches from solidification studies which can be applied to predict the weld-pool agitation and the effect of elements which influence the surface energy. The same is true for model data to describe temperature profiles, transformation, precipitation etc.

In the case of residual stress calculation in weldments we are concerned with uncertainties regarding

— distribution of heat input source,
— individual or lumped pass approach,
— thermophysical data up to 3000°C,
— heat transfer coefficient = *f*(agitation),
— microstructural changes = *f*(chemical composition, grain size),
— law of mixing, when the formation of the phases (F,P,B,M) is treated individually,
— constitutive equations for the mechanical response during cooling as a function of temperature, strain, strain rate and microstructure,

— ferrite, pearlite, bainite strength = *f*(chemical composition, cooling rate).

In order to predict the precipitation kinetics and their influence on the toughness behaviour we need

— thermodynamic data including interaction energies,
— mutual solubility of carbonitrides,
— effective diffusion coefficients for kinetics,
— size distribution effects, etc.

In order to model the flow phenomena in weld pools, it was shown[18] that the final results (weld pool geometry, surface velocity, peak temperature) depend very strongly on

— thermophysical properties and diffusivities in the liquid and solid,
— dependence of surface energy on temperature and chemical composition,
— viscosity = *f*(temperature, chemical composition) and
— energy absorption coefficient.

From these examples we can draw the conclusion that we always need experimental investigations in order to generate model specific data and to verify the model results. An excellent tool for this task is a Gleeble machine, where particular effects can be tested separately using well defined boundary conditions.[19] The high heating rate and the possibility of performing coupled thermo-mechanical tests make this machine indispensable for any modelling work in the field of weldability.

CONCLUSIONS

Weldability models are based on our current knowledge of metallurgy, derived from models for microstructural changes in the base material and adjusted to the rapidly changing thermal cycle during welding.

However, the limits of fundamentally based models remind us of phenomena which we have not yet understood. Although white spots are evident, it is necessary to concentrate our efforts on th em. In addition, several microstructural models need special material data, which are difficult to generate or rarely published, but a combined action could improve this situation. Finally, it can be concluded that any modelling progress is mainly based on an improved understanding of the underlying phenomena and, although numerical modelling may be very attractive, experimental research work should not be neglected.

REFERENCES

1. H. CERJAK and K.E. EASTERLING (eds): *Mathematical Modelling of Weld Phenomena*, The Institute of Materials, London, 1993.
2. B. BUCHMARY: Computer in der Werkstoff- und Schweißtechnik, Anwendung von mathematischen Modellen, *Fachbuchreihe Schweißtechnik*, 1991, vol. 112, DVS-Verlag Düsseldorf.
3. J. GOLDAK: in: *Recent Trends in Welding Science and Technology*, eds. S.A. DAVID and J.M. VITEK: ASM International, 1990, 71–82.
4. T. DEBROY: in: International Trends in Welding Science and Technology, eds. S.A. DAVID and J. M. VITEK, *ASM International*, 1993, 17–25.
5. Y. UEDA *et al.*: *Transactions of JWRI*, 1992, **21** (2), 251–283.
6. J. STRID and K.E. EASTERLING: *Acta Metall.*, 1985, **35** (11), 2057–2074.
7. H. KAMMERSTETTER: PhD Thesis, Graz University of Technology, 1989.
8. B. BUCHMARY and J.S. KIRKALDY: in: *Fundamentals and Applications of Ternary Diffusion*, ed. G.R. PURDY: Pergamon Press, 1990, 164–172.
9. B. BUCHMARY: in: *Mathematical Modelling of Weld Phenomena*, eds. H. CERJAK and K.E. EASTERLING: The Institute of Materials, London, 1993, 227–240.
10. B. BUCHMARY, R. PRADER and H. CERJAK: 'Thermodynamische Berechnung von Diffusionsvorgängen in Gradientenwerkstoffen', *VGB-Konf. Forschung in der Kraftwerkstechnik 1993*, 24./25.2.1993, Essen, paper C7/1–16.
11. U. DILTHEY, G. HABEDANK, T. REICHEL, W. SUDNIK and A. IWANOW: *Schweißen und Schneiden*, 1993, **45** (3), pp. 148–153.
12. H.K.D.H. BHADESHIA and L.-E. SVENSSON: in *Mathematical Modelling of Weld Phenomena*, (eds. H. CERJAK and K.E. EASTERLING), The Institute of Materials, London, 1993, 109–178.
13. H. BAUMGARDT and C. STRASSBURGER: in: Thyssen Techn. Berichte H.1/85, 42–49.
14. G.M. EVANS: *Oerlikon-Schweißmitt.*, 1991, **49**, 18–39.
15. B. BUCHMARY: *Proc. Int. Conf. Computer Technology in Welding*, Brighton, 4–7 June 1990, Paper 32.
16. W. SCHWEIGER: in: Erfolgreiche Anwendungen von Datenbanken, Expertensystemen und Simulationen in der Oberflächentechnik, VDI-Berichte 936, 1992, 197–205.
17. K. MUNDRA, T. DEBROY, T. ZACHARIA and S.A. DAVID: *Welding Research Supplement*, Sept. 1992, 313–320.
18. Proc. Conf. Physical Simulation of Welding, Hot Forming and Continuous Casting, CANMET, Ottawa 1988.

COOLING TIME PARAMETER AND HARDENABILITY ESTIMATION OF HAZ IN WELDING OF MEDIUM, HIGH CARBON MACHINE STRUCTURAL STEELS

FUKUHISA MATSUDA*

Welding Research Institute, Osaka University, 11-1, Mihogaoka Ibaraki-shi, Osaka 567, Japan

WU SHYUAN LIU**

Isuzu Motors Limited

ABSTRACT

Hardenability is one of the important factors for estimating weldability of steels, therefore a cooling time parameter ($\Delta t_{8/5}$) was used for estimating the microstructure and hardenability of lower carbon steels.

However, it is not clear whether the parameter $\Delta t_{8/5}$ is suitable or not for higher carbon steels.

Therefore this paper investigates the suitable cooling time for these steels using a Gleeble tester. Furthermore by using the new cooling parameter ($\Delta t_{8/3}$) a carbon equivalent estimation equation on hardenability has been obtained.

The main conclusions obtained are as follows:

1. The cooling time parameter of $\Delta t_{8/3}$ is better than that of $\Delta t_{8/5}$ for estimating the hardenability of HAZ for higher carbon steels whose carbon content is more than 0.3wt.%.

2. The hardness of HAZ for medium and high carbon (C≥0.3%) machine structural steels could be predicted within 15% deviation by using the analysis in combination with the parameter $\Delta t_{8/3}$ and carbon equivalent.

* Professor

** formerly Graduate Student, Osaka University.

INTRODUCTION

It is well known that the hardenability of HAZ is one of the important factors for estimating the weldability of steels. Therefore the cooling time parameter from 800 to 500°C ($\Delta t_{8/5}$) has usually been used in defining welding procedures which avoid martensite in low carbon high strength steels.

However, it was not clear if $\Delta t_{8/5}$ is also a useful parameter for higher carbon steels, because the M_s temperature of these steels is lower than that for low carbon steels.

This paper investigates the usefulness of applying the $\Delta t_{8/5}$ parameter for the HAZ of higher carbon steels.

It was demonstrated conclusively that the $\Delta t_{8/3}$ parameter is better than $\Delta t_{8/5}$ for these steels. A new equation is derived using the $\Delta t_{8/3}$ parameter and carbon equivalent.

EXPERIMENTAL PROCEDURES

STEELS USED

Medium, high carbon steel bars of JIS S25C, S35C, S40C, S55C, SKV70, SK5, SK3 and medium, high carbon low alloy steel bars of JIS SC*r*440, SUJ2, SCM435,445, SNCM420,439,447 were used as base metal. All of these are commercial steel bars whose chemical compositions are given in Table 1.

EXPERIMENTAL PROCEDURES

The shape and size of the simulated round bar specimen was 10 mm in diameter and 55 mm in length with a circular notch in the centre of the length which is the same as that shown in the report.[1]

The simulated thermal cycle is given by a dynamic testing machine (Gleeble 1500) which is based on resistance heating. The cooling time parameters are defined as $\Delta t_{8/5}$ from 800 to 500°C, and $\Delta t_{8/3}$ from 800 to 300°C. Lower and upper cooling times of $\Delta t_{8/3}$ for the given parameter $\Delta t_{8/5}$ are shown in Fig. 2, the reason for which will be described in the 'Selection of the cooling time parameter' . In this research the given $\Delta t_{8/5}$ were adopted for 3.5, 6, 21, 50, and 100 s for which $\Delta t_{8/3}$ varied individually from 9.7 to 13.7, from 17 to 28, from 70 to 135, from 187 to 310 and from 280 to 400 s. Moreover an additional $\Delta t_{8/3}$ parameter was tried to extend to 2500 s in order to introduce the equation of hardenability of the HAZ.

EXPERIMENTAL RESULTS AND DISCUSSIONS

SELECTION OF THE COOLING TIME PARAMETER

Figure 1 shows the relationship between $\Delta t_{8/5}$ and hardness for continuous cooling simulated HAZ for each steel used. Generally, increasing $\Delta t_{8/5}$ shows a decrease in hardness in each steel, although the variation of hardness is larger in higher carbon steel as the hardness of martensite is higher in higher carbon steel. It is estimated that between 3.5 and ca. 30 s of $\Delta t_{8/5}$ there is a big microstructural change from martensite to fully ferritic-pearlitic or bainitic structures.

In Fig. 1, there is a dotted line which is calculated from Okumura's equation[2] using the chemical composition of JIS S55C in Table 1. As a result, the data of dotted line are lower than the investigated data of S55C on hardness in this cooling time range. Especially, as the difference is small in the

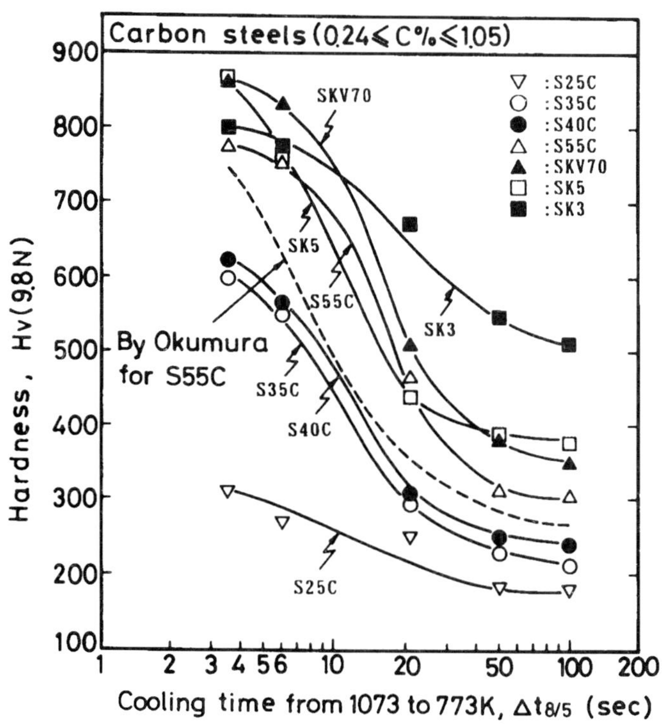

Figure 1. Effect of cooling time ($\Delta t_{8/5}$) on hardness.

Table 1 Chemical compositions of steels used (as received base metal).

	steel	Chemical Composition (wt. %)									
		C	Si	Mn	P	S	Ni	Cr	Mo	N	O
Carbon steels	S25C	0.24	0.20	0.39	0.018	0.018	0.04	0.04	0.01	—	—
	S35C	0.33	0.24	0.69	0.018	0.018	0.02	0.03	0.01	—	—
	S40C	0.39	0.25	0.70	0.026	0.028	0.02	0.12	0.01	—	—
	S55C	0.54	0.26	0.69	0.022	0.011	0.05	0.19	0.02	—	—
	SKV70	0.72	0.22	0.75	0.025	0.023	0.07	0.22	0.01	0.0096	0.002
	SK5	0.89	0.24	0.29	0.022	0.020	0.04	0.12	—	0.0043	0.001
	SK3	1.05	0.33	0.98	0.020	0.018	0.05	0.47	—	0.0130	0.001
Cr steels	SCr 440	0.40	0.29	0.76	0.025	0.012	0.07	1.09	0.02	0.0120	0.004
	SUJ2	0.96	0.26	0.46	0.015	0.009	0.05	1.46	0.01	0.0073	0.001
Cr-Mo steels	SCM 435	0.36	0.23	0.80	0.022	0.013	0.10	1.11	0.15	—	—
	SCM 445	0.45	0.24	0.78	0.017	0.018	0.07	1.07	0.15	0.0140	0.001
Ni-Cr-Mo steels	SNCM420	0.21	0.24	0.61	0.020	0.011	1.61	0.46	0.21	—	—
	SNCM439	0.39	0.28	0.73	0.012	0.007	1.72	0.69	0.17	0.0094	0.003
	SNCM447	0.48	0.26	0.81	0.025	0.012	1.65	0.72	0.15	—	—

* As received base metal.

range of ca. $\Delta t_{8/5}$: 6~21 s, and is not large at $\Delta t_{8/5}$: 3.5 s and more in the range than 50 s.

The authors have investigated the reason why there is so much difference.

Using Inagaki's monographs,[3] the authors calculated both $\Delta t_{8/5}$ and $\Delta t_{8/3}$ for welding conditions actually used (weld heat input 9~63 kJ/cm, preheating less than 573 K, plate thickness 12~34 mm) for SAW and SMAW. Figure 2 shows the relationship between $\Delta t_{8/5}$ and $\Delta t_{8/3}$ in the measured results. As a result, $\Delta t_{8/3}$ is shown in between the two lines of lower and upper cooling time with each $\Delta t_{8/5}$ parameter. Generally, increasing $\Delta t_{8/5}$ shows an increase in difference of $\Delta t_{8/3}$, e.g. the difference is ca. 65 s for 21 s of $\Delta t_{8/5}$ and ca. 123 s for 50 s. Therefore, even at same $\Delta t_{8/5}$, the cooling time from 500 to 300°C ($\Delta t_{5/3}$) is considerably different depending on the changing heat input, preheating temperature or plate thickness.

Figure 3 shows the hardness change against $\Delta t_{8/5}$ at $\Delta t_{8/5}$: 3.5, 6, 21, 50 and 100 s with various $\Delta t_{8/3}$ in S35C, S55C, SKV70. The hardness difference between the lower cooling time (solid mark) and upper cooling time (open mark) at each $\Delta t_{8/5}$ of 3.5, 6, 21, 50, 100 s is then compared. In the shortest cooling time of $\Delta t_{8/5}$, 3.5 s, there is no difference in hardness between the solid and open marks in each steel, the structure of which was almost fully martensite despite a variation of $\Delta t_{8/3}$. Moreover, in the longest cooling time of $\Delta t_{8/5}$ in each steel (S35C, 21 s; S55C, 50 s; SKV70, 50 s) there is also no difference in hardness for each steel, the structure of which was a mixture of

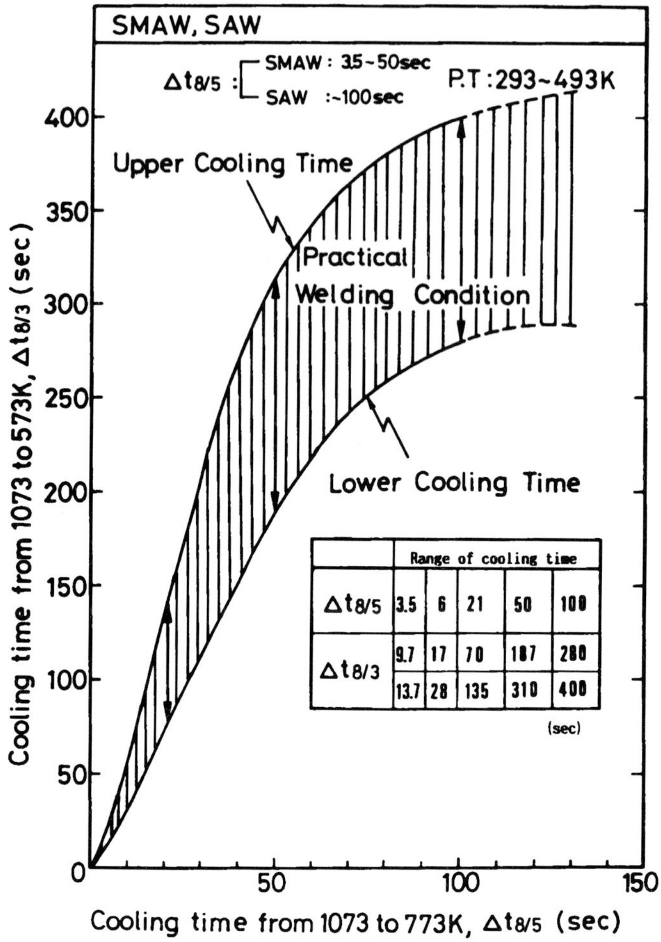

Figure 2. Relationship between $\Delta t_{8/5}$ and $\Delta t_{8/3}$ in actual welding operation.

ferrite and pearlite. However, for intermediate values of $\Delta t_{8/5}$, there is a remarkable difference in hardness for each steel between the solid and open marks. In these $\Delta t_{8/5}$, the ratios of martensite and bainitic structures were different depending on the difference of $\Delta t_{8/3}$, that is open and solid marks, even at the same $\Delta t_{8/5}$.

The relationship for $\Delta t_{8/5}$ and the hardness difference (ΔH_v) between the lower and upper cooling time of $\Delta t_{8/5}$ in S25C, S35C, S55C and SKV70 is shown in Fig. 4. As a result, the maximum hardness difference increases with C content. Moreover, the maximum hardness occurs at $\Delta t_{8/5}$: 6–8 s in S25C and S35C, at 10 s in S55C and at 21 s in SKV70.

Figure 5 shows a microstructural change of SKV70 at $\Delta t_{8/5}$: 21 s, in which (a) shows 70 s and (b) 135 s for $\Delta t_{8/3}$. In Fig. 5 (a) there is ca. 50% martensite and 50% bainite and pearlite, but in Fig. 5 (b) there is only ca. 15% martensite

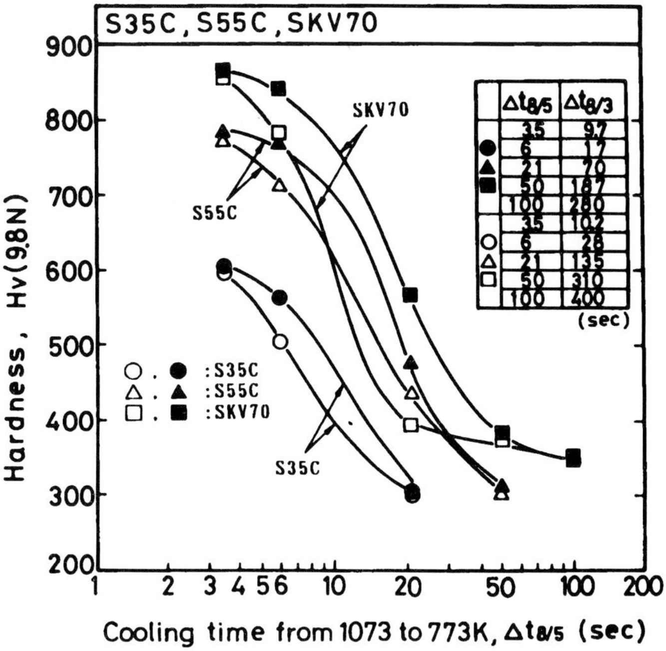

Figure 3. Effect of cooling time of $\Delta t_{8/3}$ on hardness at given $\Delta t_{8/5}$.

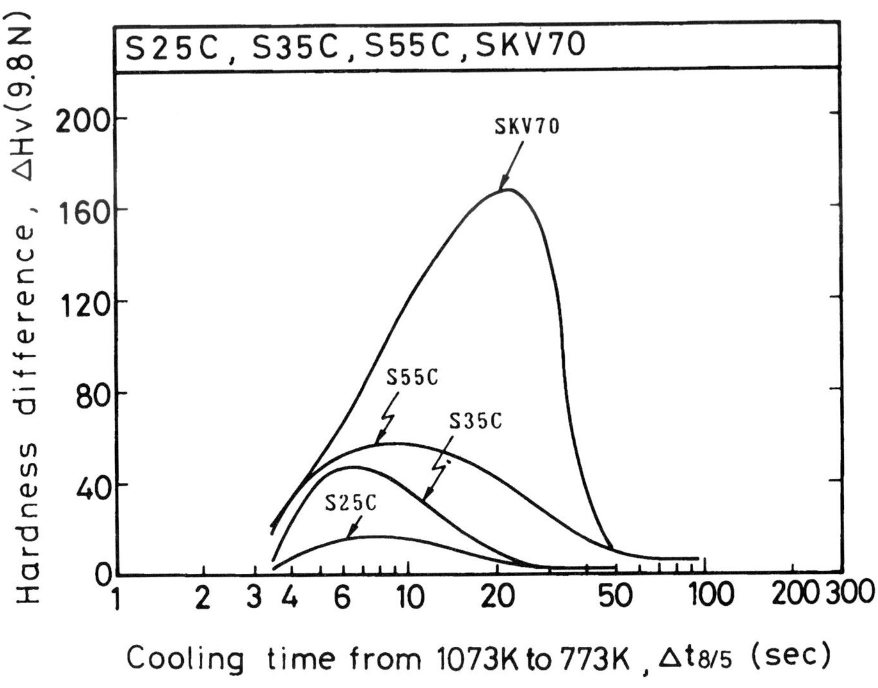

Figure 4. Effect of cooling time of $\Delta t_{8/3}$ on hardness difference (ΔH_V) at given $\Delta t_{8/5}$.

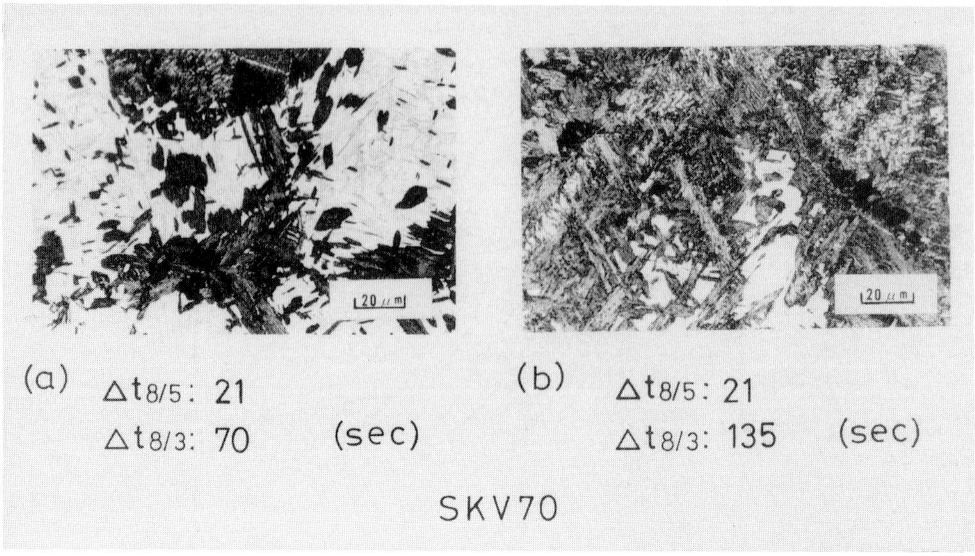

Figure 5. Microstructural change of SKV70 at 70 and 135 s of $\Delta t_{8/3}$ at $\Delta t_{8/5}$:21 s.

and 85% bainite and pearlite. Therefore, within the actual welding operation the microstructure and hardness were changed by varying $\Delta t_{8/3}$, even at the same $\Delta t_{8/5}$ in these steels.

PROPOSAL FOR THE ESTIMATION EQUATION ON HARDENABILITY

The authors have confirmed previously that the $\Delta t_{8/3}$ parameter is better than $\Delta t_{8/5}$ for estimation of the hardenability on HAZ for plain, medium and high carbon steels. We have then introduced an estimation equation for general medium and high carbon machine structural steels including low level *Cr, Mo* and *Ni* alloying elements using the $\Delta t_{8/3}$ parameter and carbon equivalent.

Estimated result of hardenability

Figure 6 shows the relationship between $\Delta t_{8/3}$ (12, 20, 71, 195, 270, 480, 640, 1280 and 2440 s) and hardness for continuous cooling simulated HAZ in medium, high carbon and low alloy steels ($0.21 \leq C\% \leq 1.05$).

Generally, increasing the $\Delta t_{8/3}$ parameter shows a decrease in hardness in each steel, and there is a remarkable decrease in hardness of more than 20 s in carbon steels. However, in comparison with the results from carbon steels, this decrease in hardness is not remarkable in low alloying steels, especially in SNCM439, 447 and SUJ2, with additions of ca. 0.7*Cr* and 1.7% *Ni,* and 1.5*Cr.*[4]

The decrease in hardness with the continuous cooling method is difficult within the actual range of welding heat input.

As a result of microscopic investigation of the structural change, on the left side of the figure, the dotted line shows the maximum cooling time in order to show 100% martensitic structure, and on the right side the other dotted line shows the minimum cooling time in order to show 0% bainitic structure, namely, 100% ferritic and pearlitic structure in these steels. However, in SNCM439, 447 and SUJ2 the minimum cooling time of 100% ferritic and pearlitic structures was estimated by CCT diagrams which were measured.[5]

Introduction to the estimation equation on hardenability

From Fig. 6 the authors have simply estimated that the relationship between hardness (H_v) and cooling time ($\Delta t_{8/3}$) is as shown in Fig. 7. As a result, the shape of the curve for the relationship between H_v and $\Delta t_{8/3}$ are only shown by the following equation:[2]

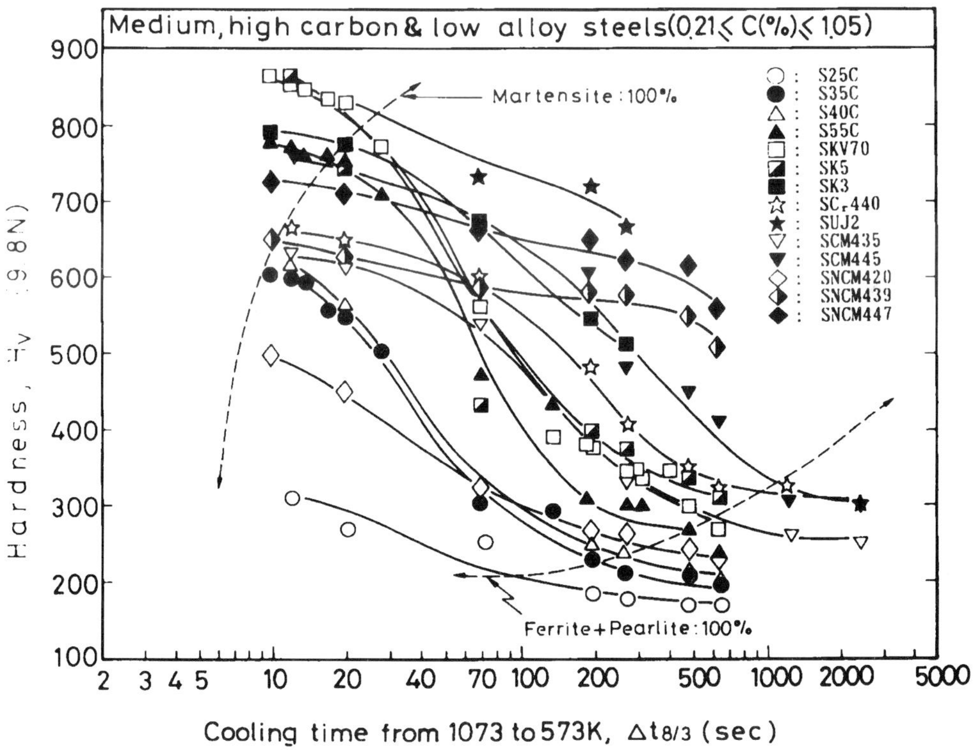

Figure 6. Relationship between $\Delta t_{8/3}$ and hardness of medium, high carbon and low alloy steels.

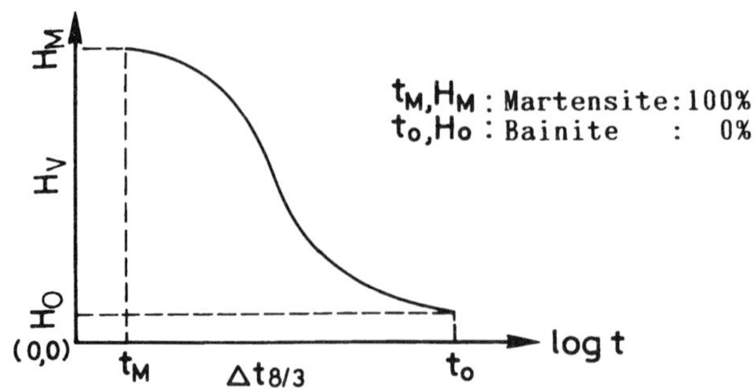

Figure 7. Shape of essential curve between 100% martensite and 0% bainite.

$$H_V = A \ . \ \tan^{-1} \Delta t_{8/3} + B \tag{1}$$

where A and B are constants.

Now if the parameters of hardnesses of 100% martensite (H_M) and 0% bainite (H_0) and the cooling times of 100% martensite (t_M) and 0% bainite (t_0) are introduced, Equation 1 will be converted to Equation 2 by using the four parameters (H_M, H_0, t_M and t_0):

$$H_V = (H_M + H_0)/2 + (H_M - H_0)/2 \times \tan^{-1}[(2 \log \Delta t_{8/3} - \log t_0 - \log t_M)/2]\tan^{-1}[(\log t_M - \log t_0)/2] \tag{2}$$

Here, the unit of cooling time is shown by the logarithm.

Therefore, the relationships between the four parameters (H_M, t_M, H_0, t_0) and their chemical composition have been individually measured by computer analysis as follows:

100% martensitic hardness, H_M

Generally, the hardness of 100% martensite are only decided by the C content. The results of the authors' measurements agree well with those of Liumura.[2] The relationship between the hardness and C carbon is shown in Fig. 8. ○ shows the data of plain carbon steels, and ● shows the data of low alloying steels. The equation for the relationship between H_M hardness and C content is shown in Equation 3:

$$H_M = 1265C - 182C^2 - 466C^3 - 10C^4 + 218 \tag{3}$$

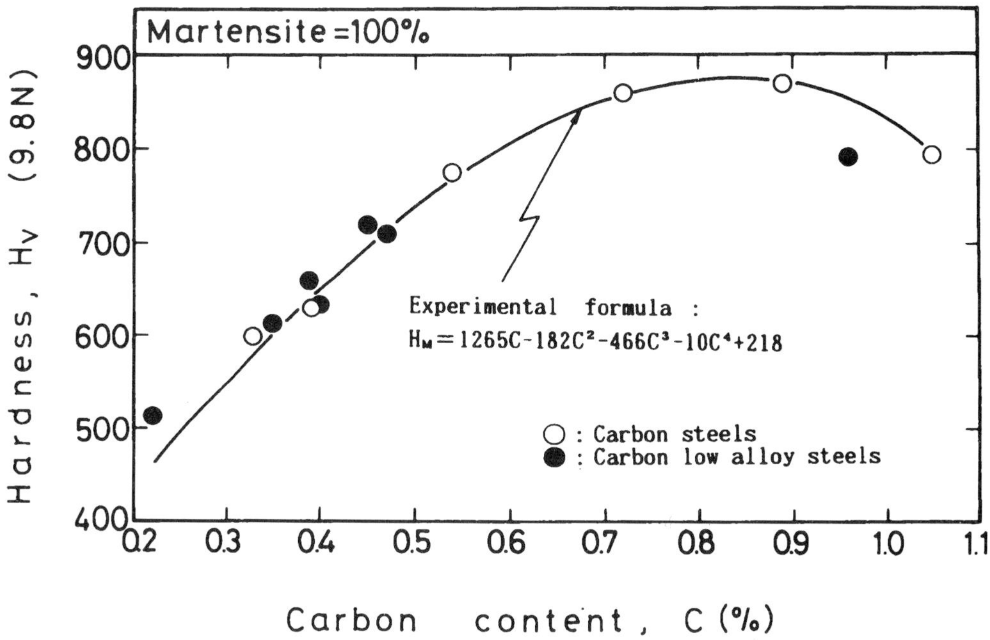

Figure 8. Relationship between carbon content and hardness on 100% martensite.

Maximum cooling time for 100% martensite, t_M

The relationships between cooling time, log t_M, and the carbon equivalent I for t_M, C_{eq} I (t_M), in multiple regression analysis are shown in Equations 4 and 5:

$$\log t_M = 0.48C_{eq}\ \mathrm{I}\ (t_M) + 0.2 \tag{4}$$

Here:

$$C_{eq}\ \mathrm{I}\ (t_M) = \mathrm{C} + M_n + (2)S_i + (1/10)N_i + (4/5)C_r + (1/5)\ M_o \tag{5}$$

Figure 9 shows the relationship between the estimated and experimental values of the maximum cooling time for 100% martensite in regression analysis with Equations 4 and 5, where the coherency which is represented by the square of the correlation coefficient (R^2) is 0.93.

100% ferrite and pearlite hardness, H_0

The relationships between H_0 and suitable carbon equivalents, C_{eq} II (H_0) are represented by multiple regression analysis which are shown in Equations 6 and 7:

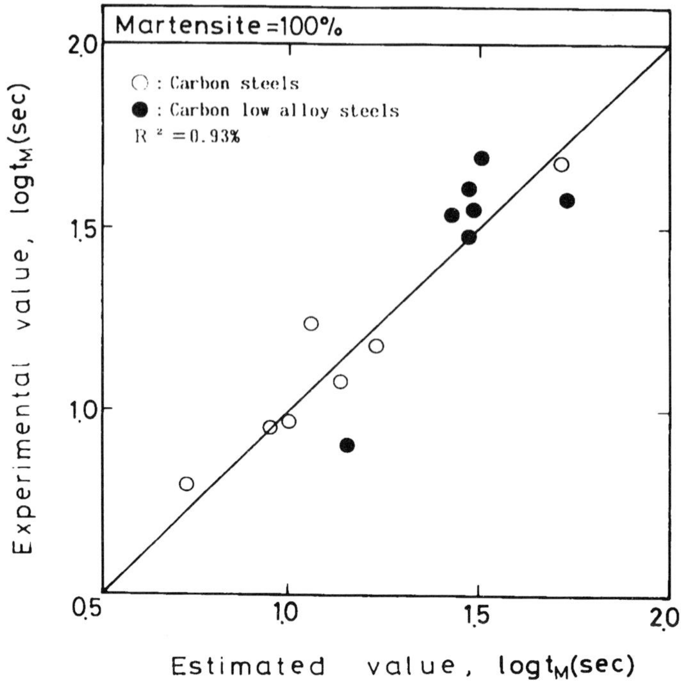

Figure 9. Comparison of estimated and experimetnal values of cooling time on 100% martensite.

$$H_0 = 173C_{eq}\ \text{II}\ (H_0) + 140 \tag{6}$$

Here:

$$C_{eq}\ \text{II}\ (H_0) = C + (1/33)\ M_n + (1/14)\ S_i + (1/20)\ N_i + (1/4)\ C_r + (8/5)\ M_0 \tag{7}$$

Figure 10 shows the relationship between estimated and experimental values of hardness (H_0) on 100% ferrite and pearlite structures in regression analysis with Equations 6 and 7, where the square of the correlation coefficient (R^2) is 0.92.

Minimum cooling time for 100% ferrite and pearlite, t_0

The relationships between the cooling time, log t_0, and the carbon equivalent C_{eq} III (t_0) in multiple regression analysis are shown in Equations 8 and 9:

$$\log\ t_0 = 0.8C_{eq}\ \text{III}\ (t_0) + 1.57 \tag{8}$$

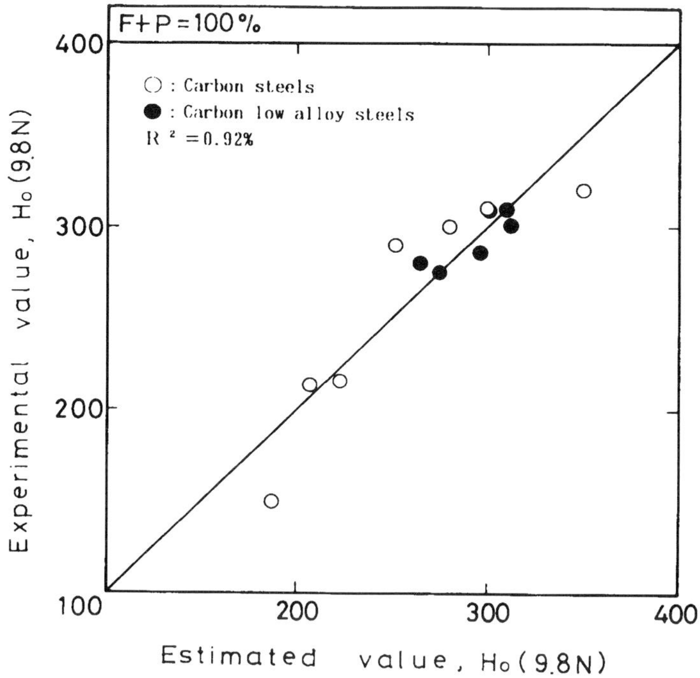

Figure 10. Comparison of estimated and experimetnal values of hardness on 100% F + P (ferrite + pearlite).

Here:

$$C_{eq}\ \text{III}\ (t_0) = C + (1/4)\ M_n + (3/4)\ S_i + (2/5)\ N_i + (11.10)$$

$$C_r + (2/5)\ M_0 \qquad (9)$$

Figure 11 shows the relationship between the estimated and experimental values of the cooling time (log t_0) for 100% ferrite and pearlite structures in regression analysis with Equations 8 and 9, where the square of the correlation coefficient (R^2) is 0.92.

Proposal for the equation on hardenability estimation

There are well characterised equations represented by computer analysis where the correlation coefficients are more than 0.92 described above.

By substituting Equations 3, 4, 5, 6, 7, 8 and 9 into Equation 2, Equation 10 is introduced which can estimate the hardenability of HAZ in the welding of medium, high carbon machine structural steels. That is:

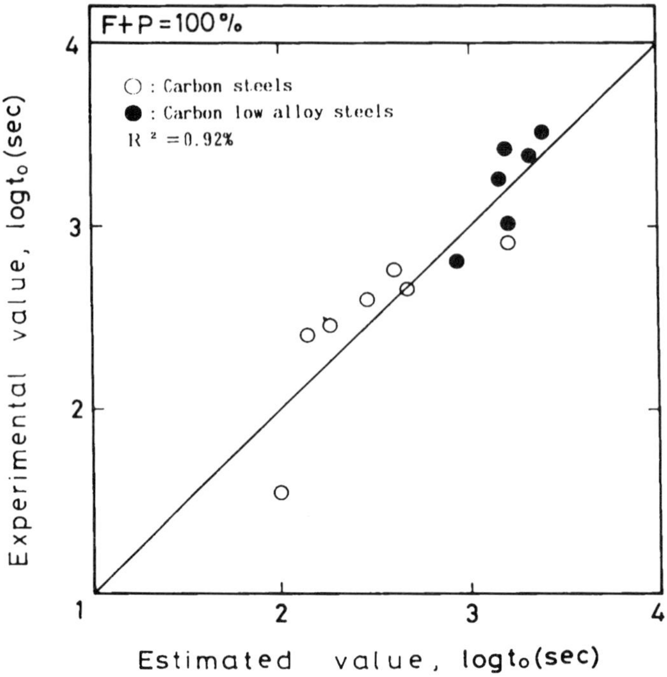

Figure 11. Comparison of estimated and experimental values of cooling time on 100% F + P (ferrite + pearlite).

$$H_V = (633C - 91C^2 - 233C^3 - 5C^4 + 178.8 + 86.5C_{eq}\,\mathrm{II}\,(H_0)) +$$

$$(633C - 91C^2 - 233C^3 - 5C^4 + 39.2 - 86.5C_{eq}\,\mathrm{II}\,(H_0)) \times$$

$$\tan^{-1}(\log t_{8/3} - 0.24C_{eq}\,\mathrm{I}\,(t_M - 0.4C_{eq}\,\mathrm{III}\,(t_0) - 0.89)\;/$$

$$\tan^{-1}(0.24C_{eq}\,\mathrm{I}\,(t_M) - 0.4C_{eq}\,\mathrm{III}\,(t_0)) \qquad (10)$$

where $0.21\% \leq C \leq 1.05\%$, $Si \leq 0.33\%$, $Mn \leq 0.98\%$, $Ni \leq 1.72\%$, $Cr \leq 1.46\%$, $Mo \leq 0.21\%$.

Figure 12 shows the relationship between the estimated and experimental value of hardness (H_V) with regression analysis as in Equation 10. As a result, part of the estimated values marked * are detached from the range within 15% of the experimental value, that is those steels which are S25*C* and SNCM420 with 0.24 and 0.21%C.

Therefore, Equation 10 seems to be of value for medium, high carbon machine structural steels with a carbon content more than 0.3%.

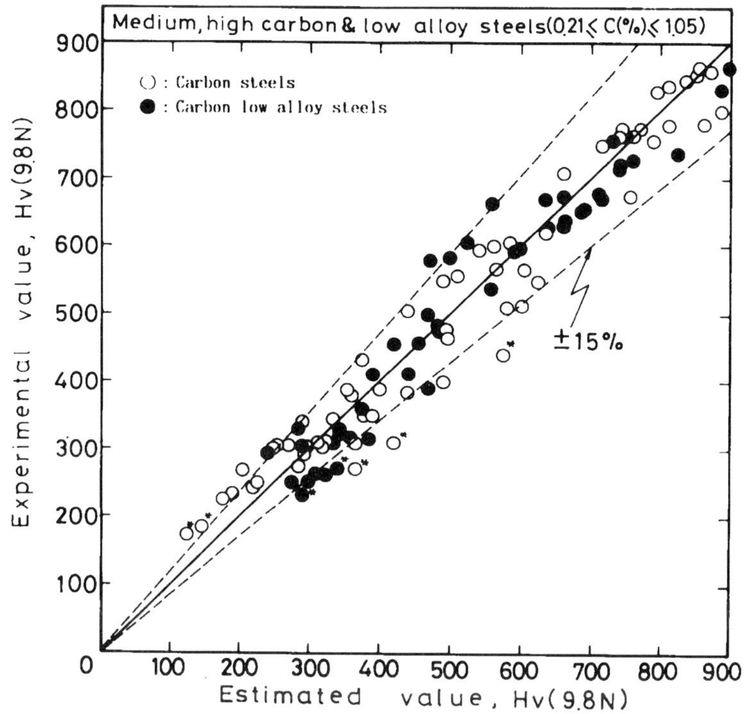

Figure 12. Relationship between estimated and experimental hardness on the estimation equation for medium, high carbon and low alloy steels.

CONCLUSIONS

In order to estimate the hardenability of HAZ in welding of medium and high carbon machine structural steels (0.2 to 1.05% *C*), this paper has treated the selection of the cooling time parameter and the proposal for the estimation equation on hardenability of HAZ using both the cooling time parameter and carbon equivalents. These experiments were done using Gleeble 1500.

The main conclusions obtained are as follows:

1. The cooling time parameter of $\Delta t_{8/5}$ has been used so far for estimation of hardenability of HAZ for conventional low-carbon, high-strength steels. However, in steels with a carbon content of more than 0.3% the cooling time parameter of $\Delta t_{8/3}$ was better to evaluate the hardenability.

   The hardness was a little differential depending on the cooling time below 500°C. The hardness difference was augmented with increasing carbon content.
2. For these steels the hardnesses of 100% martensite (H_M) and 100% ferrite + pearlite structures (H_0), and the maximum cooling time for

100% martensite (t_M) and the minimum cooling time for 100% ferrite + pearlite structures (t_0) have been estimated by individual carbon equivalents using multiple regression analysis. The estimated equation agrees with the experimental values.

3. With use of parameters of H_M, H_0, t_M and t_0 the hardenability estimation was tried for these steels. The essential equation used is as follows:

$$H_V = (H_M + H_0)/2 + (H_M - H_0)/2 \times \tan^{-1} [(2 \log \Delta t_{8/3} - \log t_0 - \log t_M)/2] \ / \ \tan^{-1} [(\log t_M - \log t_0)/2]$$

4. As a result Equation 10 was obtained as an estimation equation. The relationship between estimated and experimental values concerning hardness in these steels was in good agreement within ±15% except for steels of C:0.21%. Therefore, the authors have proposed here that Equation 10 is used as the estimation equation of hardenability of HAZ for such steels as 0.30% $\leq C \leq$1.05%, $Si \leq$0.33%, $Mn \leq$0.98%, $Ni \leq$1.72%, $Cr \leq$1.46%, Mo0.21%.

REFERENCES

1. F. MATSUDA: Toughness improvement of the HAZ for machine structural carbon and low alloys steels (Report 1), Transactions of JWRI, 1989, **18**, 2, 79.
2. M. OKUMURA: Effects of elements on the hardenability and cold crack susceptibility of HAZ of constructional steel weldments, PhD thesis of Osaka University 1990, 6 (in Japanese).
3. M. INAGAKI and T. KASUGAI: Welding of high strength steels, 1965. Cited in *Welding Metallurgy* MATSUDA, F. (ed.), 1972, 4 (in Japanese).
4. T. KASUGAI and M. INAGAKI: Effect of alloying elements on transformation behavior in synthetic weld heat-affected zone of steels (V), *J. Japan Weld. Soc.*, 1975, **44**, No. 8, 73 (in Japanese).
5. JAPAN METAL SOC: *Metal Handbook*, Maruzen, 1984, 400–404 (in Japanese).

PRACTICAL APPLICATION OF THE PUBLISHED MODEL FOR ESTIMATING THE FRACTIONS OF DIFFERENT MICROSTRUCTURE COMPONENTS IN ONE-PASS WELD METAL

A.S. KOTELCHUK and O.B. GLUSCHENKO
E. O. Paton Electric Welding Institute, Kiev, Ukraine

This report presents some results of the research conducted at the E. O. Paton Electric Welding Institute in the field of numerical analysis and prediction of the microstructures of a low-alloyed low-carbon one-pass weld metal. One of the most promising applications of the results of this work is optimising the development of welding consumables.

The required properties of the weld metal can often be described in terms of its microstructure composition. Thus, microstructure formation during the welding thermal cycle is of great importance. There are many publications where the relation composition–structure–property is presented analytically in the form of equations. (For instance, the prediction of ultimate strength and/or yield point by taking into account the strengthening effect of elements in solid solution plus the contribution of different structure components.)[1]

Many publications during the last ten years have been devoted to experimental metallographic analysis of weld metal aimed at finding out the mechanism of microstructure transformations.[2] These investigations make it possible to develop quantitative models that describe the influence of chemical composition and parameters of the welding cycle on microstructure composition of weld metal.

As far as joining materials by welding is concerned, the relations between 'What we have' and 'What we get' are quite complicated. Nevertheless they are described by the diagram shown in Fig. 1. This diagram does not pretend to include all ties, but it seems that the most significant ones are taken into

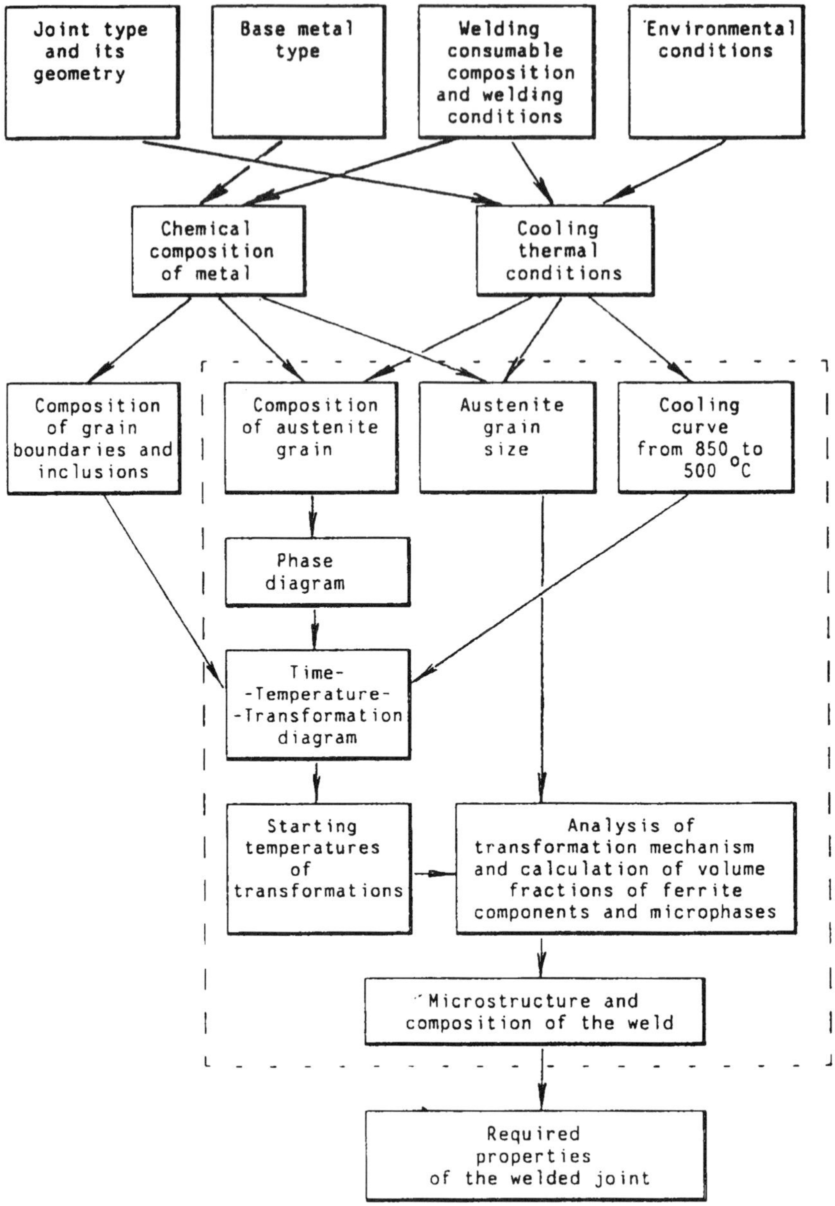

Figure 1. Main relations that should be taken into account when predicting the properties of welded joint.

account. The part of the diagram surrounded by dash lines was developed by Bhadeshia *et al.*[3–7] It was incorporated into developments at the E. O. Paton Institute. Practically it is the only model today that uses substantial physical formations. This fundamental character of the model fully compensates any losses of precision when compared with local regression models, because of its internal physical logic.

The main ideas of this model are as follows. Because of the high temperature gradients involved in arc welding, the solidification of low-alloyed steel weld proceeds with epitaxial growth of delta-ferrite at the fusion boundary in a cellular manner, the major axes following the direction of maximum heat flow. Allotriomorphic austenite then nucleates at the delta-ferrite cell boundaries and as the result of its anisotropic growth along the boundaries the columnar austenite grains closely resemble the original delta-ferrite morphology. The primary austenite grains are considered as hexagonal prisms, their length being much higher than the cross-sectional dimensions. Further primary microstructure formation continues through the growing of allotriomorphic (primary) ferrite that nucleates at the columnar austenite grain boundaries. This growth is controlled by a diffusion type mechanism. Then, as the temperature drops, the transformation by the displacement mechanism becomes kinetically more beneficial. Thus, the growth of Widmanstätten ferrite starts by mixed diffusion and through a displacement transformation mechanism controlled by the redistribution of carbon in the vicinity of the needle tip of side-plates. Two cases of growing side-plate ferrite are taken into account. The first is characterised by the following: Widmanstätten side-plate ferrite grows through the entire cross-section of the grain without 'impingement'. This case is typical for low-alloyed weld metals. The second corresponds to the growth of Widmanstätten ferrite with 'impingement' with the originating islands of acicular ferrite in the grain volume on non-metallic inclusions.[3–7]

This model was realised in the form of software for IBM PC compatible computers. The model is not closed and uses a number of parameters that cannot be estimated inside the model itself. Although the calculations are based on the fundamental relations developed by Bhadeshia,[8–13] when putting the model into practice some additions were made, aimed at a more precise determination of the parameters using the local built-in data base 'composition + thermal cycle = structure'. The additions introduced deal with the more precise estimation of the nucleus formation rate of Widmanstätten and acicular ferrite and of the kinetic parameters of primary ferrite growth using effective diffusive resistance. With this approach it seems promising to develop the model on the basis of a model for non-metallic inclusion formation to estimate the ('potential') rate of acicular ferrite nucleus formation. The prerequisites for this are provided by the physical mechanism, according to which favourable conditions for growth of acicular ferrite occur when the inclusions' surface is covered with the compounds of TiO or Al_2O_3.MnO, that have a lattice with small mis-matching to the crystallographic lattice of alpha-iron.

The main stages of parameter input, calculation and output of the results are shown in Figs. 2, 3, 4 and 5. Figure 2 presents the input of chemical composition, welding conditions and austenite grain size. Figure 3 presents the calculation of the isothermal time–temperature–transformation diagram (or

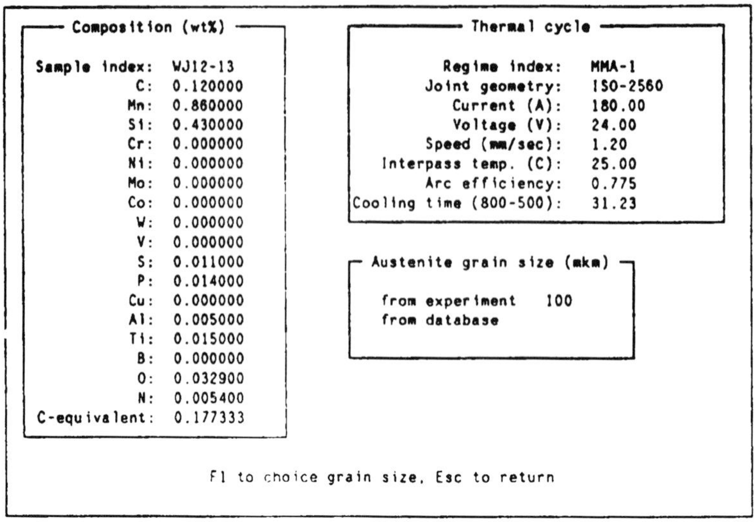

Figure 2. An example of the screen for parameter input which is required for estimation of microstructure volume fractions.

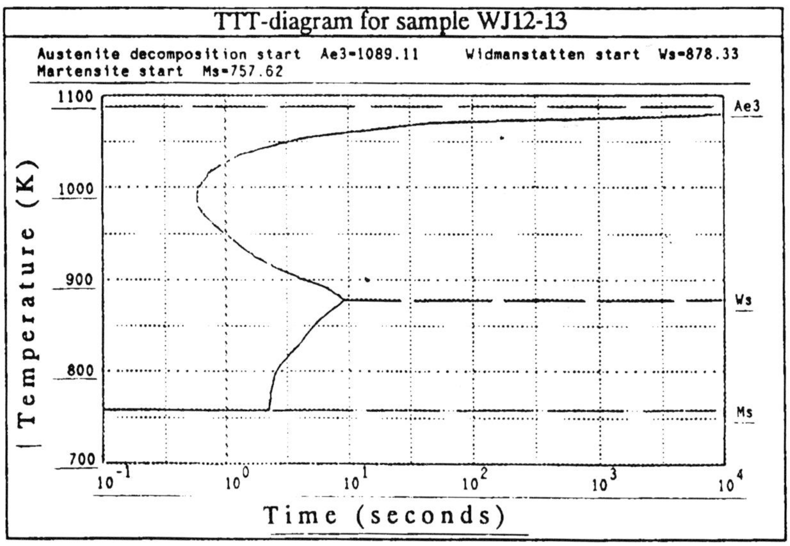

Figure 3. An example of calculated isothermal TTT-diagram screen.

use of diagram already calculated, as the program has the possibility to accumulate the results of the TTT-diagram calculation as well as input data, so that it looks like a small data base). Figure 4 shows the superimposition of the cooling curve corresponding to selected welding conditions and the determination of the austenite decomposition starting temperature under continuous cooling. Figure 5 shows the calculation of the volume fractions of microstructure components.

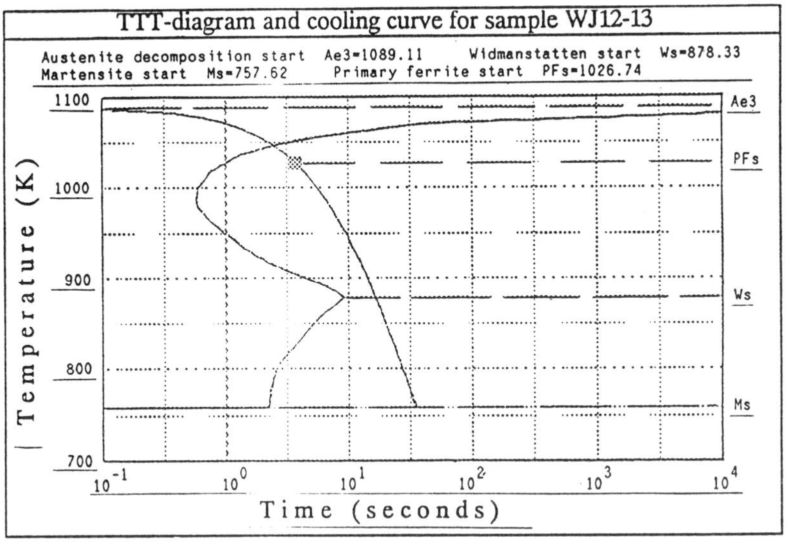

Figure 4. An example of the screen representing the TTT-diagram with cooling curve. Calculation of starting temperature of primary ferrite formation.

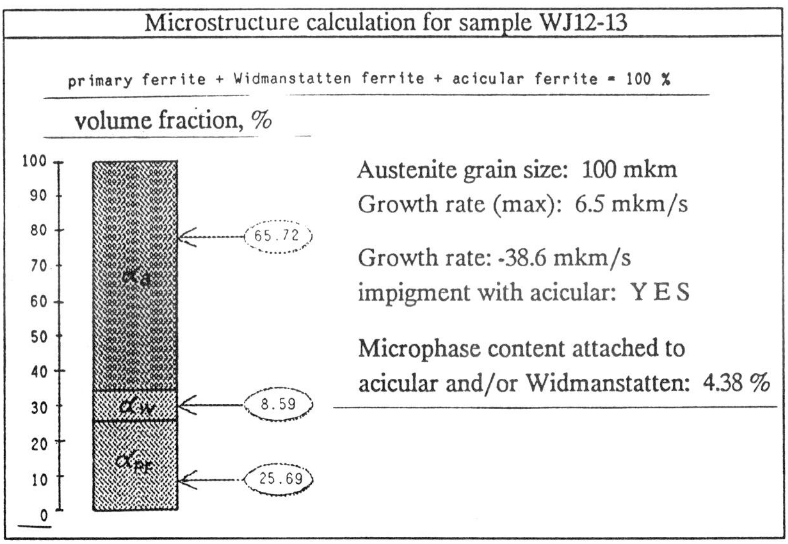

Figure 5. An example of the output screen of the microstructure volume fractions calculations

As already mentioned, these calculations cannot give the precise quantitative data on microstructure of weld metals; neither do these data seem to be so necessary. It appears that it is more significant to have the possibility to determine trends, in a simple numerical experiment, in changing the microstructure volume fractions when the chemical composition and/or welding conditions vary. Figure 6 (a and b) represent the influence of increasing heat input (in terms of increasing interpass temperature at manual arc welding) on the calculated microstructure of the weld metal of two

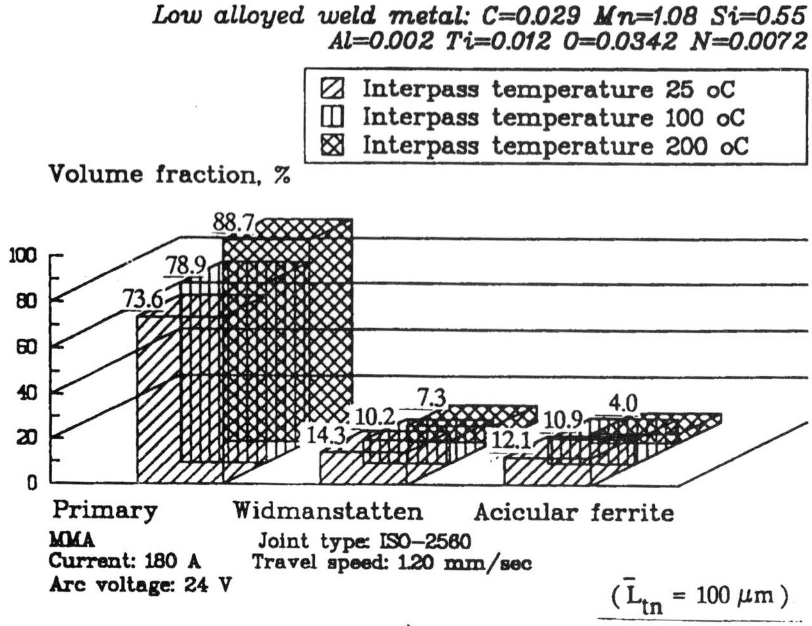

a)

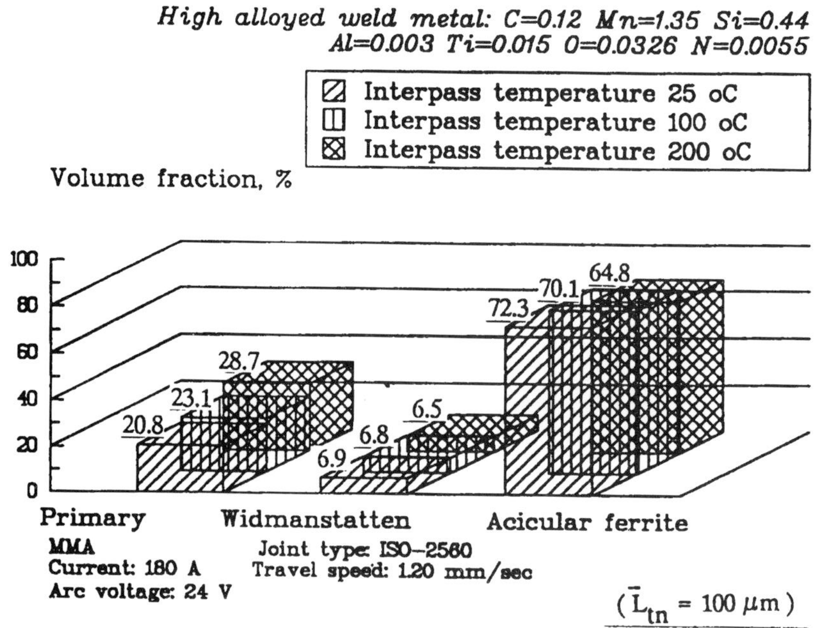

b)

Figure 6. Estimated influence of interpass temperature on microstructure composition of the weld metal.

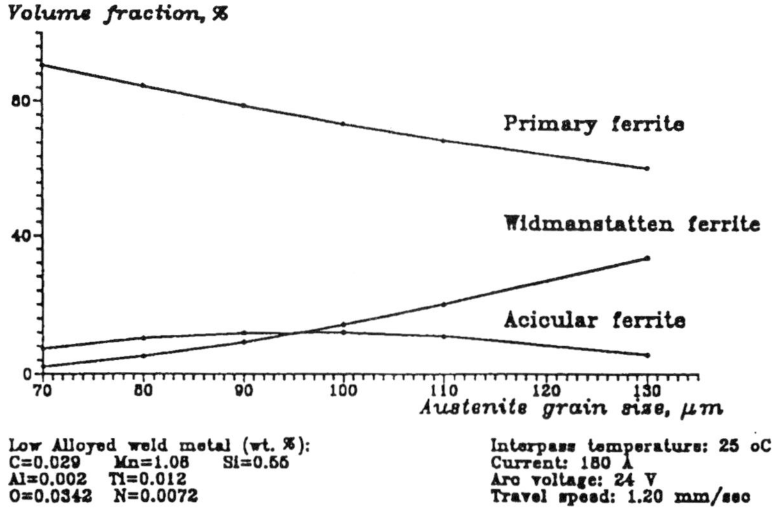

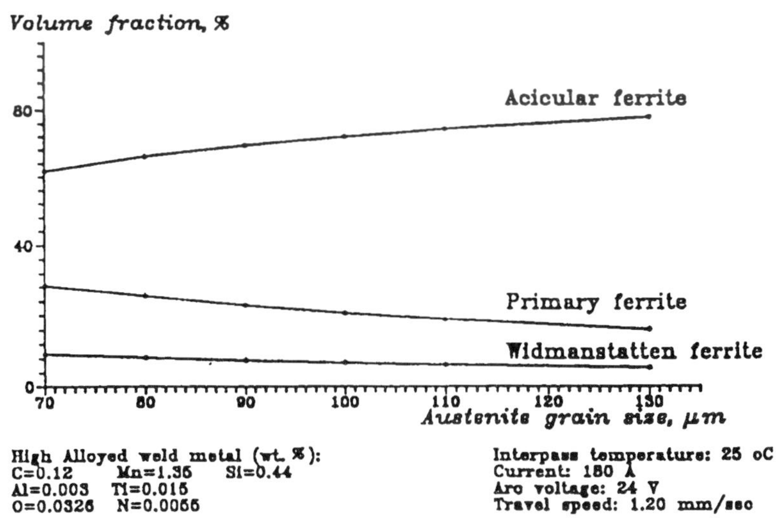

Figure 7. Estimated influence of austenite grain size (the mean lineal intercept measured on transverse section of the weld) on microstructure composition of the weld metal.

different types: low- and high-alloyed. They correspond to the two different mechanisms of transformation, mentioned above.

Unfortunately, attempts to estimate the primary austenite grain size using the approaches provided by the given model[6] have failed. This question remains open. One important parameter, which should be determined as accurately as possible, is the primary austenite grain size. Calculations in Fig. 7, a and b show the influence of this parameter on the volume fraction of different microstructure components. It is seen that the influence is more significant for low-alloyed weld metals in comparison with high-alloyed ones. In some cases it is even possible to determine the optimal grain size by reaching the maximum volume fractions of beneficial microstructure components.

The following conclusions show that the model based on ideas developed by Bhadeshia is applicable for the prediction of microstructure changes with varying outer conditions and parameters. As the result of its practical application the software is thus described.

Nevertheless the model has the possibility of further improvement. It seems that this can be done by taking into account the process of acicular ferrite nucleation using the same approaches that were used in the present model.

REFERENCES

1. A.A.B. SUGDEN and H.K.D.H. BHADESHIA: A model for the strength of the as-deposited regions of steel weld metals, *Metallurgical Transactions*, 1988, **19A**, 6, 1597–1602.
2. D.J. ABSON and R.J. PARGETER: Factors influencing the as-deposited strength, microstructure and toughness of MMA welds suitable for Mn–C steel fabrications, *International Metals Reviews*, 1986, **31**, 141–194.
3. H.K.D.H. BHADESHIA, L.E. SVENSSON, and B. GRETOFT: A model for the development of microstructure in low-alloy steel (Fe–Mn–Si–C) weld deposits, *Acta Metallurgica*, 1985, **33**, (7), 1271–1283.
4. H.K.D.H. BHADESHIA, L.E. SVENSSON and B. GRETOFT: Prediction of the microstructure of submerged arc linepipe welds, *Proc. of the 3rd Int. Conf. 'Welding and Performance of Pipelines'*, London, 18–21 November, 1986, The Welding Institute, Cambridge, UK. Part 1. 1986. pp. 17.1–17.10.
5. H.K.D.H. BHADESHIA and L.E. SVENSSON: Effect of alloying additions on microstructure and properties of vertical-up MMA welds. 1. Method of theoretical assessment, *Joining and Materials*, 1989, pp. 182R–187R.
6. H.K.D.H. BHADESHIA and L.E. SVENSSON: Effect of alloying additions on microstructure and properties of vertical-up MMA welds. 2. Results and discussion, *Joining and Materials*, 1989, pp. 236R–238R.
7. H.K.D.H. BHADESHIA: Modelling the microstructure in the fusion zone of steel weld deposits, in *Recent Trends in Welding Science and Technology*, TWR'98 Gatlingburg, Tennessee, USA, 14–18 May, 1989. ASM International, 1990. pp. 189–198.

8. H.K.D.H. BHADESHIA: Shear transformations in steels, *Acta Metallurgica*, 1981, **29**, (6) 1117–1130.
9. H.K.D.H. BHADESHIA: The diffusional formation of ferrite in iron and its alloys, *Progress in Materials Science*, 1985, **29**, 321–386.
10. H.K.D.H. BHADESHIA: Driving force for martensitic transformation in steels, *Metal Science*, 1981, **15**, (4), 175–177.
11. H.K.D.H. BHADESHIA: Thermodynamic extrapolation and martensite-start temperature of substitutionally alloyed steels, *Metal Science*, 1981, **15**, (4), 178–180.
12. H.K.D.H. BHADESHIA: A thermodynamic analysis of isothermal transformation diagrams, *Metal Science*, 1982, **16**, (3), 159–165.
13. H.K.D.H. BHADESHIA: Application of first-order quasichemical theory to transformations in steels, *Metal Science*, 1982, **16**, (3), 167–169.

GRAIN GROWTH AND EMBRITTLEMENT OF STEEL WELDS

IVAN HRIVNAK
Technical University, Kosice, Slovakia

ABSTRACT

The most important changes which result in the mechanical properties when welding steels are the grain growth and microstructural transformation. The grain in the underbead zone is considerably coarser when compared to the grain size of the parent plate.

GRAIN GROWTH IN THE UNDERBEAD ZONE

Following a complete transformation $\alpha \rightarrow \gamma$ (or following recrystallisation in metals without any polymorphous transformation), grains fail to be in equilibrium with regard to the surface energy γ_{gb} of grain boundaries. Larger grains swallow up smaller ones, resulting in grain growth. If $\overline{d}$ designates the mean grain diameter, then the mean driving force per unit volume may be expressed as

$$\Delta G = 2\frac{\gamma_{gb}}{\overline{d}} \tag{1}$$

i.e. the surface energy of grain boundaries (γ_{gb}) divided by the mean grain curvature $\overline{d}/2$. Ashby and Easterling[1] designated the mean growth rate of grains as $\overline{v}$:

$$\overline{v} = \frac{d\overline{d}}{dt} = M\,2\,\frac{\gamma_{gb}}{\overline{d}} \tag{2}$$

where M is the so-called mobility of grain boundaries, strongly dependent on temperature. By integrating this expression, we have

$$\bar{d}^2 = d_0^2 + Kt \tag{3}$$

where $K \approx M\gamma_{gb}$ and d_0 is the original size of grain. Experimental results show the dependence of the expression.

$$\bar{d} = K' t^n \tag{4}$$

where again K' is the proportionality constant and n an exponent whose value depends on whether the steel contains particles hindering mobility of grain boundaries (precipitates) or inclusions. According to Ashby and Easterling,[1] in the presence of second phase particles (Fig. 1), the driving force for a migrating boundary 2 γ_{gb} over d is balanced by the growth of the boundary surface due to its interaction with the particles. If we designate the volume share of particles as V_f of radius r, the total number of integrating particles is $\frac{3V_f}{2r^2}$, so that the braking force P per unit of the boundary surface may be expressed as

$$P \approx \frac{3V_f}{2\pi r^2} \pi r \gamma_{gb} = \frac{3V_f \gamma_{gb}}{2r} \tag{5}$$

For equilibrium, i.e. when the particles block grain growth, we must have $P = \Delta G$, hence,

$$2\frac{\gamma_{gb}}{\bar{d}} = \frac{3V_f \gamma_{gb}}{2r} \tag{6}$$

and for a maximum grain size in such a case (hence $d = d_{max}$)

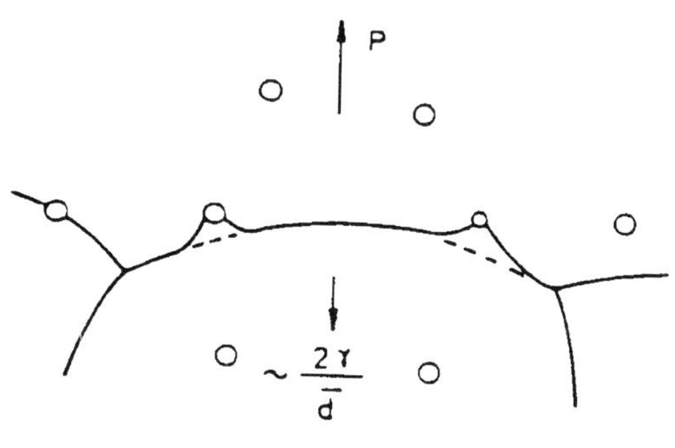

Figure 1.

$$\bar{d}_{max} = \frac{4r}{3V_f} \tag{7}$$

This relation was first deduced by Zener (quoted in reference 1). It shows that carbonitride particles (e.g. of micro-alloying elements) present in the underbead zone may effectively hinder grain growth, if they are not dissolved. However, their role may also be partly taken over by particles of inclusions (especially of oxides) and partly by segregates (e.g. P, S). Ikawa *et al.*[2] studied grain growth in the heat-affected zone. They assumed that this growth was controlled by diffusion, directed by surface energy and required no nucleation. A change in grain size could then be expressed as

$$\frac{\mathrm{d}d}{\mathrm{d}t} = A_f\,(d)\,\exp\left(-\frac{Q}{RT(t)}\right) \tag{8}$$

where $T(t)$ expresses that part of the thermal cycle which lies above the critical temperature of precipitate solution, and therefore of grain growth. A is the kinetic constant and $f(d)$ is the function of d. The resultant grain size can be obtained by integrating the last expression

$$\int_{d_i}^{d_f} \frac{\mathrm{d}d}{A_f(d)} = I \tag{9}$$

where d_i is the original, and d_f is the final grain size. The left-hand side of the equation depends solely on d_i and d_f, and I is the constant for the final change of the grain size d. Therefore,

$$I = \int_0^x \exp\left(-\frac{Q}{RT(t)}\right)\mathrm{d}t \tag{10}$$

In this expression, Q is the activating energy of auto-diffusion. It can be determined by experimentally estimating the grain size as a function of the welding thermal cycle $T(t)$, i.e. T_{max} and $t_{8/5}$, e.g. with the aid of a simulation of welding cycles. We can thus set up the so-called welding diagram as the dependence of grain size on the maximum temperature of the cycle and heat input which characterizes the cooling time $t_{8/5}$. Figure 2 illustrates such a diagram for austenitic stainless steel 18/8 CrNi (type 316). The figure shows contours of equal-sized grains in the heat-affected zone, also the temperature–time zone of precipitate dissolutions–carbides $Cr_{23}C_6$.

Matsuda and Okumura[3] followed grain growth in the underbead zone in micro-alloyed structural weldable steels in the presence of titanium nitride particles, TiN. According to these authors, grain size in the underbead zone may be expressed as

$$\bar{d} = 0.75\,\frac{r}{V_f} \tag{11}$$

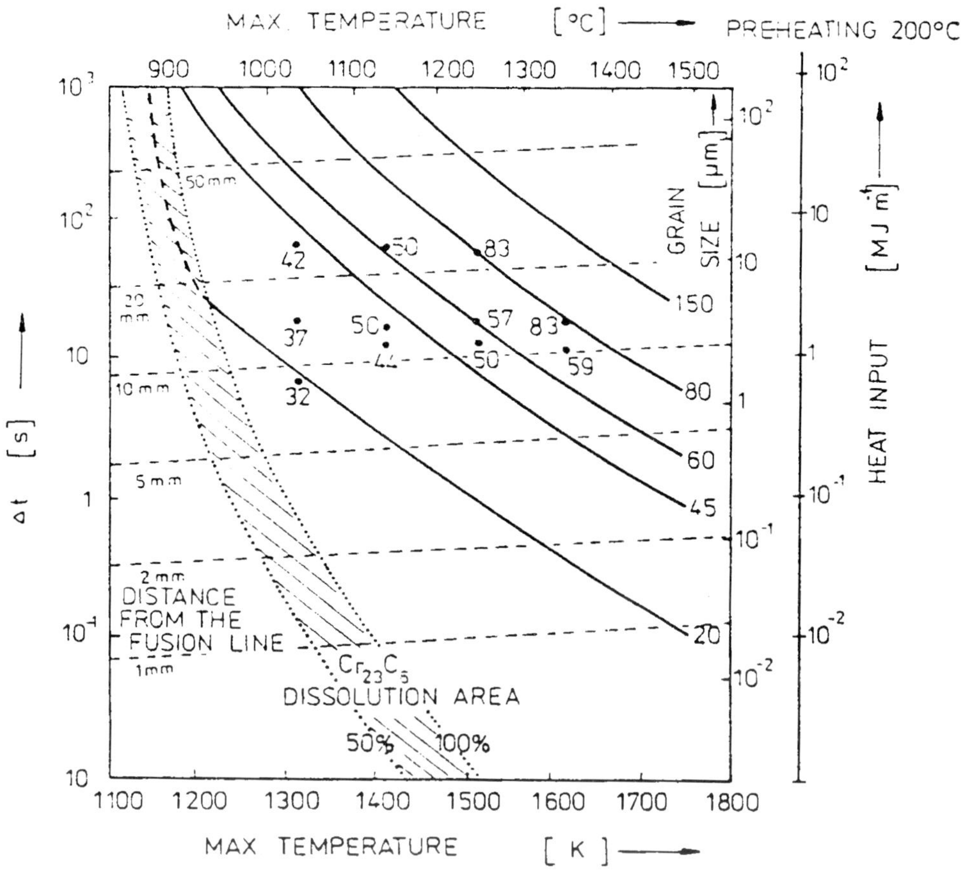

Figure 2. Weld diagram showing dependence of grain size on the maximum temperature and heat input for austenitic stainless steel 18/8 Cr Ni (type 316).

or generally as $\bar{d} = K(r/V_f)$, where K is a dimensionless constant, r the mean size of the precipitate (in this case of TiN) and V_f is the volume fraction of the precipitate. If $V_f << 1$, then the mean distance of particles x may be expressed as $x = \bar{r}/V_f$

Generally the grain size of the underbead zone is proportional to the maximum temperature and cooling time:

$$d = kt^{n}_{8/5(12/8)} \tag{12}$$

where k and n are constants. The constant k exhibits and Arrhenius temperature dependence. Normally, the maximum value of 0.5 for n is achieved only for ultra-high-purity metals at temperatures approaching their melting points. Departures from $n = 0.5$ behaviour were generally thought to result from solute drag, interactions with particles or the effect of textures.

The grain size can also be expressed by the relationship[4]

$$d = \left[\int_{t_{T_S}}^{t_{T_f}} \sum A \exp\left(-\frac{Q}{RT(t)} t_{8/5} \right) \right]^{1/n} \tag{13}$$

where t_{T_S} is the temperature at which transformation $\alpha \rightarrow \gamma$ starts, t_{T_f} temperature at which this transformation is completed.

Burges *et al*[5] use the parameter P_a for the growth of austenitic grains,

$$P_a = \left(\frac{1}{T} - \frac{R}{\Delta H} \ln \frac{t}{t_0} \right)^{-1} \tag{14}$$

where T is the temperature in K, t is time, R is the gas constant and H is the activation energy of γ-grain growth. Parameter P_a can also be used to characterize the thermal cycle of welding

$$P_a = \frac{1}{T_M} - \frac{R}{\Delta H} \ln \left[\frac{\alpha Q}{\lambda (T_M - T_0)} \right]^{-1} \tag{15}$$

where T_M is the maximum temperature of the cycle in K, $\alpha = 0.03$ for carbon steels, λ is the thermal conductivity, T_0 the preheating temperature (K), and Q the specific weld input (kJ cm^{-1}). In carbon structural steels, an activation energy $\Delta H = 240.8$ k J mol^{-1} was found by measurement, which corresponds to the self-diffusion of iron.

Grain growth in the underbead zone presents one of the major weldability problems in all modes of welding for all metals and alloys. A coarse 'primary' grain immediately causes several drawbacks:

- according to the Hall–Petch equation, it decreases yield strength ($\sim kd^{-1/2}$);
- a coarse grain slows down the kinetics of austenite decomposition, and consequently favours the production of Widmannstätten ferrite (at slow cooling times $t^{8/5}$) or diffusionless transformations;
- it increases steel's susceptibility to liquation, cold and stress relief cracking;
- it lowers the strength of zone ($\sim kd^{-1/2}$) in metals without a polymorphous transformation.

With current structural steels, the critical temperature of grain growth is ~1050°C; in the case of microalloy steels, this temperature may rise to 1350°C.

The degree of grain growth depends on the chemistry of the steel and on the shape of the weld thermal cycle and, hence, on the heat input of welding.

Let us investigate the width of the underbead zone (i.e. the part of the heat-affected zone in which grains are coarsened considerably) in narrow gap submerged arc welding. Let us designate the width of the gap prior to welding as a_0 (Fig. 3). After welding the edges of the steel plates to be welded are partially remelted to the width a, where a–a_0 is that part of the metal which is remelted and becomes a part of the weld metal.

It is known[6] that the distance between the heat source and point of the HAZ in which temperature ceases (T_{max}) can be expressed by equation

$$r = \left(\frac{2}{\pi e} \frac{q}{\rho c v} \frac{1}{T_{max} - T_0} \right)^{\frac{1}{2}} \quad \text{for thick plates}$$

$$r = \left(\frac{2}{\pi e} \right)^{\frac{1}{2}} \frac{q}{2 d \rho c v} \frac{1}{T_{max} - T_0} \quad \text{for thin plates} \qquad (16)$$

where q is the heat generated by the arc $q = E\,I$, where E is the arc voltage, I the arc current (A), c is the specific heat, ρc is the volume thermal capacity, d is the thin plate thickness, e basis of Neperian logarithm, v is the welding speed and T_0 is the preheating or interpass temperature.

The heat input Q can be expressed as

$$Q = \eta \frac{q}{11\,000\ v} \ (\text{kJ cm}^{-1})$$

where η is the thermal efficiency (0.90–0.99 for submerged arc welding, 0.66–0.85 for manual metal arc welding, 0.50–0.70 for welding in gas shielding).

Let us investigate the case of three-dimensional heat flow, that is the welding of thick plates. For structural steel $Qc = 4.5 \times 10^6$ J m^{-3} K^{-1}, $\alpha = 9.1 \times 10^{-6}$ M^2 s^{-1}, $\lambda = 41.0$ W s^{-1} K^{-1}). For the remelted zone $T_{max} = T_m$, T_m = melting temperature, which can be calculated from the chemical composition of the steel,

$$T_m = 1537 - 90C \ [7]$$

$$\text{or } T_m = 1537 - (88\,C + 8\,Si + 5\,Mn + 5\,Cu + 1.5\,Cr + 4\,Ni + 2\,Mo + 2\,V + 30\,P + 25\,S) \ [^\circ C] \qquad (17)$$

and inserting $T_m = T_{max}$ into equation[16] we obtain $r = r_m = a/2$.

The width of the remelted zone is then $\dfrac{2r_m - a_0}{2} = r_m \dfrac{a_0}{2}$

Let us take the temperature above in which the grains start to coarsen as $T_{crit} = 1150^\circ C = 1423$ K. The distance of that point from the heat source is

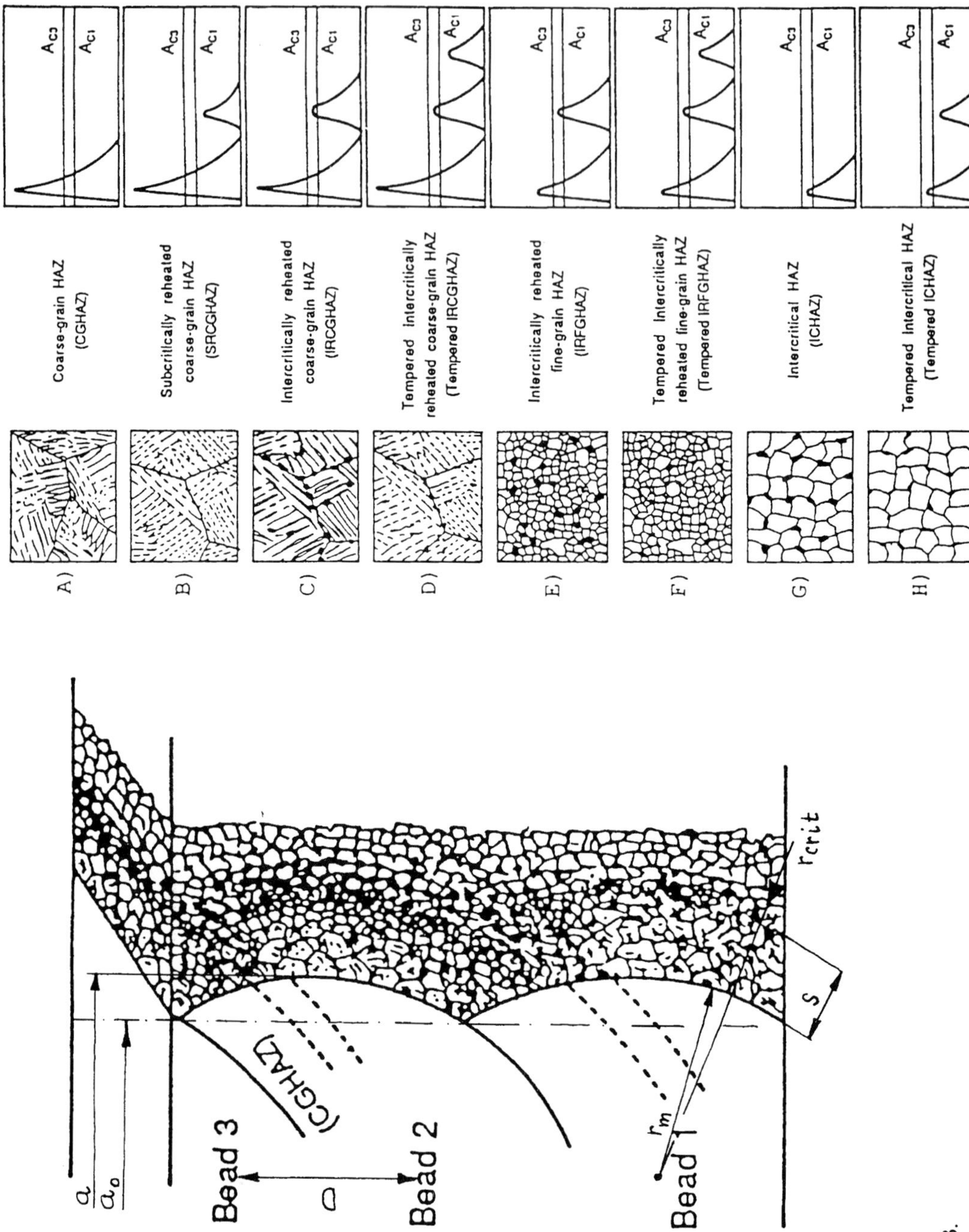
A) Coarse-grain HAZ (CGHAZ)
B) Subcritically reheated coarse-grain HAZ (SRCGHAZ)
C) Intercritically reheated coarse-grain HAZ (IRCGHAZ)
D) Tempered intercritically reheated coarse-grain HAZ (Tempered IRCGHAZ)
E) Intercritically reheated fine-grain HAZ (IRFGHAZ)
F) Tempered intercritically reheated fine-grain HAZ (Tempered IRFGHAZ)
G) Intercritical HAZ (ICHAZ)
H) Tempered Intercritical HAZ (Tempered ICHAZ)
Ac3
Ac1
Bead 3
Bead 2
Bead 1
(CGHAZ)
a
a/ao
rm
rcrit
S

Figure 3.

$$r_{\text{crit}} = \left(\frac{2}{\pi e} \frac{q}{\rho c v} \frac{1}{1423} - T_0 \right)^{\frac{1}{2}} \tag{18}$$

and the width of underbead (coarse grained heat affected zone, SG HAZ) zone is

$$s = r_{\text{crit}} - r_{\text{m}}$$

$$s = \left(\frac{2}{\pi e} \frac{q}{\rho c v} \right)^{\frac{1}{2}} \left[\left(\frac{1}{1423 - T_0} \right)^{\frac{1}{2}} - \left(\frac{1}{T_{\text{m}} - T_0} \right)^{\frac{1}{2}} \right] \tag{19}$$

EMBRITTLEMENT OF THE HAZ

Recently Devillers *et al.*[8] proposed a model to predict toughness in HAZ. As mentioned above, the CGHAZ may be brittle especially when unfavourable microstructures prevail in it. In modern microalloyed steels the chemical composition dependence of the notch toughness transition temperature is negligible at the faster cooling rates (lower $t_{8/5}$ values) but is important for the higher heat inputs. At higher $t_{8/5}$ values (over ca. 80 s) the side-plate structure is observed with carbide colonies and/or M–A (martensite–austenite) islands in between, termed upper bainite or granular bainite. Another zone exhibiting poor toughness in multipass welds is the intercritically reheated coarse grain zone (ICCGHAZ), which generally yields high carbon martensite islands. This is often considered the LBZ (local brittle zone) in CTOD tests. It is therefore important to predict the percentage of the LBZ in the fracture path taking into consideration the dependence of the welding parameters (heat input, preheating or interpass temperature). The distance of the Ac_1 isotherm affected by the next pass, r_{Ac_1}, can be expressed by equation [16] taking into account $T_{max} = T_{Ac1} = 723°C = 996$ K

$$r_{Ac_1} = \left(\frac{2}{\pi e} \frac{q}{\rho c v} \frac{1}{996 - T_0} \right)^{\frac{1}{2}} \tag{20}$$

Similarly for the Ac_3 isotherm:

$$r_{Ac_3} = \left(\frac{2}{\pi e} \frac{q}{\rho c v} \frac{1}{T_{Ac} - T_0} \right)^{\frac{1}{2}} \tag{21}$$

where T_{Ac_3} can be calculated from the chemical composition (wt.%) of steel:

$$Ac_3 = 910 - 203\,C^{1/2} - 15.2\,Ni + 44.7\,Si + 10.4\,V + 31.5\,Mo + 13.1\,W - (30\,Mn + 11\,Cr + 20\,Cu - 700\,P - 400\,Al - 120\,As - 400\,Ti) \quad [^{\circ}C]$$

$$Ac_1 = 723 - 10.7\,Mn - 16.9\,Ni + 29.1\,[SC]+ 16.9\,Cr + 290\,As + 6.38\,W$$

And the width of the intercritically reheated coarse grained HAZ is then

$$1 = r_{Ac_1} - r_{Ac_3} \tag{22}$$

The surface area of ICCGHAZ is then

$$F \approx s \times 1$$

For 1 pass and in the case of N passes deposited at the same preheating and interpass temperature, T_0, the total area of ICCGHAZ is

$$A = (N - 1)\, s \times 1 \tag{23}$$

Concerning the vertical spacing between two subsequent deposits, *D*, results in the dependence

$$D = 1.8 + 0.1Q \ [\text{mm}]$$

where Q is in kJ/cm.

A slight restoration of toughness may be observed in the tempered CGHAZ or highly tempered ICCGHAZ. Measurements using high strength steel weldments have revealed[9] that the lowest decomposition temperature for the M–A constituent is 523–573 K where the heat of the third pass which tempers the ICCGHAZ near the first pass or not, depends not only upon the heat input but also on the spacing between the passes.

CONCLUSION

In this paper some approaches to predict the austenite grain size in the underbead zone of steel weldments were given. In general it was assumed that the grain coarsening temperature in modern structural steels lies at 1150°C. Based on this approach the width of the coarse grained heat-affected zone was calculated for the simple case of narrow gap submerged arc welding. The fracture area of intercritically reheated CGHAZ as well as the expression to calculate the tempered ICCGHAZ were given. As stated in reference 8, it is then possible to evaluate the effect of steel chemistry and welding parameters on the width, geometry and portion of individual parts of the heat affect zone.

REFERENCES

1. M.F. ASHBY and K.E. EASTERLING: 'A first report on diagrams for grain growth in welds', *Acta Metall,* 1969, **30**, 1982.
2. H. IKAWA, H. OSHIGE and S. NOI: 'Study on the grain growth in weld heat affected zone' (Report 5). *J. Welding (Japan)*, 1977, **7**, 396.
3. S. MATSUDA and N. OKUMURA: 'Effect of TiN precipitate particles on the austenite grain size of low carbon low alloy steels.' *Trans. I.S.I.J.*, **18**, 198.
4. P.J. ALBERRY: 'Computer model for multipass repair welds in SA 508 class 2 alloy', *Welding J.*, 1989, **Oct**, 410.
5. P. BOURGES, P. MAYNIER, and R. BLONDEAU: 'Prevision de l'evolution des characteristiques de la zone affectée et du metal fondu lors du detensiounement', *Proc. Ing. Conf. Stress Relieving Heat Treatments of Welded Constructions,* 6–7 July, 1987, Sofia (IIW/IIS) Pergamon Press, 1987.
6. I. HRIVNAK: *Theory of weldability of metals and alloys,* Elsevier, Amsterdam, 1992.
7. L. SMRHA: *Solidification and crystallization of steel ingots,* SNTL, Prague, 1983 (in Czech).
8. L. DEVILLERS, D. KAPLAN and R. TESTARD: 'An approach for predicting microstructures and toughness properties in HAZs of multipass welds of structural steels', Doc. IIW - IX-1672-92 (IIS/IIW-1164-92).
9. I. HRIVNAK, F. MATSUDA and K. IKEUCHI: Investigation of M-A constituent in high strength steel welds. *Trans. JWRI,* 1992, **21** (2), 9.

A MATHEMATICAL MODEL AND PROGRAM FOR IBM PC/AT TO OPTIMISE THE PARAMETERS OF DIFFUSION BONDING

T.G. TARANOVA, A.V. TAYANOVSKAYA
and
G.M. GRIGORENKO
E.O. Paton Electric Welding Institute of the Academy of Sciences of Ukraine, Kiev, Ukraine

ABSTRACT

It is a known fact that in diffusion welding a joint forms as a result of the deformation of the materials being joined. There exist a number of hypotheses that explain the joint formation process in different ways. One describes the nature of the formation of a joint between solid materials. This work discriminates between three main stages of the joint formation process: formation of a physical contact, activation of contact surfaces and volumetric interaction. In welding of similar metals the first and second stages practically merge, since activation of the two contact surfaces starts during the process of their convergence in the crushing of individual microprojections due to mutual plastic deformation. Recrystallisation that causes formation of common grains in the contact zone can serve as a criterion for the completion of the third stage.

METHOD

We consider the first stage of joint formation which involves contact. Diffusion bonding can be described in a number of stages. The physical contact is a state in which electron interaction becomes possible between the reacting materials. This state is a necessary condition for any reaction. In diffusion welding the physical contact forms due to plastic deformation of the near-contact volumes of the materials being joined.

With such methods as diffusion bonding the stage of formation of the physical contact can impose some limitations on the formation of a sound joint. Naturally, the model of formation of the physical contact will dictate the process, but only if the plastic deformation mechanism is taken into account and if this mechanism is used to determine the kind of relationship between the deformation rate and the welding parameters. It is necessary to determine the relationship between the plastic deformation rate, temperature, pressure and time. Analysis of the data in the literature reveals at least two basic approaches to the study of kinetics of contact formation. In reference 2, the authors proceeded from an equation of the deformation rate at a stage of the steady creep (J. Weertman's equation). On the other hand, it may be that the complete contact will form before the beginning of the steady creep state.[3] This latter approach seems to be in better agreement with our experiments.

Thus, to construct the calculation models for the physical contact formation we used an equation of the deformation rate for the unsteady creep stage allowing for the fact that there is a gradual decrease in stress.

At the unsteady creep stage at a constant stress the deformation rate, depending on a time, is usually described by the following expression:

$$\dot{\varepsilon} = At^{-n} \tag{1}$$

where A is the coefficient depending on the temperature and the applied stress, n is the factor which can have values from 0 to 2.

Based on the kinetic theory of dislocations, the deformation rate can be written as follows:

$$\dot{\varepsilon} = \nu L_0 b \rho \exp\left(-\frac{E_{\kappa}(P)}{RT}\right) \tag{2}$$

where ν is the frequency of the atomic oscillations, L_0 is the path of the dislocations to an obstacle, b is the modulus of Burger's vector, ρ is the density of the mobile dislocations, E_{κ} is the activation energy of plastic deformation at the set stress P, R is the universal gas constant and T is the temperature.

To use equation 2 to describe the plastic deformation rate at the unsteady creep stage, we assume the following. A change in the unsteady creep rate depending on the time, according to equation 1, is caused by a decrease in the density of the mobile dislocations, i.e. $\rho = \beta t$, where β is a constant.

Equation 2 can then be rewritten in the following form:

$$\dot{\varepsilon} = \eta t^{n} \exp\left(-\frac{E_{\kappa}(P)}{RT}\right) \tag{3}$$

where $\eta = \nu L_0\, b\, \beta$.

In equation 3 the $\dot{\varepsilon}(t)$ dependence is as in Equation 1. When applying equation 3 for the calculation models of physical contact formation, one should bear in mind that the stress in the contact zone is $P = P(t)$, i.e. $E_{\kappa} = E_{\kappa}[P(t)]$. To develop the model further we assume the following:

1. One of the surfaces is assumed to be absolutely rigid and ideally smooth.
2. Each microprojection at the second surface is represented as a right circular cone which is transformed into a right truncated cone in the process of deformation. Naturally, at $t = 0$ the volume of the microprojection is $V(0) = F_H h_H/3$ (where h_H is the initial height of the cone), while at $t > 0$:

$$V(t) = X(t)\frac{F_H + F_\kappa(t) + \sqrt{F_H \cdot F_\kappa(t)}}{3} \tag{4}$$

where F_H is the area of the lower base of the cone, F_K is the area of the upper base of the truncated cone and X is the variable height of the cone.

3. According to the law of conservation of mass, for all $t > 0$ it holds:

$$V\,(0) = V\,(t) \tag{5}$$

4. Introduce the dimensionless values:

$$\overline{X}(t) = \frac{X(t)}{h_H}; \quad \overline{F}_\kappa(t) = \frac{F_\kappa(t)}{F_H} \tag{6}$$

5. Stress in the contact:

$$P(t) = \frac{P_P}{F_\kappa(t)}, \tag{7}$$

$$P_P < P < \infty, \quad 0 < t < \infty$$

6. According to Equation 3:

$$\frac{\mathrm{d}\overline{X}(t)}{\mathrm{d}t} = \eta t^n \exp\left(-\frac{E_\kappa(P)}{RT}\right) \tag{8}$$

Then from equation 5 and from the dependencies for $V(0)$ and $V(t)$, using equation 6 and going over to the relative values, we will find the following:

$$\overline{X}(t) = \frac{\overline{F}_\kappa(t) + \sqrt{\overline{F}_\kappa(t)}}{1 + \overline{F}_\kappa + \sqrt{\overline{F}_\kappa(t)}} \tag{9}$$

By differentiating equation 9 with respect to t, we obtain:

$$\frac{\mathrm{d}\overline{F}_\kappa(t)}{\mathrm{d}t} = \frac{\left(1 + \overline{F}_\kappa(t) + \sqrt{\overline{F}_\kappa(t)}\right)^2 \cdot \sqrt{\overline{F}_\kappa(t)}}{0.5 + \sqrt{\overline{F}_\kappa(t)}} \cdot \frac{\mathrm{d}\overline{X}}{\mathrm{d}t} \tag{10}$$

Then, using equation 8, from equation 10 we find the differential equation that defines function $F_K(t)$:

$$\frac{d\bar{F}_{\kappa}}{dt} = \eta t^{-n} \frac{(1 + \bar{F}_{\kappa}(t) + \sqrt{\bar{F}_{\kappa}(t)})^2 \cdot \sqrt{\bar{F}_{\kappa}(t)}}{0.5 + \sqrt{\bar{F}_{\kappa}(t)}} \exp\left(-\frac{E_{\kappa}(P)}{RT}\right) \tag{11}$$

where the kind of function $E_K(p)$, as earlier, is unknown, while $P = P(t)$. Therefore, it is impossible to integrate equation 11. However, we will be interested in the behaviour of $\bar{F}_K(t)$, when $\bar{F}_K$ is close to 1. Assume, if $t \to \infty$, then $P \to P_P$ (certain calculated values of the pressure) and, hence, $E_K(P) \to E_K(P_P)$. The initial condition for equation 11 will have the form:

$$\bar{F}_{\kappa}(0) = 0 \tag{12}$$

Assuming $\bar{F}_K = 1$ in the right part of equation 11, we obtain:

$$\frac{d\bar{F}_{\kappa}}{dt} = 6\,\eta\, t^{n} \exp\left(-\frac{E_{\kappa}(P_P)}{RT}\right) \tag{13}$$

By integrating equation 13 with respect to the initial condition equation 12, we find that in the first approximation:

$$\bar{F}_{\kappa}(t) = \frac{6\,\eta\, t^{1-n}}{1-n} \exp\left(-\frac{E_{\kappa}(P_P)}{RT}\right) \tag{14}$$

If we again replace $E_K(P_P)$ by $E_K(P)$ in equation 14, then, with the value of $\bar{F}_K$ close to 1 and by making some approximation, we will obtain the following:

$$\bar{F}_{\kappa}(t) = \frac{6\,\eta\, t^{1-n}}{1-n} \exp\left(-\frac{E_{\kappa}(P)}{RT}\right) \tag{15}$$

Thus, to determine from equation 15 the area of the physical contact formed at different values of temperature, pressure and time of the process, it is necessary to know the values of parameters $E_K(P)$, n and η for a particular material.

The present work studies 36 x 18 mm samples of nickel NP-2 annealed in vacuum at 1123 K for an hour. Prior to bonding they were washed in acetone and alcohol. Bonding was performed at temperatures from 1073 to 1273 K under compression pressure from 5 to 40 MPa; the bonding time varied from 1 to 30 min. Surface preparation was done by electrolytic polishing. Experimental studies of the contact formation kinetics were conducted within the range of temperatures and pressures which were used to study creep of nickel.

According to the model developed, we will be interested in the $\dot{\varepsilon}(t)$ dependence. To determine this dependence it is usually suggested[3] that the method of the graphical differentiation should be used which, as a rule, leads to a significant error. And we suggest an approximation of the experimental curves by polynomial of the type $a_0 + a_1x + a_2x^2 + a_3x^3 + \ldots$, then an analytical calculation of the derivative and plotting of the $\dot{\varepsilon}(t)$ dependence. Further on, we think it necessary to plot the $\ln \dot{\varepsilon} = f(1/T)$ dependence. At different values of P the tangent of the slope of the straight lines obtained will express the activation energy of the plastic deformation at the preset stress, since, from equation 2,

$$\ln \dot{\varepsilon} = \ln \eta - n \ln t - \left(\frac{E_\kappa(P)}{RT} \right) \quad (16)$$

$$\frac{\delta (\ln \dot{\varepsilon})}{\delta \left(\frac{1}{T} \right)} = -\frac{E_\kappa}{R} = tg\alpha \quad (17)$$

For nickel the activation energy is estimated as $E_\kappa = 125$ kJ/mole, this being in agreement with the data from reference 3.

To estimate n and η we will use the initial equation 2 for the plastic deformation rate. Taking the logarithm of this equality will yield equation 16.

Then, introduce the following designations:

$$y = \ln \dot{\varepsilon}$$

$$x = \ln t$$

$$a_1 = \ln \eta - \frac{E_\kappa(P)}{RT}$$

$$a_2 = -n \quad (18)$$

With these designations equality equation 18 will have the form:

$$y = a_1 + a_2 x \quad (19)$$

We determined the values of parameters n and η in the equation for the physical contact area. The practical value of the developed physical contact formation model consists in the fact that, if used, it will be sufficient to obtain the creep curves for a material under investigation to determine a kind of dependence (equation 15).

The pressure enters into this equation in the implicit form. According to the data of references,[1, 3] it is the kind of $E_\kappa(P)$ dependence that is the subject of the discussion. From the data of Zhurkov[4] and Osipov,[5] function $E_\kappa(P)$ is linear with respect to P, i.e.:

$$E_\kappa = \gamma_1 - \gamma_2 P \quad (20)$$

where γ_1, γ_2 are certain coefficients.

To determine the parameters of equation 20, it is necessary to plot the $E_K(P)$ dependence for different T. Then the three parameters, P, T and t, will enter into equation 15 in the explicit form. We used the direct optimisation method and obtained the optimum values of the welding parameters, P, T, t, the physical contact area achieved being 100%,: $P = 15$ MPa, $T = 1243$ K, $t = 4.5$ min. The mathematical treatment was carried out using IBM PC/AT-286. Programming was performed using Turbo PASCAL.

Metallographic and fractographic studies using the scanning electron microscope JSM-840 were conducted to experimentally estimate the physical contact area.

Fractographic studies showed the high quality of the welded joint produced at $T = 1273$ K, $P = 15$ MPa, $t = 10$ min.

CONCLUSIONS

1. The mathematical model developed and software written in the Turbo PASCAL for IBM PC/AT-286 allow an optimisation of the diffusion bonding parameters on the basis of a model using physical contact.
2. Analysis of the results obtained provides a satisfactory agreement between experimental and calculated values of P and T.
3. As far as the bonding time is concerned, it should be noted that it takes 4.5 min to ensure the formation of the complete physical contact at $T = 1273$ K and $P = 15$ MPa, this being in agreement with the calculated values.
4. An increase of up to 10 min in the holding time leads to activation of the combined recrystallisation and to providing the full-strength welded joint. The further increase in the holding time makes the grains in the welding zone much coarser, thus causing a degradation of the welded joint performance.
5. Equation 15 can be used in practice for engineering calculations to design particular pieces. All the unknown parameters in equation 15 are determined by the creep curves; therefore, this makes it possible to essentially reduce the amount of experimental research required for the development of welding technology.

REFERENCES

1. YU.L. KRASULIN: Interaction between metal and semiconductor in solid phase, *M. Nauka*, 1971, 119.

3. E.S. KARAKOZOV: Joining of metals in solid phase, *M. Metallurgiya*, 1976, 260.
4. B.V.I. ZHURKOV and A.N. BAKHTIBAEV: Time and temperature dependence of strength of single crystals, *Solid State Physics*, 1969, **11** (3), 690–692.
5. K.A. OSIPOV: Some activated processes in solid metals and alloys, *M. nauka*, 1972, 131.

INVESTIGATION OF RELATIONSHIPS BETWEEN CHEMICAL COMPOSITION AND MECHANICAL PROPERTIES USING THE INTERACTIVE GRAPHIC SYSTEM 'SHEFFLER'

A.V. TAYANOVSKAYA and G.M. GRIGORENKO
E.O. Paton Electric Welding Institute of the Academy of Sciences of Ukraine, Kiev, Ukraine

Present-day alloys are characterised by complex chemical compositions, because they have to meet numerous service requirements. Therefore, in order to develop new compositions and to forecast their service behaviour the application of advanced computing facilities must be used.

The process of developing new grades of steels is known to be semi-empirical in nature. Premises and assumptions made by experimenters are most often based on the results of single-factor experiments. This requires many experiments where no allowance is made for the mutual effect of all the factors. Lately, experimental design methods have achieved acceptance. Such methods allow multifactorial experiments to be considered and estimated to a certain degree of accuracy.[1, 2]

Another method for optimisation of an alloy composition is also well-known. This method involves the estimation of various structural and phase characteristics in the selection of compositions that would meet the preset criteria. This method for optimisation of alloy compositions is described in detail in reference 3, and uses the graphic representation of the results in ternary diagrams. Such an approach allows a series of heat-resistant compositions to be developed,[4, 5] the compositions of the known IN100, 713C alloys susceptible to precipitation of embrittling phases to be corrected and the service behaviour of heat resistant alloys[6] to be predicted.

The amount of the strengthening phase and the temperature of its dissolution, the lattice parameters of the γ' phase and matrix, the number of electron vacancies, etc. are usually used as the criteria for calculations.

The present work describes the interactive graphic system 'Sheffler' developed by the authors that is designed to investigate the relationships between chemical composition and technological properties of steels and alloys. The system uses both methods for optimisation of chemical compositions of steels and alloys.

The effect of chemical composition of steel on its structure, the phase transformations taking place in it, as well as the relationship of these parameters with basic mechanical, physical and technological properties have been the subject of investigations of long standing. However, the extensive experimental materials available in the present literature are inadequately systematised. The authors suggest that the Sheffler structural diagram that provides the graphic representation of a relationship between the chemical composition of alloyed steels and their structure should be used as a data model to systematise experimental materials. This can be achieved by adding the nickel and chromium equivalents and by conventionally subdividing all alloying elements into austenite- and ferrite-forming ones. Despite some known drawbacks, the Schaeffler diagram has been used widely by technologists to forecast the structural-phase state of both steels and alloys.

The system utilises the graphic dialog. It is used to display steels and alloys under investigation in the Sheffler diagram taken from the relational database and to choose those required for a further analysis. Thus, by selecting steel from an appropriate structural characteristic in the Sheffler diagram, a user can call in the database to retrieve the information on its chemical composition and mechanical, physical and technological properties.

As an example, consider some known alloys (Table 1). When displayed in the Sheffler diagram, the majority of these alloys fall into the martensitic–ferritic region, this being associated with properties of the alloys. At present, available in industry, particularly in boiler and turbine manufacture, are wide ranges of heat-resistant 12% chromium steels, which can be utilised as materials for pipes, blades, fasteners and other components. These steels have been developed as a result of much research carried out at different times, and in various countries and laboratories in order to meet diverse requirements of customers. Such circumstances made it difficult to estimate and compare the extensive materials accumulated concerning high-chromium steels correctly. Even if the steels compared are very close or similar to each other in chemical composition, it is not always possible to obtain comparable results because of differences in metal melting, hot working and heat treatment, as well as the uniform approach to determine such properties, especially at high temperatures. Among heat-resistant high-chromium steels, those with 12 wt.% chromium hold the lead. Modern modifications of heat-resistant steels based on 12 wt.% chromium are complex alloys containing a great number of alloying elements, that sometimes exert different effects on the steel structure and properties.

Table 1. Chemical composition of the 8–12 wt.% chromium steels

Steel	Chemical composition (wt.%)									
	C	Si	Mn	Cr	Ni	Mo	V	W	Nb	Co
08Kh13	0.08	<0.8	<0.8	12	–	–	–	–	–	–
12Kh13	0.09 – 0,15	<0.8	<0.8	12	–	–	–	–	–	–
20Kh13	0.16 – 0,25	<0.8	<0.8	12	–	–	–	–	–	–
15Kh11MF	0.12 – 0,19	<0.5	<0.7	11.5	–	0.6 – 0.8	0.25 – 0.4	–	–	–
15Kh12VNMF (EI802)	0.12 – 0,18	<0.4	<0.9	12	0.4 – 0.8	0.5 – 0.7	0.15 – 0.30	0.7 – 1,1	–	–
13Kh12N2V2MF (EI961)	0.1 – 0,16	<0.6	<0.6	12	1.5 – 1.8	1.35 – 0.5	0.18 – 0.30	1.6 – 2,0	–	–
18Kh11MNFV (EP291)	0.15 – 0,21	<0.6	<1.0	11.5	0.5 – 1.0	0.8 – 1.1	0.2 – 0.4	–	0.2 – 0.45	–
14Kh12V2MF (EI756)	0.1 – 0,17	<0.5	<0.8	12	–	0.6 – 0.9	0.15 – 0.3	1.7 – 2,2	–	–
18Kh12VMBFR (EI993)	0.15 – 0,22	<0.5	<0.5	12	–	0.4 – 0.6	0.15 – 0.3	0.4 – 0,7	0.2 – 0.4	–
15Kh12VM1BFR (EP752)	0.1 – 0,15	<0.5	<1.0	12	0.5 – 1.0	0.7 – 1.0	0.15 – 0.3	0.55 – 0,85	0.15 – 0.3	–
12Kh8M1VFBR (EP505)	0.1 – 0,15	<0.5	<1.0	6 – 8	0.5 – 1.0	0.7 – 1.0	0.3 – 0.5	0.55 – 0,85	0.15 – 0.4	–
12KhK6MFB	0.10	–	–	10.5	0.6	–	0.7	0.2	045	6.0
12KhK6MVFBR	0.07	–	–	10.5	–	0.85	0.5	0.8	–	6.5

When analysing the data of numerous investigations conducted on alloyed 12 wt.% chromium steels, one should note that simultaneous alloying with several elements at a time, i.e. complex alloying, is more favourable for providing the high heat resistance, than the equivalent alloying with one element or with one element but in a much greater amount. With the higher requirements for heat resistance (temperature and stresses) in steels, chemical compositions of the 12 wt.% chromium steels become more complex. In addition to chromium, these steels can contain various combinations and quantities of molybdenum, tungsten, niobium, vanadium, boron and other elements.

The data of Table 1 are indicative of the complex alloying of steels and of their more complex compositions. Noticeable is the fact that the list comprises the 6 wt.% Co steels (steels 12KhK6MFB and 12KhK6MVFBR). Like carbon, cobalt decreases the content of the structurally free ferrite in 12 wt.% chromium steels, and this, in turn, affects their properties. Thus, the long-term strength tests of the EI756 steel specimens revealed the more than 10 times increase in time to fracture in the presence of 2–3 wt.% Co. Alloying the 12 wt.% Cr–Mo–V–W–Nb steel with 4–5 wt.% Co has a positive effect on the heat-resisting properties, improves long-term strength and relaxation resistance.

As seen in Table 1, along with 12 wt.% chromium steels now in use are also 8 wt.% Chromium steels, but such steels are few. It can be seen from comparing some foreign patents that the lower chromium steels hold promise. Apparently, the interest in 7–9 wt.% chromium steels, which are inferior to 12 wt.% chromium steels in heat resistance, can be explained by the higher workability of the former during heat treatment and welding.

Table 1 gives a number of steels with higher nickel content. These steels, being less heat-resistant, are employed at lower temperatures.

Steels 08Kh13, 12Kh13 and 20Kh13 differ in carbon content, which determines their structure and properties. Steel 08Kh13 containing less than 0.08 wt.% carbon belongs to the ferritic class. Due to the high chromium content, the steel possessses high corrosion resistance. It is used to manufacture shapes, sheets, pipes and billets that must have oxidation resistance at up to 750–800°C and are then employed for making low-strength weldments. Steel 12Kh13 belongs to the martensitic–ferritic class, while steel 20Kh13 to the martensitic class. Because of the higher carbon content it possesses a higher ability to quenching and has higher strength properties and hardness than steel 12Kh13 and, moreover, steel 08Kh13. Steel 12Kh13 has the highest heat resistance among this group of steels.

Compared to steels 12Kh13 and 20Kh13, steel 15Kh11MF has a lower chromium content. Due to this fact, this steel is characterised by a lower susceptibility to ferrite formation, which it acquires when adding some extra molybdenum and vanadium to it. The steel belong to the martensitic class. Steel 18Kh11MNFB differs from steel 15Kh11MF in the extra content of niobium, which is added to improve heat resistance. This steel belongs to the martensitic-ferritic class, i.e. 18Kh11MNFB has the higher heat resistant properties.

Steel 15Kh12VNMF also belongs to the martensitic–ferritic class. Compared to steel 15Kh11MF, it is alloyed with an extra amount of tungsten to improve heat resistance.

Steel of the EI756 grade is the only one of 12 wt.% chromium among those under consideration which is alloyed mostly with one element, but in a relatively larger quantity, with about 2 wt.% tungsten, in particular. This steel belongs to the martensitic–ferritic class.

Steel 15Kh12VM1BFR (EP752) is among the multi-alloyed ones and is a modification of steel EI993. Unlike steel EI993, it contains a higher amount of tungsten, molybdenum and boron. Due to this alloying, steel EP752 possesses the highest heat-resistant properties of all 12 wt.% chromium steels. It also belongs to the martensitic–ferritic class.

Unlike all the other above-mentioned steels, steel EP505 contains 7–8 wt.% chromium. It also belongs to the multi-alloyed steels containing molybdenum, tungsten, niobium, vanadium and boron. The steel is purely martensitic because of the lower chromium content.

To illustrate, we chose the steels which have the same structural characteristics in the Sheffler diagram, i.e. within the martensitic–ferritic class. We derived the regression equations for the index of strength F and G for long-term strength as an index that characterises heat resistance (the data on these indices for the steels chosen were taken from references[7, 8]). Due to the multiple regression analysis realised in the system which employs methods of direct optimisation and, thus, imposes no restrictions on the number of variable parameters, construction of the regression equations of any complexity required by a user is allowed. The authors chose the optimum chemical composition of 12 wt.% chromium steel so as to maximise the integral criterion:

$$I = | F | \quad | G |$$

where: $F = a_1 C_1 + a_2 C_2 + \ldots + a_n C_n$

$G = b_1 C_1 + b_2 C_2 + \ldots + b_n C_n$

C_i are the concentrations of the alloying elements.

For optimisation we used the minimum function search criterion:

$$K = -I + \sum_{i=1}^{6} A_i \; t_i$$

where I is the integral criterion, t_i the positive coefficients and A_i are the penalty functions:

$$A_1 = \begin{cases} 0, & \text{if } F(\vec{C}) > F1 \\ F1 - F(\vec{C}), & \text{if } F(\vec{C}) < F1 \end{cases}$$

$$A_2 = \begin{cases} 0, & \text{if } F(\vec{C}) < F2 \\ F(\vec{C}), - F2 & \text{if } F(\vec{C}) > F2 \end{cases}$$

$$A_3 = \begin{cases} 0, & \text{if } G(\vec{C}) > G1 \\ G1 - G(\vec{C}), & \text{if } G(\vec{C}) < G1 \end{cases}$$

$$A_4 = \begin{cases} 0, & \text{if } G(\vec{C}) < G2 \\ G(\vec{C}), - G2 & \text{if } G(\vec{C}) > G2 \end{cases}$$

$$A_4 = \begin{cases} 0, & \text{if } \forall C_i > 0 \\ /\min C_i, & \text{if } \exists\ C_i < 0 \end{cases}$$

$$A_6 = \begin{cases} 0, & \text{if } \sum_i C_i < \text{const} \\ /\max C_i, & \text{if } \sum_i C_i > \text{const} \end{cases}$$

The penalty functions A_1, A_2, A_3, A_4 allow for the restrictions on the strength and heat resistance indices

$$F \in [F1,\ F2], \qquad G \in [G1,\ G2],$$

A_5, A_6 – for the physical values of the concentrations

$$\sum_i C_i < \text{const},\ C_i > 0$$

The calculated values were found to in good agreement with the experimental data.

The system is realised in the TURBO PASCAL programming language for IBM PC/AT 286 and is a set of the specialised modules controlled from the interactive graphic interface.

In the future the interactive graphic system 'Sheffler' will be supplemented with the new program modules to realise the estimations of various structural and phase characteristics, such as the number of electron vacancies, the lattice parameters of the γ phase, etc.

REFERENCES

1. C. SIMS and V. HAGEL Heat-resistant alloys, *M. Metallurgiya*, 1976, 568.
2. B.B. GULYAEV and L.F. PAVLENKO, Modelling of the computer-aided search for optimum alloying, *M. Nauka*, 1975, 11–14.
3. TAKEHIRO ONO *et al.*, Design of single-crystal heat-resistant nickel-base superalloys containing molybdenum, *Tetsu to hagane*, 1988, (11), 133–140 (in Japanese).
4. H. HARADA and M. YAMAZAKI, Development of Ti, Ta and W containing heat-resistant phase-strengthened nickel alloys, *Tetsu to hagane*, 1979, (7), 1059–1068 (in Japanese).
5. R. WATANABE and T. KUNO, Development of precipitation hardened heat-resistant nickel alloys, *Tetsu to hagane*, 1975, (11), 2274–2294 (in japanese).
6. V.V. RTISHCHEV., Calculation methods for prediction of phase composition, structural characteristics and long-term strength limits by chemical compositions of heat-resistant nickel–chromium-base alloys. Heat-resistant and high-temperature nickel-base steels and alloys. *M. Nauka*, 1984, 144–150.
7. A.F. SILAEV, G.P. FEDORTSOV-LUTIKOV, and M.F. SHESHENEV, Heat-resistant chromium steels for power engineering, *M. Metallurgizdat*, 1963, 184.
8. K.A. LANSKAYA, Heat-resistant high-chromium steels, *M. Metallurgiya*, 1976, 215.

COMPUTERISED ANALYSIS OF HYDROGEN MASS TRANSFER IN WELDS AND STEELS

DIPL.-ENG. VITALI PAVLIK

Welding Institute, Technical University of Aachen, Pontstrasse 49, D-52062 Aachen, Germany

1. INTRODUCTION

One important aspect of high strength steel weldability is the risk of cold cracks during or after welding. Cold cracking can occur both in the weld metal and in the heat affected zone. Their formation is caused by stress and hydrogen.

Information about hydrogen distribution in a welded joint is of interest in preventing cold cracking. Because of this, a number of publications have been devoted to the numerical calculation of hydrogen distribution in weld joints.[1–4]

Some of the methods are based on Fick's law that defines the concentration gradient as the motive force of diffusion, others which take into account the chemical potential gradient are formulated in terms of the potential function. However, in any case the basic parameter which determines the hydrogen mass transfer is the diffusion coefficient. Thus, to carry out a calculation applied for the real weld joints it is necessary to have information about the temperature and concentration dependence of hydrogen diffusion coefficient in the deposited metal, HAZ and in the base metal.

Hydrogen diffusion depends to a large extent on the steel composition, the metal structure and on the plastic deformation. Hydrogen can be ideally dissolved in the metal lattice or entrapped in various structural defects, such as point defects, dislocations, grain and phase boundaries, micro and macro cavities etc. These traps differ from each other with the binding energy of hydrogen. Strongly trapped hydrogen is often called residual hydrogen, whilst the hydrogen which can effuse from the specimen at room temperature is described as diffusible. But this is an arbitrary division, because hydrogen can be untrapped by heat.

The interaction of hydrogen with traps becomes especially strong at low temperatures (<~200°C). At higher temperatures the diffusion coefficient tends to that for the perfect metal.

In the present report the mathematical model implemented in a computer program for determining the hydrogen diffusion coefficient in metals is suggested. Some practical results of its application for investigation of the hydrogen mass transfer in welds and steels are described.

2. METHOD OF MEASURING THE HYDROGEN EFFUSION

The determination of hydrogen diffusion coefficient can be carried out using the experimental data of the mass transfer kinetics in a specimen of a regular geometric shape, for which it is possible to solve the diffusion problem in an analytic form.

In this work the measurements of the isothermal hydrogen effusion from cylindrical specimens charged with hydrogen during welding or by electrolytic polarisation were used. The laboratory equipment, which is used has made it possible to measure the hydrogen effusion at the temperature from 20 to 250°C. For this purpose the specimen is placed in a hermetic chamber, which is heated in a resistance furnace made of copper. The temperature is measured by thermocouples at the copper wall of the furnace and maintained constant during the test. To determine the effused quantity of hydrogen the test chamber can be opened by means of two solenoid valves. By means of this all the hydrogen contained in the chamber is transferred to a thermal conductivity detector by the carrier gas argon, Fig. 1. The effused quantity of hydrogen is calculated on the basis of the measured signal from the detector. This process is repeated periodically so that the sum of the individual measured values gives the total quantity of hydrogen effused from the specimen. The dependence of evolution rate from hydrogen content **V(Q)** can be obtained by numerical differentiation of **Q(t)**.

3. MATHEMATICAL MODEL FOR DETERMINING THE HYDROGEN DIFFUSION COEFFICIENT

Let us consider the Fick's equation for the hydrogen effusion from the specimen of radius R and length L, suggesting that in the initial stage hydrogen is uniformly distributed in a metal and the rate of the gas desorption from the surface into ambient atmosphere is much higher than the rate of diffusion that means the zero boundary condition. Thus the diffusion problem can be written as follows

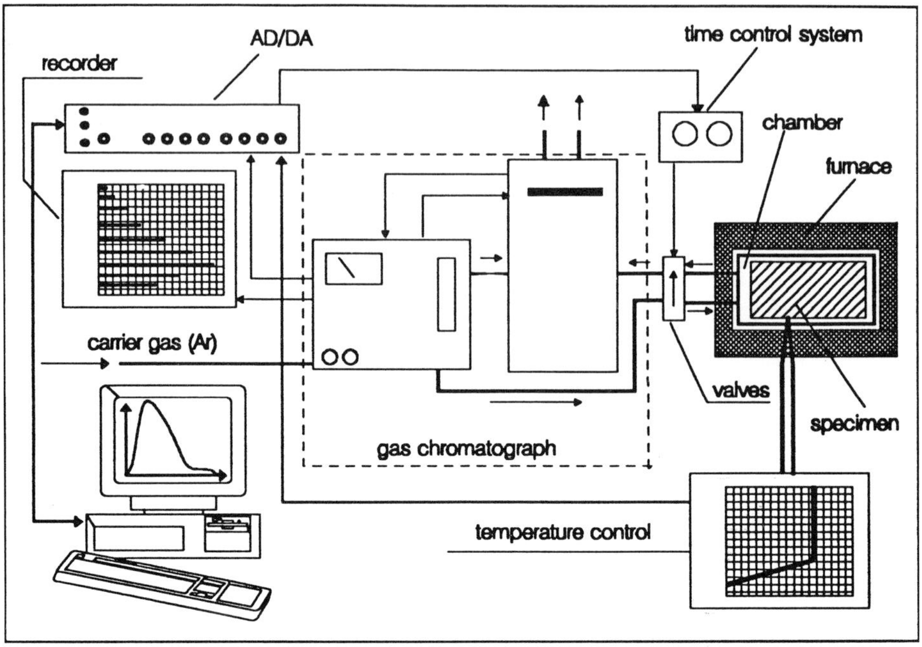

Figure 1. Analytical unit for measuring effusion of hydrogen.

$$\frac{\delta C}{\delta t} = \frac{1}{r}\frac{\delta}{\delta r}(rD\frac{\delta C}{\delta r}) + \frac{\delta}{\delta y}(D\frac{\delta C}{\delta y}) \tag{1}$$

$$C(0,r,y) = C_0 \tag{2}$$

$$C(t,r,0) = C(t,r,L) = C(t,R,y) = 0$$

where C is the concentration of hydrogen; D – diffusion coefficient; t – time; r and y – cylindrical coordinates; C_0 – initial hydrogen concentration.

If the diffusion coefficient is constant the solution of the problem (1)–(3) can be obtained with the Fourier method of variable separation. The equation for the concentration is written as follows

$$C(t,r,y) = \frac{32C_0}{\pi}\sum_{m=0}^{\infty}\sum_{n=1}^{\infty}\frac{J_0\left(\frac{\mu_n}{R}r\right)\sin\left(\frac{2m+1}{L}\pi y\right)}{\mu_n J_1(\mu_n)(2m+1)}\exp\left\{-\left[\frac{\mu_n^2}{R^2}+\frac{(2m+1)^2}{L^2}\pi^2\right]Dt\right\} \tag{4}$$

where J_0 and J_1 are the Bessel functions of the first kind of order zero and first respectively, μ_n are the roots of equation $J_0(x) = 0$.

From Equation 4 it is possible to obtain the time dependence of the hydrogen content in the specimen Q by integrating this equation through the volume of the specimen. The rate of hydrogen effusion V is the derivative of Q on time:

$$Q(t) = 2\pi \int_0^R r\delta r \int_0^L C(t,r,y)\delta y$$

$$= \frac{32 C_0 L R^2}{\pi} \sum_{m=0}^{\infty} \sum_{n=1}^{\infty} \frac{1}{(2m+1)\mu_n^2} \exp\left\{-\left[\frac{\mu_n^2}{R^2} + \frac{(2m+1)^2}{L^2}\pi^2\right] Dt\right\} \quad (5)$$

$$V(t) = \frac{\delta Q(t)}{\delta t}$$

$$= \frac{32 C_0 L R^2 D}{\pi} \sum_{m=0}^{\infty} \sum_{n=1}^{\infty} \left[\frac{1}{(2m+1)^2 R^2} + \frac{\pi^2}{\mu_n^2 L^2}\right] \exp\left\{-\left[\frac{\mu_n^2}{R^2} + \frac{(2m+1)^2}{L^2}\pi^2\right] Dt\right\} \quad (6)$$

After following linear substitution of variables in the equation (1)

$$\tilde{r} = \frac{r}{R}, \quad \tilde{y} = \frac{y}{R}, \quad \tilde{C} = \frac{C}{C_0}$$

the problem (1)–(3) takes form

$$\frac{\delta \tilde{C}}{\delta F_0} = \frac{1}{\tilde{r}} \frac{\delta}{\delta \tilde{r}} \left(\tilde{r} \frac{\delta \tilde{C}}{\delta \tilde{r}}\right) + \frac{\delta}{\delta \tilde{y}} \left(\frac{\delta \tilde{C}}{\delta \tilde{y}}\right) \quad (7)$$

$$\tilde{C}(0, \tilde{r}, \tilde{y}) = 1 \quad (8)$$

$$\tilde{C}(F_0, \tilde{r}, 0) = \tilde{C}(F_0, \tilde{r}, \frac{L}{R}) = \tilde{C}(F_0, 1, \tilde{y}) = 0 \quad (9)$$

where $F_0 = \dfrac{Dt}{R^2}$ is dimensionless time called Fourier number.

For the effusion rate $V(t)$ and hydrogen content $Q(t)$ the dimensionless equivalents also exist

$$\tilde{V} = \frac{V}{C_0 R D} \quad (10)$$

$$\tilde{Q} = \frac{Q}{C_0 R^3} \quad (11)$$

The values $\tilde{V}$ and $\tilde{Q}$ are unambiguously determined as the functions of Fourier number by solving Equations 7–9.

The determination of the hydrogen diffusion coefficient can be carried out by following procedures:

(a) From the experimental value of Q it is necessary to calculate the dimensionless equivalent $\tilde{Q}$ using the equation (11), then to determine the corresponding value $\tilde{V}$ from $\tilde{Q}$ by solving equations (7)–(9) (numerical method), and finally to calculate D from equation (10) taking into account $\tilde{V}$ and the experimental results for the effusion rate V.

(b) A longer time from the beginning of effusion, results in a low hydrogen content in the specimen, the regular regime of effusion at which the effusion rate is proportional to the hydrogen content reaches, and determination of D can be done using the equation

$$D = \left[\frac{\mu_1^2}{R^2} + \frac{\pi^2}{L^2}\right]^{-1} \frac{V}{Q} \qquad (12)$$

which results from (5) and (6) at $t \to \infty$

(c) The third possibility of diffusion coefficient determination is based on the expression for the Fourier number $F_0 = \dfrac{D\,t}{R^2}$ and the graph $F_0(t)$, which is calculated by numerical method from the experimental data $Q(t)$.

The methods (a) and (c) are based on the exact solution of the diffusion problem and enable to determinate the diffusion coefficient at the whole hydrogen concentration range, which takes place during effusion. The calculations in accordance with the first of them are performed for each experimental point $V(Q)$ separately so any disturbances of the measuring procedure or deviation of effusion behaviour from the theoretically expected one influence the results obtained. In a contrast with this, the diffusion coefficient in the method (c) is calculated from the slope of the graph $F_0(t)$ which uses many experimental points and difficulties mentioned above can be eliminated. Besides, this method directly shows whether the hydrogen effusion follows to the accepted mathematical model, in accordance with which the calculated dependence $F_0(t)$ must be straight line, or not.

The dependencies of the evolution rate from the hydrogen content, calculated according to the equations 5–6 are shown on the Fig. 2. The initial hydrogen content was taken to be equal 5 ml/100g. With the decreasing of hydrogen content the curves tend to the straight lines. The slope of graphs is determined by diffusion coefficient and specimen dimensions according to equation 12. The accuracy of this equation does not depend on the diffusion coefficient, initial hydrogen content and specimen sizes but is determined by the relative hydrogen content $Q(t)/Q_0$. The vertical dotted lines on the Fig. 2

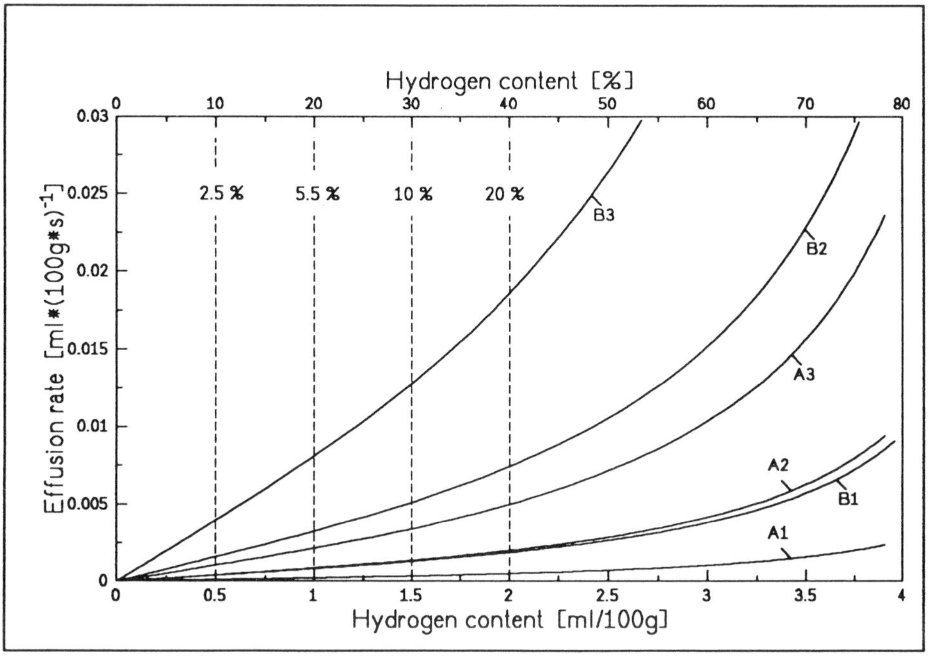

Figure 2. The kinetic curves of hydrogen effusion from cylindrical specimens of different dimensions for different diffusion coefficients calculated from equations 5 and 6. Specimen sizes: "A" – R = 4 mm, L = 15 mm; "B" – R = 2 mm, L = 10 mm. Diffusion coefficients (10^{-5} cm^2/s): "1" – 0.5, "2" – 2.0, "3" – 5.0. The numbers on the vertical dotted lines indicate the precision of the formula (12).

indicate the accuracy of the formula (12) calculated for single point. But if we use a number of experimental points which correspond to the hydrogen content less than the level indicated by dotted line and utilise the regression procedure, the theoretical accuracy of this method of hydrogen diffusion coefficient determination is higher.

To check whether the effusion of hydrogen from the metal of welded joints charged with hydrogen during welding or by electrolytic polarisation corresponds to the considered mathematical model the measurements of hydrogen effusion from the specimens prepared with multilayer welding, have been carried out. The basic coated electrode UONI 13/55 was used, and the effusion temperature was equal to 100°C. The electrolyte for charging was composed from 95% distilled water and of 5% H_2SO_4 (concentrated). Furthermore 0.033% of natriumthiosulphur was added to the solution. The current density was equal to 5 mA/cm^2. On the Fig. 3 the graphs of Fourier number multiplied by square of specimen radius calculated in accordance with the method (c) are shown. As it is possible to see, the results of measurements and subsequent calculations are located very good on the same straight line. The slope of this line is equal to the diffusion coefficient. The deviation of the

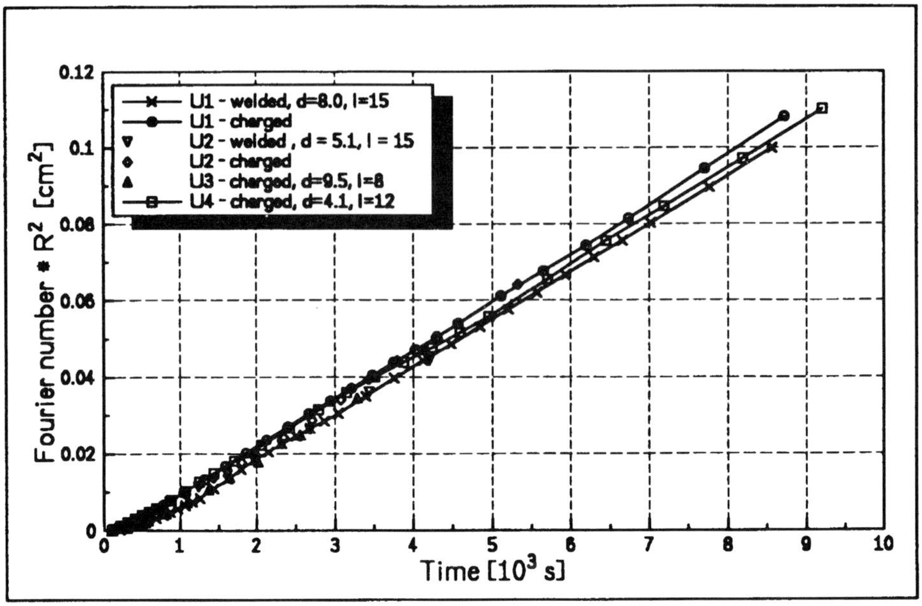

Figure 3. Dependence of Fourier number on time for different specimen sizes (d – diameter l – length) and methods of charging with hydrogen : during welding – "welded" and by electrolitic charging – "charged". Basic coated electrod UONI 13/55. Temperature is equal to 100 °C. Diffusion coefficient is equal to 1.1 ± 0.05 cm^2/s.

results from the straight line which occurs at the beginning of effusion is caused by the finite time, necessary for the specimen to reach the temperature of the final value. The results obtained do not depend on the specimen sizes. It proves, that effusion of hydrogen is described by the accepted mathematical model. Besides, it possible to use electrolytic charging of metal with hydrogen for the subsequent effusion measurements.

4. INFLUENCE OF TRAPS ON HYDROGEN EFFUSION

The measurements of hydrogen effusion from steels and, especially, from the weld metal deposited by rutile coated electrodes show the deviation of kinetic curves $V(Q)$ from theoretically expected ones with the decreasing of the hydrogen content in the specimen to a certain level. This effect is demonstrated in Fig. 4 and Fig. 10 for the rutile coated electrode ANO-4. Instead of the solid straight lines which follow from the equations (5) and (6)

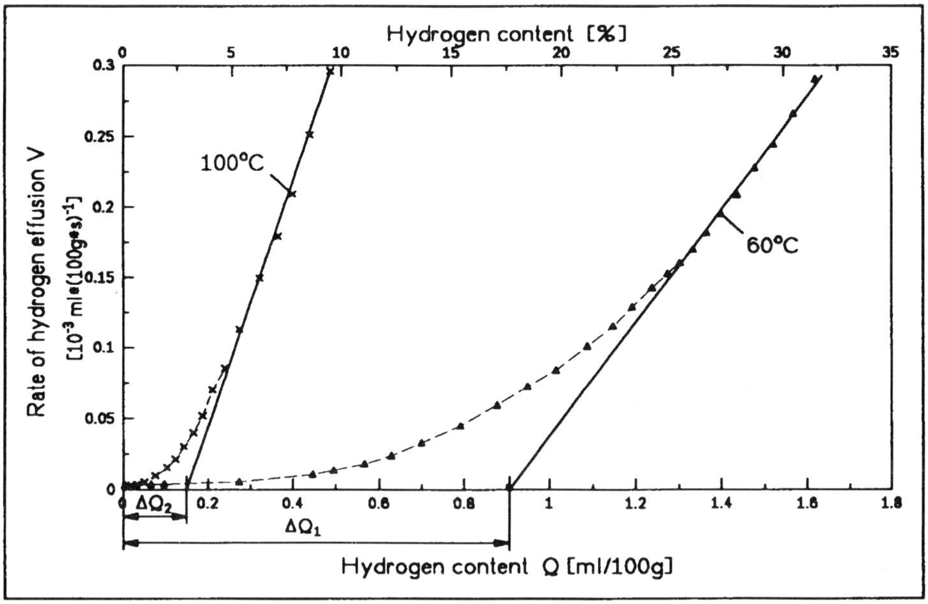

Figure 4. Influence of traps on the behavior of effusion curves under low hydrogen content in the specimen for analysis temperature 60 and 100°C. Rutile coated electrode ANO-4.

under low hydrogen content (Fig. 2) the data of measurements are located as shown in Fig. 4. It results in an increase of time which is necessary to allow the escape of hydrogen from the specimen because the effusion rate decreases very slowly.

This phenomenon is caused by the evolution of hydrogen from traps which is possible to measure when the level of diffusible hydrogen is low enough and can be described using McNabb and Foster's theory of hydrogen diffusion in non perfect metal.[5] The diffusion equation in the presence of traps can be written as follows:

$$\frac{\delta C_1}{\delta t} + \frac{\delta C_t}{\delta t} = D\Delta C_1 \tag{13}$$

where C_l designate the concentration of diffusible hydrogen, and C_t-hydrogen concentration in traps. This equation suggests that the flow is supported by concentration gradient of diffusible hydrogen only, that is traps are located so far from each other that diffusion jumps of "trap–trap" are impossible.

The variation of hydrogen concentration in traps can be expressed as the difference between the rate of capture and release by the following equation:

$$\frac{\delta C_t}{\delta t} = kC_l(C_{tm} - C_t) - pC_t \tag{14}$$

where k is the rate constant of trapping, p – rate constant of detrapping, C_{tm} is the capacity of traps. After dividing equation 14 by C_{tm} the following equation for the trap occupancy fraction $n = \frac{C_t}{C_{tm}}$ is obtained:

$$\frac{\delta n}{\delta t} = k(1 - n)C_l - pn \tag{15}$$

If the rate of diffusible hydrogen effusion from the specimen is much higher then the rate of hydrogen evolution from the traps the solution of equation 14 can be obtained in the approximation that the trapped hydrogen does not influence the evolution of diffusible hydrogen. To take into account the influence of hydrogen in the diffusible zone the rate of evolution from traps the following approximation for the concentration C_l can be accepted

$$C_l(t) = C_0 \exp(-at) \tag{16}$$

which follows from the solution of Fick's equation 5. Here the constant a is determined by the diffusion coefficient and specimen sizes:

$$a = D\left(\frac{\mu_1^2}{R^2} + \frac{\pi^2}{L^2}\right) \tag{17}$$

The charging of the specimen with hydrogen takes time so we can suggest that equilibrium between diffusible and trapped hydrogen is attained. Thus the initial condition for equation 15 is following

$$n_0 \equiv n(0) = \frac{kC_0}{kC_0 + p} \tag{18}$$

which follows from equation 15 when $\frac{\delta n}{\delta t} = 0$.

After substitution of equation 16 to equation 15 the linear differential equation of first order is obtained, the solution of which is written as follows

$$n(t) = \exp\left(-pt + \frac{kC_0}{a}\exp(-at)\right)\left(F(t) + \frac{kC_0}{kC_0 + p}\exp(-\frac{kC_0}{a})\right) \tag{19}$$

where $F(t) = dC_0\int_0^t \exp(-a\tau)\exp\left(p\tau - \frac{kC_0}{a}\exp(-a\tau)\right)d\tau$.

The equation 19 contains three constants – a, p and kC_0. The constant a relates to the evolution of diffusible hydrogen and can be determined from diffusible part of effusion curve $V(Q)$. The constant p which represents the hydrogen evolution from traps can be calculated as the slope of graph $V(Q)$ when this graph corresponds to the trapped hydrogen. Such parts of curves

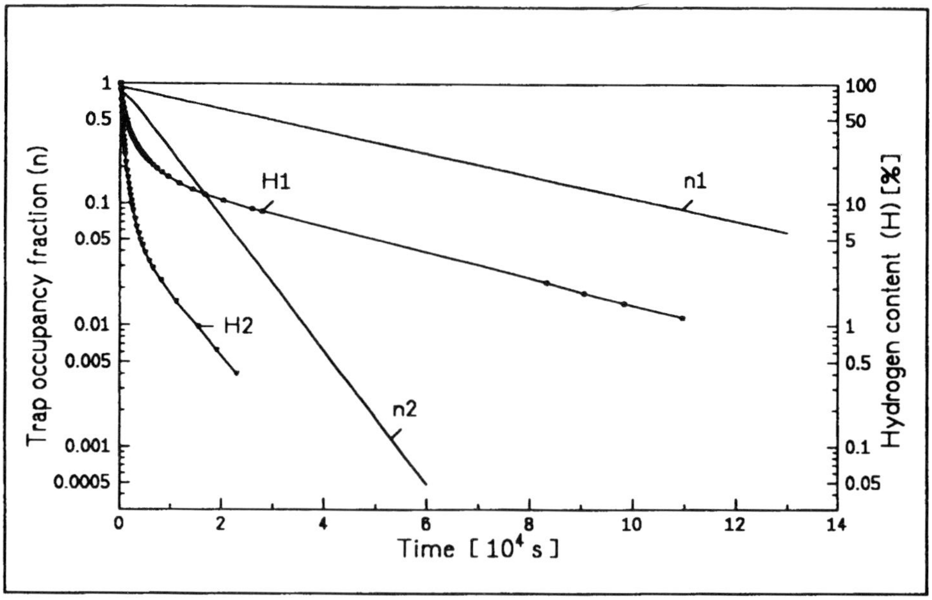

Figure 5. Time dependence of the traps occupancy fraction calculated on equation 19 and experimentally measured hydrogen content in the specimens. Rutile coated electrode ANO-4. Indices n1,H1 correspond to the temperature 60°C; n2,H2 – 100°C.

V(Q) correspond to the hydrogen content in the specimen ΔQ_1 and ΔQ_2 in Fig. 4.

The results of trap occupancy fraction calculations in accordance with equation 19 for the experimental data represented in Fig. 4 are shown in Fig. 5. The constant kC_0 was suggested to be equal $10.p$ and $5.p$ for the temperature 60 and 100°C respectively. On this figure the experimentally measured hydrogen content in the specimen vs. time is also presented. As it is possible to see the decrease rate of measured hydrogen at the beginning period of effusion is much higher than the change of trap occupancy fraction, so the traps do not essentially influence the measured hydrogen effusion. But after a long time when the diffusible hydrogen escapes from the specimen the measured effusion corresponds to the hydrogen evolution from traps and the slope of graphs *H(t)* and *n(t)* are the same.

To compare more accurately the values of diffusible hydrogen evolution rate and the rate of detrapping the calculations of hydrogen concentration in traps and rate of hydrogen evolution from traps under 100°C were carried out. If we suggest that 6 percent of the totally measured hydrogen represent the hydrogen in traps the calculated result corresponds to the measured one Fig. 6. The measured rate of hydrogen effusion at the time which corresponds to 10% of hydrogen content is ten times higher than that for hydrogen evolution

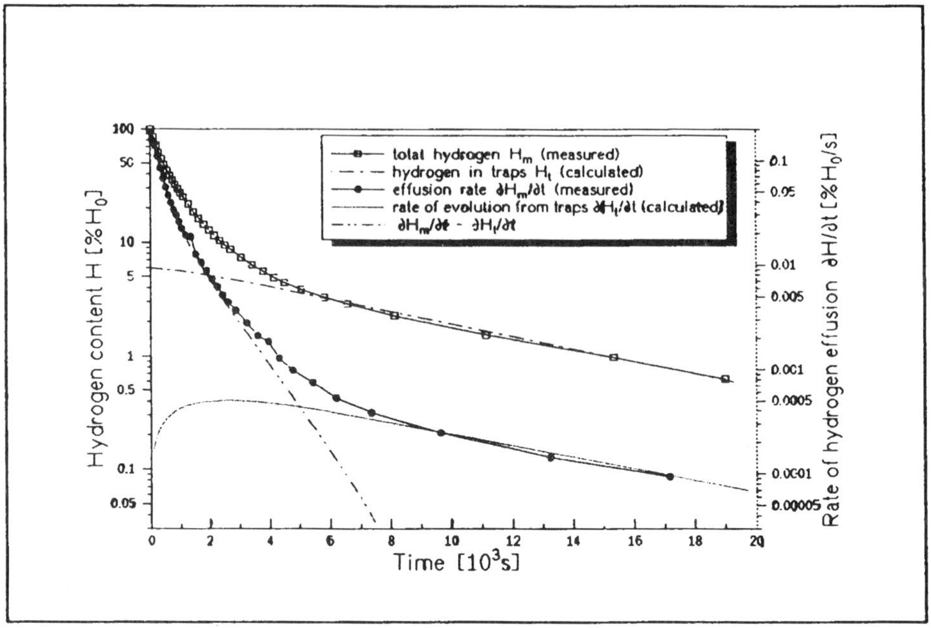

Figure 6. The hydrogen evolution from traps. The comparison of calculations with experiment.

from traps. Besides the difference between measured effusion rate and effusion rate from traps is presented by a straight line on a semilogarithmic scale as it follows from the solution of Fick's equation 6 for a long time after the beginning of effusion.

5. THE FINAL PROCEDURE OF DETERMINING THE HYDROGEN DIFFUSION COEFFICIENT

The previous analysis show that evolution of hydrogen from traps does not essentially effect the effusion of diffusible hydrogen until the level of hydrogen is sufficiently high. It enables the avoidance of the influence of traps on the measured hydrogen diffusion coefficient. With this purpose it is necessary to extrapolate the straight part of effusion curve *V(Q)* to the point of zero effusion rate and take away the value of trapped hydrogen. Such values are represented by ΔQ_1 and ΔQ_2 in Fig. 4 and by ΔQ_3 on Fig. 10 for the hydrogen effusion from the metal made of a rutile coated electrode.

In the computer program the automatic search for the straight part of the effusion curve is realised on the basis of regression analysis. Then the corrected data are treated in accordance with methods (a) and (c). The results of calculations consist of the values of diffusion coefficient obtained by using all methods described in the Chapter 3. The most acceptable is the value of D obtained from the slope of the graph $F_0(t)$ because it is based on many measured points and describes hydrogen diffusion with exact solution of Fick's equation. The program uses up to 40 values of μ_n and 40 values of **m** in the equations 5 and 6 so the mistake of calculations is negligible in comparison with the measurements.

The program also enables the treatment of the origin experimental data. In this case the graph $F_0(t)$ at a long time from the start of effusion represents the hydrogen evolution from traps. It is possible to calculate the constant of detrapping rate p by multiplying the apparent diffusion coefficient which corresponds to the slope of this graph on the geometric factor $\left[\frac{\mu_1^2}{R^2} + \frac{\pi^2}{L^2}\right]$.

In addition, the solutions of Fick's equation for the hydrogen evolution from an infinite cylinder, hydrogen permeation through plate and hydrogen diffusion in a thick wall hole cylinder are implemented in the program. In the case of hydrogen permeation the determination of diffusion coefficient can be carried out by calculating both the Fourier number dependence from time and by "time lag" method.

6. INFLUENCE OF DEFORMATION ON THE DIFFUSION COEFFICIENT OF HYDROGEN

The results described in this and next chapters were published before.[6] So in this paper the main attention will be paid to illustrating the application of the computing procedure involved.

For investigating the influence of plastic deformation on hydrogen diffusion a fine-grained structural steel with the tensile strength of 800 MPa was used. The steel composition is listed in reference 6. The standard specimens for mechanical testing with diameter of 6 mm were electrolytically charged with hydrogen under 25°C. Then the specimens were placed into liquid nitrogen and stretched in a tensile testing machine. The elongations were 0.2%, 4.7% and 11.2% for the specimens **E1, E2** and **E3** respectively. The central part of the specimens, with the length of ~15 mm, was cut for the measurements of hydrogen effusion. The measurements were made at 100°C for all specimens.

The hydrogen evolution curves are shown in Fig. 7. The dependencies of effusion rate vs. hydrogen content made by numerical differentiation are shown in Fig. 8. As it possible to see the effusion of hydrogen from the specimens **E2** and **E3** starts with a certain delay. Besides, the total amount of hydrogen which effused at the temperature of analysis decreases with increasing deformation. These effects can be explained by influence of traps

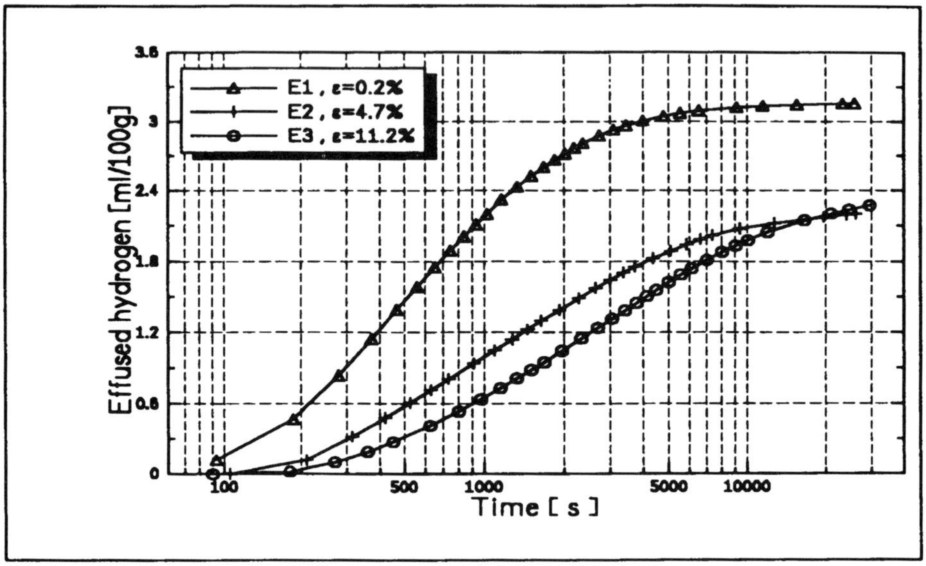

Figure 7. Influence of plastic deformation of steel on the effusion of hydrogen..

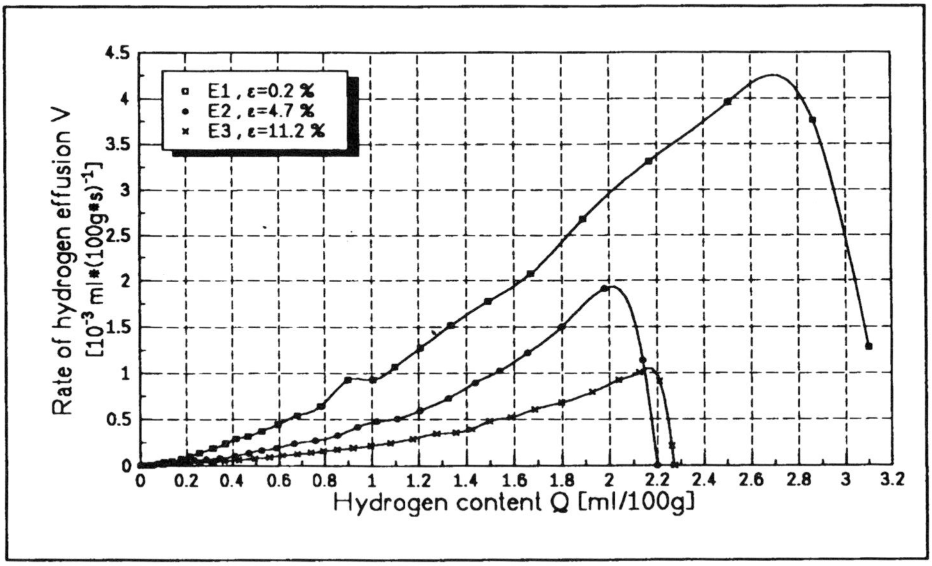

Figure 8. Influence of deformation on the kinetic of hydrogen effusion.

which were created during deformation. Since the temperature of deformation was so low (–196°C) that hydrogen diffusion was practically stopped it takes time for hydrogen to fill newly created traps during subsequent heating in the furnace. So at the beginning of effusion there is no output flow of hydrogen. The difference in total hydrogen shows that deformation, produced traps with high bonding energy for hydrogen. These traps cannot be activated at the utilised analytical temperature.

After the temperature reached the final value the kinetic curves **V(Q)** (Fig. 8) demonstrate good agreement with the theory (Fig. 2) up to the low hydrogen level. Then it is possible to see the insignificant deviations analogous to those shown in Fig. 4 for effusion from weld metal. The treatment of data in accordance with method (3) demonstrates very well the influence of deformation on hydrogen diffusion coefficient which is equal to the slope of graph **E1, E2** and **E3** in Fig. 9 ($1.2 \cdot 10^{-5}$, $0.4 \cdot 10^{-5}$ and $0.25 \cdot 10^{-5}$ cm^2/s respectively). Besides, the fact that these graphs are represented by straight lines proves that diffusion coefficient remains constant in the considered range of hydrogen concentration.

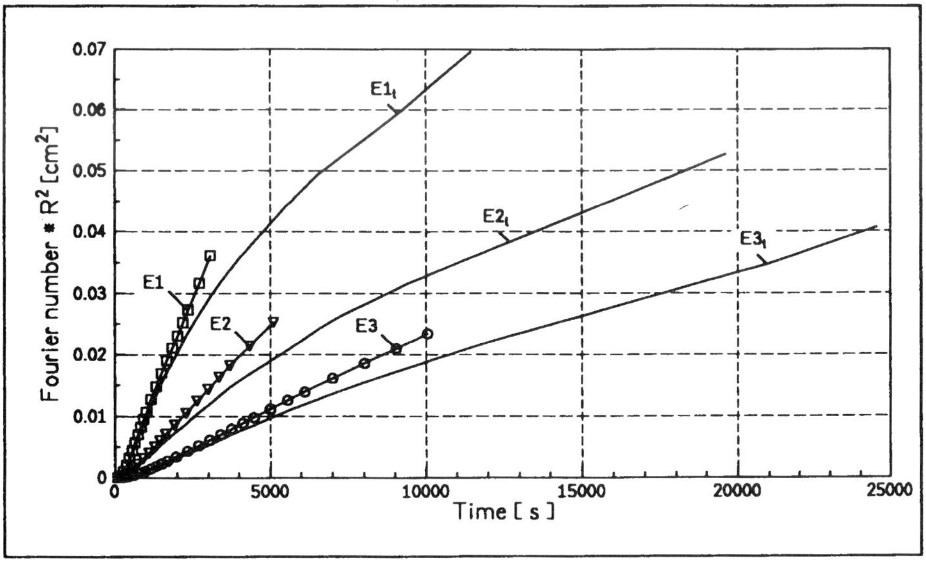

Figure 9. Influence of deformation on the dependence of Fourier number on time. The curves E1, E2 and E3 represent the hydrogen diffusion in accordance with the Fick's equation 1 and correspond to the specimen elongation 0.2, 4.7 and 11.4% respectively. The curves $E1_t$, $E2_t$ and $E3_t$ were calculated for the original effusion data and reveal the influence of traps on hydrogen effusion.

7. HYDROGEN DIFFUSION IN THE WELD METAL DEPOSITED BY BASIC AND RUTILE COATED ELECTRODES

It was shown earlier in the literature that hydrogen escape from the metal deposited by rutile coated electrode takes a much longer time than the effusion from the metal deposited by electrode with the basic coating. This effect usually is explained by the hinder of diffusion due to a large amount of non-metallic inclusions in rutile metal. In the present report the analysis of hydrogen diffusion in these two types of weld metal on the basis of developed mathematical procedure is described.

The cylindrical specimens for these experiments were prepared from the centre of a weld joints made with basic and rutile coated electrodes. All specimens have been charged with hydrogen simultaneously. The alloy composition was almost identical for both weld metal.[6] The structure was in both cases ferritic–perlitic with similar grain sizes and lamellar distances. The difference between the two weld metals consist in the silicon and oxygen content and the quantity of non-metallic inclusions. The measurements of effusion were carried out at the temperature of 60, 100 and 150°C.

In Fig. 10 the kinetics curves *V(Q)* for rutile coated electrode are shown. The amount of hydrogen measured at 150°C is higher than that measured at more low temperatures. It is caused by the activation of another kind of traps than

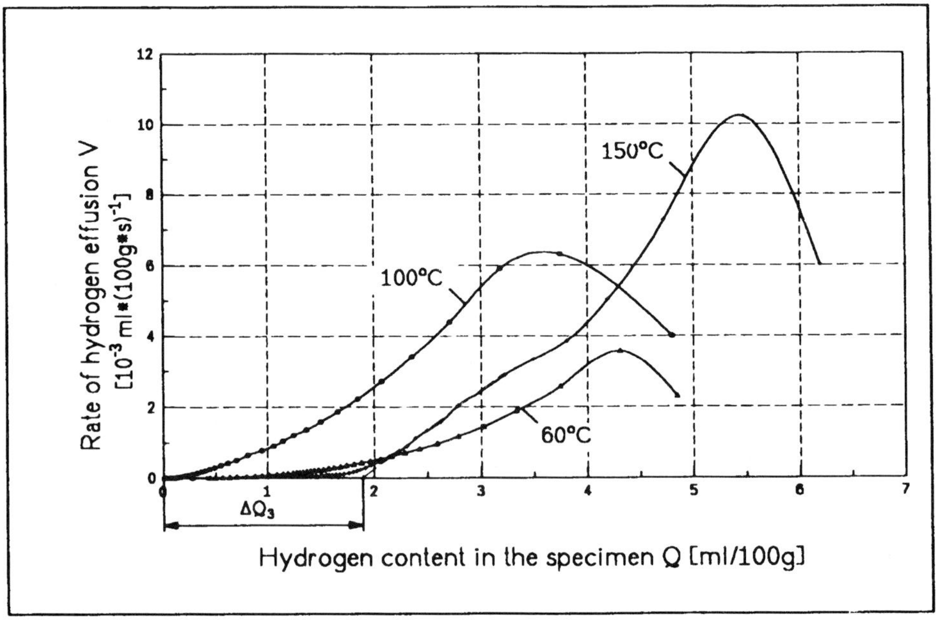

Figure 10. Influence of temperature on the kinetic of hydrogen effusion from the specimens made of rutile deposited metal. The value ΔQ_3 represents the evolution of hydrogen from traps activated at 150°C.

those which supplied the hydrogen ΔQ_1 and ΔQ_2 at 60 and 100°C (Fig. 4). We can see that $\Delta Q_2 < \Delta Q_1$. This is explained by a higher increase in detrapping rate constant p in comparison with k (14) with increasing temperature. As a result of this the initial trap occupancy fraction n_0 decreases (18). At a temperature of 150°C it is difficult to detect the influence of these traps on hydrogen effusion because p is the same order of magnitude as a (17).

The measurements of hydrogen effusion from steels and weld metals deposited with basic electrodes never show such values of residual hydrogen activated at 150°C like ΔQ_3 for rutile coated electrode (Fig. 9). It proves that this effect is caused by non-metallic inclusions which are present in rutile deposited metal in a great quantity.

The calculations of Fourier number for the two weld metals at effusion temperature of 100°C are shown in Fig. 11. For the basic electrode the influence of traps is slightly detectable. The evolution of diffusible hydrogen from rutile deposited metal occurs with a diffusion coefficient which is equal to the slope of graph "rutile (diff.)" in Fig. 11. The analogous results were obtained at other temperatures.[6]

To compare the hydrogen diffusion in the considered weld metals the temperature dependencies of diffusion coefficient are shown in Fig. 12. Here the logarithm of D is plotted vs. reciprocal temperature so the slope of the graph is equal to activation energy of diffusion as it follows from the Arhenius

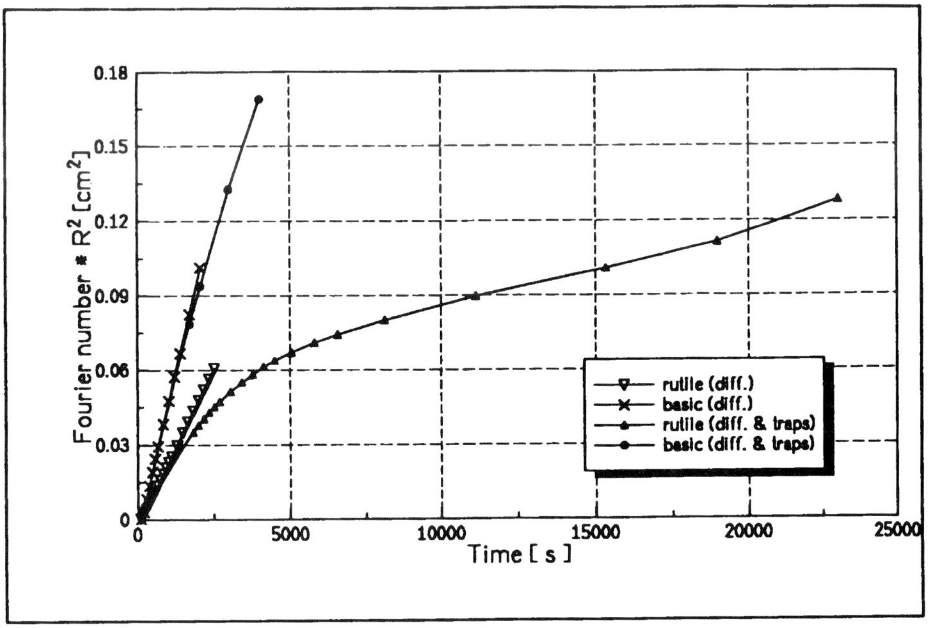

Figure 11. Influence of the type of electrode coating on the time dependence of Fourier number at the temperature of 100°C.

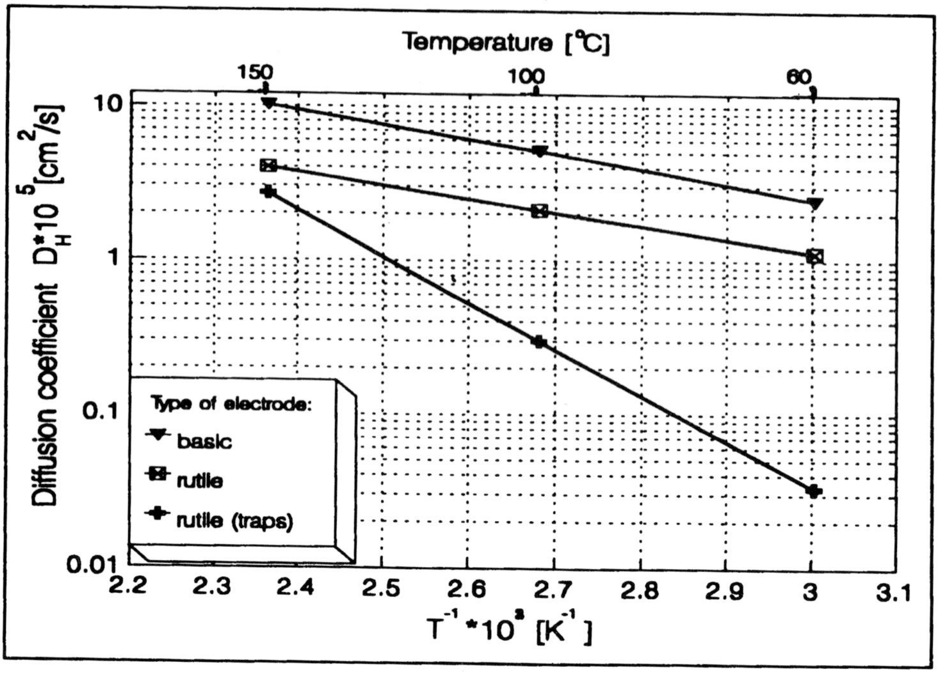

Figure 12. Temperature dependence of diffusion coefficients of hydrogen in deposited metal from basic and rutile coated electrodes and temperature dependence of the rate constant of hydrogen evolution from traps (rutile-(traps)).

equation $D = D_0 \exp\left(-\frac{E_d}{RT}\right)$. The activation energies for diffusion in both weld deposits are virtually identical. It means that diffusion transport of hydrogen occurs in both metals on the same mechanism. In the case of the rutile electrode the non-metallic inclusions act like obstacles in the way of diffusible hydrogen decreasing the value D_0. In this figure the temperature dependence of the product of detrapping constant on geometric factor $p\left[\frac{\mu_1^2}{R^2} + \frac{\pi^2}{L^2}\right]^{-1}$ for rutile electrode is shown also. It makes it possible to compare the diffusion rate with the rate of evolution from traps. This graph shows that activation energy for hydrogen evolution from traps is higher than activation energy of diffusion.

8. CONCLUSIONS

The method for determination of the hydrogen diffusion coefficient described in the present report enables us to analyse the diffusion in the whole concentration range which takes place during experiment because it is based on the exact solution of Fick's equation. The many experimental results published in the literature were obtained using approximate solutions for the diffusion problem which includes only first member of sum in equation 5. Such approximation describes the hydrogen effusion only at a sufficiently large time from the beginning of the process. But, as it was shown, in this case the measured value of D can represent the hydrogen evolution from traps. Perhaps this fact is one of the reasons for a very small diffusion coefficient obtained from the effusion measurements described in the literature.

The experimental results presented in this paper show that mass transfer of hydrogen in the tested materials is controlled by a constant diffusion coefficient in the hydrogen concentration range which is usual for welding. So this value of D can be used for calculations of hydrogen distribution in weld joints. The influence of traps, if it is sufficient, can be estimated by means of determination of parameter p and initial hydrogen accumulated by this traps.

REFERENCES

1. I.K. POCHODNJA *et al.*: Mathematical simulation of gases behaviour in weld joints (Russian). Publishing house Naukova Dumka, Kiev, 1979.
2. V.A. KARCHIN *et al.*: Kinetics of hydrogen distribution in weld joints (Russian). *Autom. Svarka*, **33** (1980), (6), 28–32.
3. G. DICKEHUT and J. RUGE: Wasserstoffverteilung in der Schweißnaht – Theorie zur Berechnung. *Schweißen und Schneiden*, 1988, **40**, (6), 289–292.
4. V. MICHAILOV, J. RUGE and K. THOMAS: Berechnung der Wasserstoffverteilung beim Schweißen, *Schweißen und Schneiden*, 1991, **43**, (11), 655–658.
5. A. MCNABB and P.K. FOSTER: A New Analysis of the Diffusion of Hydrogen in Iron and Ferritic Steels. *Transactions of AIME*, 1993, **227**, 618–627.
6. U. DILTHEY, S. TRUBE, I.K. PKHODNYA and V.A. PAVLYK: Untersuchung des Diffusionskoeffizienten von Wasserstoff in deformiertem Stahl und in Schweißgütern aus basischen und rutilumhüllten Elektroden, *Schweißen und Schneiden*, 1992, **44**, (12), 668–671.

ACKNOWLEDGEMENTS

The author wishes to thank Professor Horst Cerjak and Professor Ivan Grivnjak for their sincere invitation and support, which made this presentation possible. My thanks are also expressed to Professor Igor Pochodnja, under whose guidance the presented work has been done.

SECTION III

RESIDUAL STRESSES

COMPUTATIONAL WELD MECHANICS OF THE STEADY STATE

J. GOLDAK and M. GU
Carleton University, Ottawa, Ontario, Canada

ABSTRACT

The steady state is a fundamental state of the welding process. With appropriate symmetry, all welds rapidly approach the steady state. Since the steady state reduces the dimension from $(x, y\ z, t)$ to (x, y, z), it can lead to substantial savings in computing costs. Even more important, it enables significantly higher accuracy and resolution to be achieved than is possible in a full transient analysis. Of course, the steady state does not resolve the starting or stopping transients.

An analysis of steady state temperatures, microstructures and thermal stresses in a weld is reviewed. For thermal analyses, a finite element method using an Eulerian formulation is described. The method incorporates the addition of filler metal in a natural manner. Strategies that were developed to analyse the evolution of microstructures in the steady state thermal field are reviewed. History-dependent parameters are integrated along streamlines. A data structure to assist in creating those streamlines and setting starting conditions for each streamline is used. The method can also be applied to the temperature field at a specific step in the Lagrangian formulation. A significant gain in computing speed and/or spatial resolution over transient analyses has been achieved.

INTRODUCTION

To the author's knowledge, Westby's[1] 1968 finite difference thermal analysis and Ueda's thermal stress analysis[2] began the development of modern computational weld mechanics. Today, finite element and finite difference methods have largely replaced the series solutions developed by Rosenthal[3] and Rykalin[4] in the 1930s to solve the heat equation for moving heat sources. Their solution was a steady state solution and it used an Eulerian reference frame. It was used for computing temperature fields in welds until the 1980s and is still used as the solution with the least computational cost. In the 1980s, the finite element method (FEM) displaced the Rosenthal and Rykalin method because FEM could deal with complex finite geometries, nonlinear effects such as temperature-dependent thermal properties and complex boundary conditions including heat sources. The FEM formulations were usually transient Lagrangian formulation rather than steady state Eulerian formulations. Many successful 3-dimensional transient analyses on welds have been reported that use such Lagrangian FEM formulations.[5] FEM approximations to the steady state temperature field were obtained by running a transient analysis until it approached the steady state.

Today the transient Lagrangian formulation for computational weld mechanics analysis of large weldments faces three major limitations: (1) Temperatures, microstructures and stress and strain vary rapidly near the heat source and slowly farther from the heat source. This requires a fine mesh near the heat source and a coarse mesh far from the heat source. In a Lagrangian reference frame, the mesh must be refined dynamically.[6] The size of weld that can be analysed, the resolution in space and time that can be achieved, is limited by the computer speed and memory available. (2) Short time steps are required to lessen the effect of 'spot welding.' Since so many time steps are needed to reach the steady state, an analysis of a medium sized weld can require a day or longer on a modern workstation. (3) The addition of filler metal is usually ignored because of the excessive complexity and cost of changing the geometry and mesh between time steps to accommodate the added filler metal. However, the added filler metal is particularly important in a subsequent thermal stress analysis.

In the first part of the paper, an FEM Eulerian formulation is reviewed for the steady state temperature field in welds.[7] Like the Rosenthal solution, it is limited to constant thermal properties and to nearly prismatic domains with constant cross-section and prismatic axis in the direction of welding. Being a steady state solution, it provides no information on the transients on starting and stopping the weld. However, the steady state does allow a much higher spatial resolution in a single computational step than a transient analysis. Adding filler material becomes natural in the steady state analysis.

In the second part of the paper, the evolution of microstructure in a weld under steady state conditions is reviewed.[8] The evolution is computed by integrating along flow lines. A data structure to assist in creating flow lines and setting starting conditions of the flow lines is used. The method has proved to be fast and efficient. A significant gain in computing speed and/or spatial resolution over transient analyses has been achieved.

A steady state stress analysis for welds is presented in reference. 9.

EULERIAN STEADY STATE FORMULATION

In a Lagrangian reference frame, the conservation of energy or heat equation may be written as

$$\rho c_p \dot{T} - \nabla \cdot q - S = 0 \qquad or \qquad T_{,xx} + T_{,yy} + T_{,zz} + \frac{S}{\kappa} = \alpha \dot{T} \tag{1}$$

where the comma subscripts indicate partial derivatives. The temperature is T, the thermal conductivity is κ, the volumetric specific heat is c_p, $\alpha = \chi\pi/\kappa$ is the reciprocal of diffusivity, the thermal flux is $q = -\kappa \nabla T$ and S is a distributed heat source.

Suppose a heat source is moving at a constant speed v in the positive x-direction. We define an Eulerian (moving) frame with origin at the centre of the source, and coordinates (ξ, y, z). The transformation from (x, y, z) to (ξ, y, z) is given by: $\xi = x - vt$, where t is time. replacing the material time derivative, $\dot{T}$, with

$$\dot{T} = T_{,t} + \xi_{,t} T_{,\xi} = T_{,t} - v T_{,\xi}$$

in Equation 1, the equation of conservation of energy in the Eulerian frame is

$$T_{,\xi\xi} + T_{,yy} + T_{,zz} + \frac{S}{\kappa} = -\alpha v T_{,\xi} + \alpha T_{,t} \tag{2}$$

$T_{,t}$ is the time derivative of temperature at a point fixed in the Eulerian frame. This derivative has a value of zero in the steady state and so Equation 2 becomes:

$$T_{,\xi\xi} + T_{,yy} + T_{,zz} + \frac{S}{\kappa} = -\alpha v T_{,\xi} \tag{3}$$

If an FEM method is based on Equation 3, it would require a non-standard FEM code to deal with the resulting asymmetric operator. When α and v are constant, Equation 3 can be made self-adjoint by the transformation

$$T = T_0 + \exp\left\{\frac{-\alpha v \xi}{2}\right\} \varphi(\xi, y, z) \tag{4}$$

Substituting this into equation (3), we have

$$\varphi_{,\xi\xi} + \varphi_{,yy} + \varphi_{,zz} + \frac{S}{\kappa} \exp\left\{\frac{\alpha v \xi}{2}\right\} = \left\{\frac{\alpha v}{2}\right\}^2 \varphi \tag{5}$$

or

$$\nabla^2 \varphi - \left\{\frac{\alpha v}{2}\right\}^2 \varphi = -\frac{S}{\kappa} \exp\left\{\frac{\alpha v \xi}{2}\right\} = \tilde{S}$$

Equation 5 is symmetric negative definite. After properly transferring the boundary conditions from the T function to the ϕ function, this equation can easily be solved with a standard Lagrangian FEM code for the heat equation. Equation (5) only applies to linear systems (i.e. with α independent of T) because of Equation 4. In the following examples, the diffusivity is taken as a constant.

BOUNDARY CONDITIONS

Prisms with constant cross-sections such as the one shown in Fig. 1 are the most convenient domains for Equation 5. The boundary between the weld pool and the solid can be prescribed to the solidus temperature. The flux due to radiation and convection on the surface of the weld pool, i.e. the weld heat input, can be found from the thermal reactions.[10]

From experimental measurements, we know that at a sufficient distance, ξ_n, in front of weld pool, temperatures remain at ambient temperature; $T = T_0$ when $\xi > \xi_n$ and consequently, the flux $q = 0$ when $\xi >> \xi_n$.

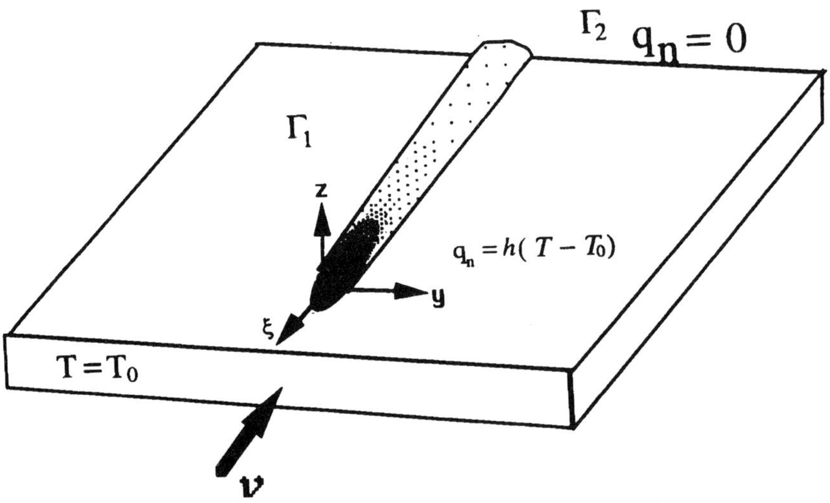

Figure 1. The Eulerian frame and boundary conditions. Γ_2 is the downstream boundary that is normal to the velocity of the source. On this surface the gradient of the temperature has been prescribed to zero. Γ_3 is the surface of the liquid solid interface of the weld pool. The temperature is prescribed on this surface which is smooth but the orientation and curvature is arbitrary. Γ_4 is the surface of the upstream end with outward normal parallel to the velocity of the source. On this surface the ambient temperature of the plate has been prescribed. Γ_1 is the remainder of the boundary surface with outward normal orthogonal to the velocity of the source. On this surface convection has been applied.

A boundary flux can exist on the prism boundaries when they are not insulated. For the case of a constant convective coefficient, h, the boundary condition is

$$q_n = -\kappa \nabla T \cdot n_\Gamma = h(T - T_0) \tag{6}$$

where q_n is the scalar external flux and n_Γ is the outward normal vector on the boundary. We use Equation 4 to express Equation 6 in terms of ϕ

$$\kappa \nabla T \cdot n_\Gamma = \kappa \exp\left\{\frac{-\alpha v \xi}{2}\right\}\left[\nabla \xi \cdot n_\Gamma - \left\{\frac{\alpha v}{2}\right\}\varphi \cdot n_\Gamma\right] = h \exp\left\{\frac{-\alpha v \xi}{2}\right\}\varphi \tag{7}$$

On the side walls of the prism, n_Γ is orthogonal to the weld path,

$$n_\Gamma = \begin{bmatrix} 0 \\ n_y \\ n_z \end{bmatrix}$$

Equation 7 for the side walls becomes

$$\nabla \varphi \cdot n_\Gamma = -\frac{h}{\kappa}\varphi \tag{8}$$

The heat source term has been set to zero. If one wished to include the weld pool, a heat source term could replace the prescribed temperature boundary condition on the weld pool solid–liquid interface.

At the outlet, the temperature will drop to room temperature because of the heat dissipation from the surfaces. Either essential or natural boundary conditions can be applied, i.e. $T = T_0$ or $\nabla T . n_\Gamma = 0$ when $\xi \Rightarrow -\infty$. Transforming temperature T to the potential function ϕ, when $\xi \Rightarrow -\infty$,

$$\nabla T \cdot n_\Gamma = \exp\left\{\frac{-\alpha v \xi}{2}\right\}\left[\nabla \varphi \cdot n_\Gamma - \left\{\frac{\alpha v}{2}\right\}\varphi \cdot n_\xi\right] = 0$$

so,

$$\nabla \varphi \cdot n_\Gamma = \left\{\frac{\alpha v}{2}\right\}\varphi \cdot n_\xi \tag{9}$$

FEM FORMULATION

Applying Galerkin's method to Equation 5 and assuming $S = 0$ gives

$$\int_\omega w \cdot \left\{\nabla^2 \varphi - \left\{\frac{\alpha v}{2}\right\}^2 \varphi\right\} d\Omega = 0 \tag{10}$$

for all w in the space of test functions.

For infinite domains the test functions must be specified carefully but for a finite domain the usual Galerkin test functions can be used. Applying Green's identity to Equation 10, it becomes

$$\int_{\omega}\left\{\nabla w \cdot \nabla \varphi - \left\{\frac{\alpha v}{2}\right\}^{2} w\,\varphi\right\} d\Omega + \int_{\Gamma} w\,\{\nabla \varphi \cdot n_{\Gamma}\} d\Gamma = 0 \tag{11}$$

Applying the boundary condition in Equation 8 to the side walls,

$$\int_{\Gamma_1} w\{\nabla \varphi \cdot n_{\Gamma}\} d\Gamma = -\int_{\Gamma_1} \frac{h}{\kappa} w \varphi \, d\Gamma \tag{12}$$

If the boundary condition in Equation 9 is applied on the downstream end,

$$\int_{\Gamma_2} w\{\nabla \varphi \cdot n_{\Gamma}\} d\Gamma = \int_{\Gamma_2} \frac{\alpha v}{2} w\,\varphi\, n_{\xi} d\Gamma$$

When the end faces are perpendicular to x, $n_x = -1$ at the rear end,

$$\int_{\Gamma_2} w\{\nabla \varphi \cdot n_{\Gamma}\} d\Gamma = -\int_{\Gamma_2} \frac{\alpha v}{2} w\,\varphi\, d\Gamma \tag{13}$$

Putting Equations 12 and 13 into Equation 11,

$$\int_{\Omega}\left\{-\nabla w \cdot \nabla \varphi - \left\{\frac{\alpha v}{2}\right\}^{2} w \varphi\right\} d\Omega - \int_{\Gamma_1} \frac{h}{\kappa} w \varphi \, d\Gamma - \int_{\Gamma_2} \frac{\alpha v}{2} w \varphi \, d\Gamma = 0 \tag{14}$$

for all test functions w. This has the same structure as a transient FEM code and can be easily solved with a standard FEM code.

RESULTS

Two test cases were analysed with the same basic mesh as shown in Figure 2. The weld pool surface was measured from a weld in which the arc was abruptly stopped and the weld pool blown out with an air jet. The empty pool is shown on the mesh in Figure 2. The pool surfaces were prescribed to the melting temperature of the base metal. Both surfaces of the weld joint had been prepared with V grooves.

TEST CASE 1

The steady state temperature field of an arc weld on an aluminium plate was computed using the method described. The Eulerian frame is created on a mesh 218 mm long, 165 mm wide, and 19 mm thick. The thickness of the frame was the same as the plate on which the weld pool was measured. The welding speed was 6 mm per second. The length of the weld pool was 323.5 mm. The centre of the pool was 10 mm from the leading edge. The maximum pool width was 15 mm. The deepest point was 10.13 mm. For aluminium the value of the parameter α was chosen as

$$\frac{c_p}{\kappa} = 1.76\times10^{4}\text{s m}^{-2}$$

Radiation and natural convection were allowed on the top and bottom surfaces and the two side edges. Mass flows in the front edge at room temperature. The computed results indicate that the diffusion rate was slower than the weld speed. Not far ahead of the weld pool, the temperature dropped quickly to room temperature and the inlet boundary condition was satisfied.

The boundary condition at the rear edge could be either essential or natural. In case we want to limit the domain size, the boundary condition becomes rather difficult to choose. It is usually not true that the rear edge is at room temperature, nor is the gradient of temperature zero. however, setting the temperature gradient equal zero in the x direction, $T_\xi = 0$, is obviously closer to the real solution. When $T_\xi = 0$ at the rear edge, heat input into the weld pool is balanced with the heat loss through surface radiation and convection and hot material flowing out from rear end.

If the downstream length is too short, the boundary condition on the downstream end will cause an error. if the downstream length is too long, the problem can become ill-conditioned. To choose an appropriate length, the domain length was changed dynamically, i.e. the rear part was expanded or shrunk. Keeping a fixed and very fine mesh near the weld pool, material at the rear end was added until the temperature field in the area of the weld pool was stable. Numerical experiments proved that with sufficient material behind the weld pool, convergence was rapid.

Figure 2(a) shows the computed isothermal contours. Figure 2(b) shows the isothermal contours on the half mesh cut along the weld path. The contours show the expected results.

Figure 3 compares the FEM 3-D steady state analysis with Rosenthal's 1-D and 2-D analytical model. The sample points for the FEM analysis were picked parallel to x and tangent to the weld pool edge on the top surface. The curves indicate that too much heat has entered the material in the 1-D or 2-D model. However, the influence of the thickness dimension is much less than that of width in this case.

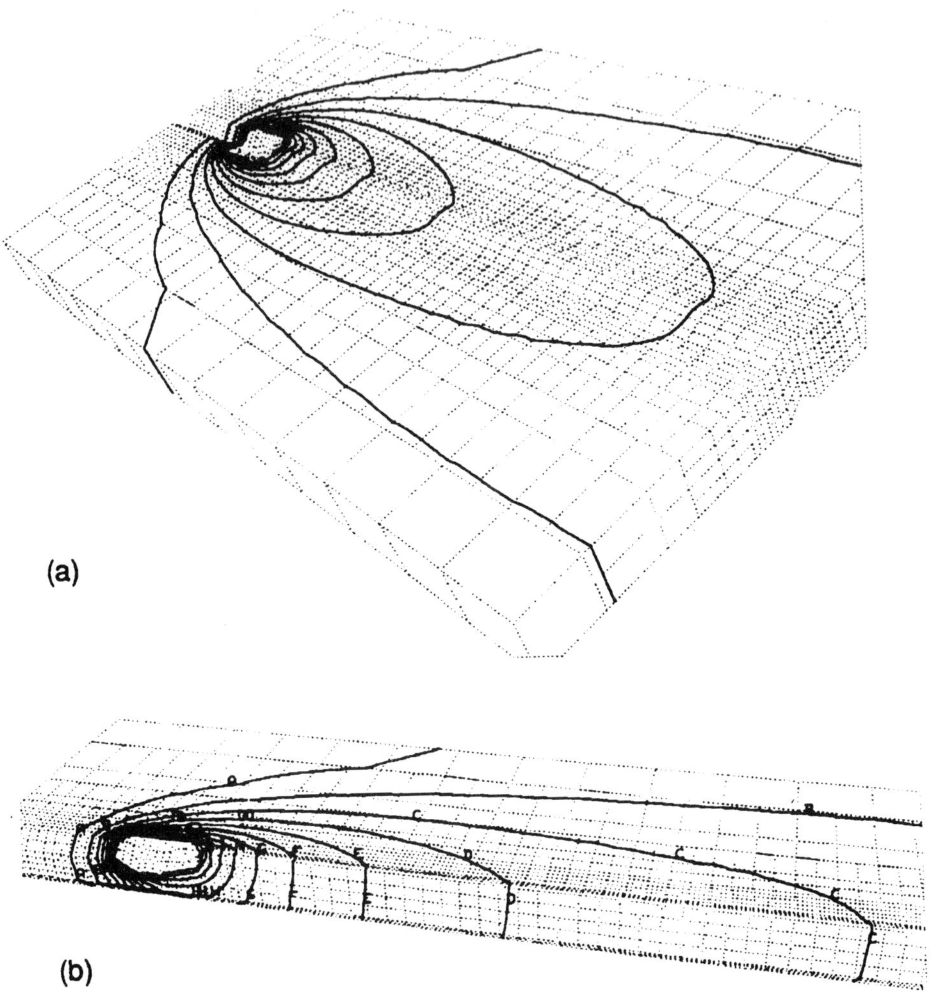

Figure 2. The FEM mesh is built on an Eulerian frame. The finest mesh is in the front of weld pool. The element sizes after the weld pool can be changed in the program, so that more or less material is included. The isothermal contours in steady state of an arc weld on aluminium. It was computed by a transformed Eulerian formulation. Because of high diffusivity, the density of the mesh around the weld pool could be lower than that for steel. (a) is the whole mesh, (b) is the mesh cut along the weld path. The temperature contours are lebelled in (b). The values of the contours are: A 54.8; B 101.0' C 148.0' D 194.0; E 240.0; F 287.0; G 333.0; I 426.0; J 472.0 ; K 519.0; L 565.0.

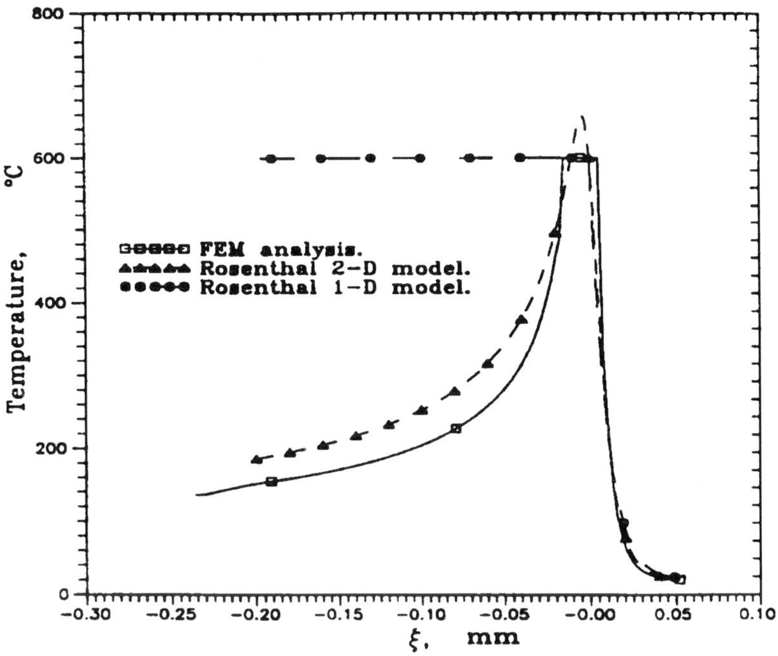

Figure 3. Comparison of thermal cycles in steady state of arc weld on aluminium. Note the 1-D and 2-D analytical solutions were calculated by Rosenthal's model. The power inputs for the analytical models were calculated according to the temperatures of the weld pool wall. Since the 1-D model used a sectional source the energy input is much higher than the 2-D model and the FEM model.

TEST CASE 2

The second test was carried on a low carbon steel plate (Fig. 4). The diffusivity of steel is smaller than that of aluminium. At the same welding speed, the Eulerian domain for steel can be smaller than that for aluminium. However, the mesh around the weld pool should be finer in order to describe the sharp gradient.

To assess the efficiency, an analogous analysis was performed by Lagrangian formulation (Fig. 5). 55 time steps were taken to reach a relatively stable temperature field around weld pool. The same value of α was used, $\alpha = 10^5$ s/m^2. The meshes for both analyses consisted of about 12 000 elements. Since the elements have to form a dense region along the whole weld path in the Lagrangian frame, the spatial resolution is obviously higher in the Eulerian frame. Compared with 3 CPU minutes for steady state analysis, 6 hours was used in the 55 steps in the transient analysis. Both analyses were performed on a Sun Sparc II workstation. It is estimated that an analysis using Equation 3

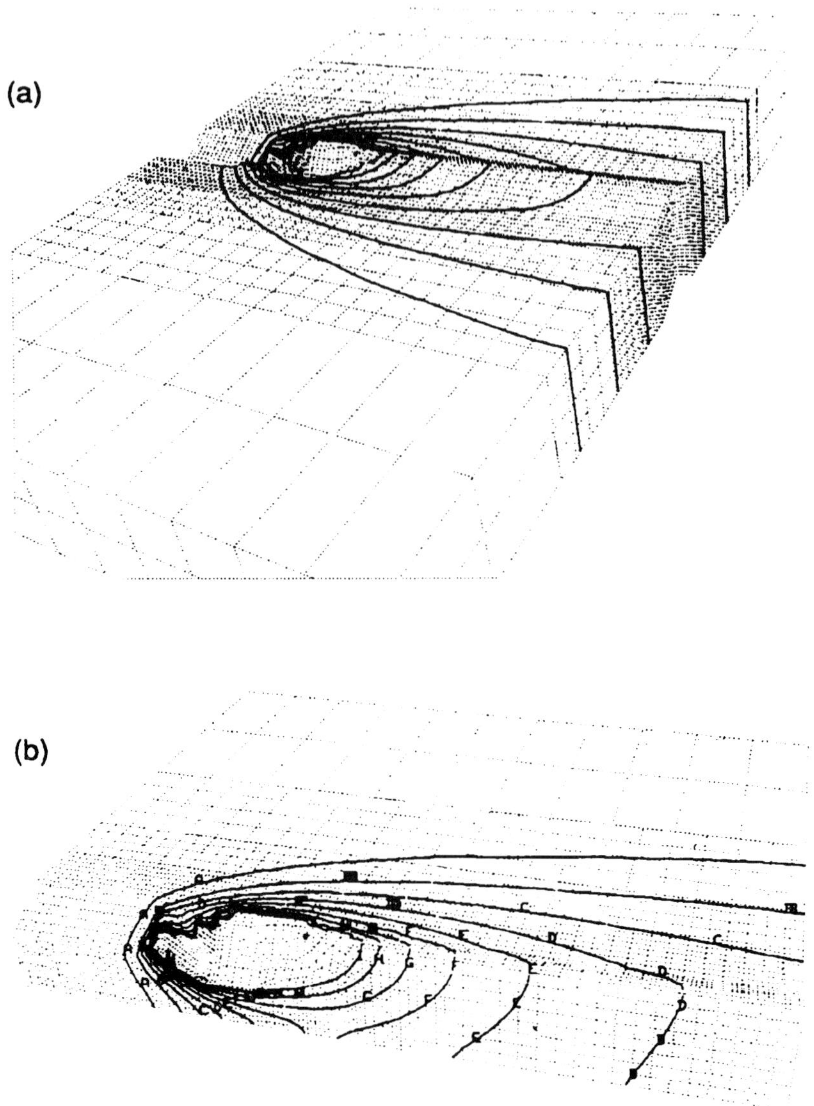

Figure 4. The isothermal contours in steady state of an arc weld on low carbon steel. It was computed by a transformed Eulerian formulation. Because of low diffusivity, the density of the mesh around the weld pool should be high, especially around the first half ot he weld pool. (a) is the whole mesh, (b) is the mesh cut along the weld path. The temperature contours are lebelled in (b). The values of the contours are: A 144.0; B 299.0; C 453.0; D 607.0; E 762.0; F 916.0; G 1070.0; H 1230.0' I 1380.0.

with Gaussian elimination and an asymmetric solver, for same mesh size, would cost approximately 17 minutes on a Sun Sparc I workstation. Such a formulation has the advantage that it can be used when α is temperature dependent, with a suitable nonlinear solver.

Figure 6 compares the thermal history of points in the vicinity of the weld pool. The millimetres indicate the distance from the weld path. On the Eulerian frame, the thermal cycle is converted from the temperatures on a flow line. The thermal cycle from steady state has a sharp gradient in time space, i.e. the temperature cools down faster after the weld pool. It was observed that the ellipsoids used to simulate the weld pool in Lagrangian frame was larger than the real size pool used in Eulerian frame. Although the length, width and depth are same for the weld pools in the two frames, the real pool is in fact thinner toward the end of the pool. The ellipsoids used in Lagrangian frame thus resulted in an incorrect higher energy input. As a result, the simulated temperature field around the weld pool given by the current method is more accurate than the previous method.

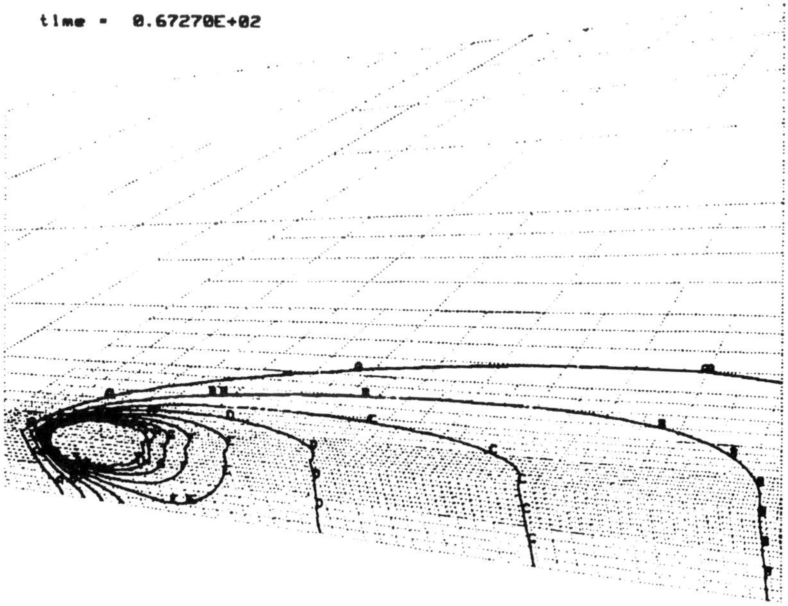

Figure 5. The isothermal contours computed by a Lagrangian formulation. The contours represent the solution after 55 time steps, 67 seconds after striking the weld torch. The velocity and diffusivity used were same as those used for Figure 4. Note without the addition of filler material the temperature changes are smaller in the direction of thickness. The values of the contours are: A 150.0; B 304.0; C 458.0; D 611.0; E 765.0; F 919.0; G 1070.0; H 1230.0; I 1380.0

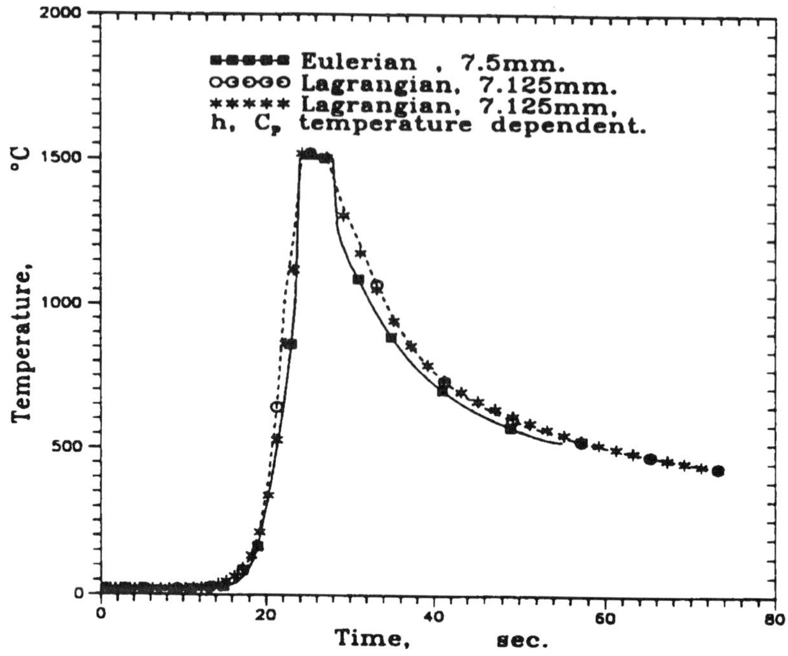

Figure 6. Comparison of thermal cycles in the point next to the weld pool between the Eulerian formulation and the Lagrangian formulation in steady state of arc weld on low carbon steel. The distance from the weld path is shown in mm.

The method can be used immediately to compute the transient temperature on stopping the arc. One need only perform a lagrangian transient analysis with the steady state temperature distribution as the initial condition.

MICROSTRUCTURE EVOLUTION IN THE HEAT-AFFECTED ZONE OF STEADY STATE WELDS

At a macroscopic level, the metal physics can be described by scalar fields specifying the fraction of each phase present, the composition of each phase and the grain size of each phase at each spatial point in the heat-affected zone (HAZ). In low alloy steels, the phases of greatest interest are ferrite, pearlite, austenite, bainite, martensite, carbide and liquid. Most steels used in welded structures are produced with a ferrite–pearlite microstructure or a ferrite and austenite–martensite constituent. When these steels are heated above the A_{e1} or eutectoid temperature, the pearlite or austenite–martensite constituent rapidly transforms to austenite with the same composition. Even though the heating rates are high, superheating only amounts to a few tens of degrees; approximately 80°C is common. So the assumption that the transformation occurs at equilibrium can be justified. However, the homogenisation of this high carbon austenite region is much slower. It is a function of the temperature, time and the distance between high carbon regions. Easterling, Ashby and Li have studied this in detail for laser heat treating surfaces.[11]

The growth in the grain size of the austenite that forms can be described as a function of temperature and time. However, growth will not begin until carbo-nitrides such as V and Nb dissolve. Easterling *et al.* used the equilibrium dissolution temperature for V and Nb to begin grain growth of austenite. TiN does not dissolve but it does coarsen. Ion *et al* applied Lipschitz, Slyozow and Wagner's theory of coarsening to TiN and the Zener theory of grain growth with pinned boundaries to estimate grain growth with TiN.[12]

On cooling, the decomposition of austenite begins at the A_{e3} temperature. The first phase to form is ferrite. If possible, pearlite, bainite and martensite will follow. Several strategies have been tried for the prediction of isothermal phase transformation kinetics. These parameters can be estimated from CCT diagrams but a distinct CCT diagram is needed for each grain size in each alloy.

In a steady state temperature field, each material point moves along a flow line. All material points on a given flow line experience the same thermal cycle. Thus the evolution of microstructure everywhere in the weld can be determined by integrating the points in a cross-section over a time interval that captures the thermal cycle. Essentially this reduced the dimensionality from four dimensions (x, y, z, t) to three (x, y, z). Previous simulations[11,12] were often based on the thermal prediction of Rosenthal's model[3], which assumed a steady state but to the author's knowledge none used an FEM steady state thermal model. Since an FEM model of a steady state temperature field around the weld pool is a true three-dimensional model, it can investigate different welds in detail taking into account differences in their unique temperature distributions. Such distributions are sometimes critical to the simulation of microstructures. Compared to a time marching Lagrangian formulation for a transient thermal analysis, the cost of computing a steady state temperature field has been reduced from hours or days to minutes. The difficulty in coupling is the FEM steady state temperature field with a microstructure prediction en extracting the thermal cycle of a material particle from the steady state temperature field, especially when the mesh is complex, i.e. unstructured.

In the method adopted by the authors the kinetics of austenite decomposition are described by ordinary differential equations (ODE) of the form:

$$\frac{df}{dt} = B(G,T)f^m[1-f]^p \tag{15}$$

The term $B\ (G,T)$ is a function of grain size, undercooling, the dependence of the carbon diffusivity on the alloy and temperature, and the phase fractions present. The exponents, m and p, were determined by experiment by Kirkaldy. Austenite grain size growth is computed with the ODE

$$\frac{dG}{dt} = \frac{1}{2G} \exp\left[-\frac{Q}{RT}\right] \tag{16}$$

This is essentially the model presented by Watt *et al.*[13].

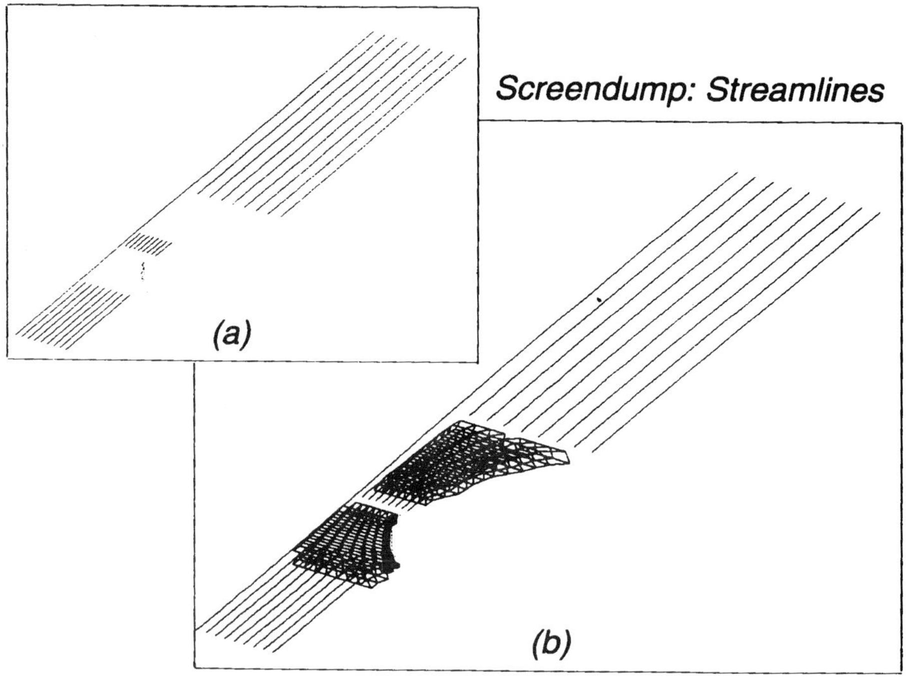

Figure 7. Numerical representation of all flow lines (streamlines) on a $y = 0$ plane. The lines are the links of points in the x direction. (a) Only the bodies of flow lines are shown. The unstructured mesh with refinement around the weld pool results in short lines. A controller prevents flow lines jumping over any element. (b) Those flow lines starting inside an element obtain their initial data by interpolating from upstream elements that are stored in heads.

The microstructure algorithm used here was motivated by the work of Kirkaldy, and originally developed for welding by Watt *et al.*[13] and Henwood *et al*[14] and enhanced by Khoral[15]. This microstructure model is limited to the heat affected zone (HAZ) in low alloy steels. A special data structure is used for the coupling of a steady state thermal field to the formation of

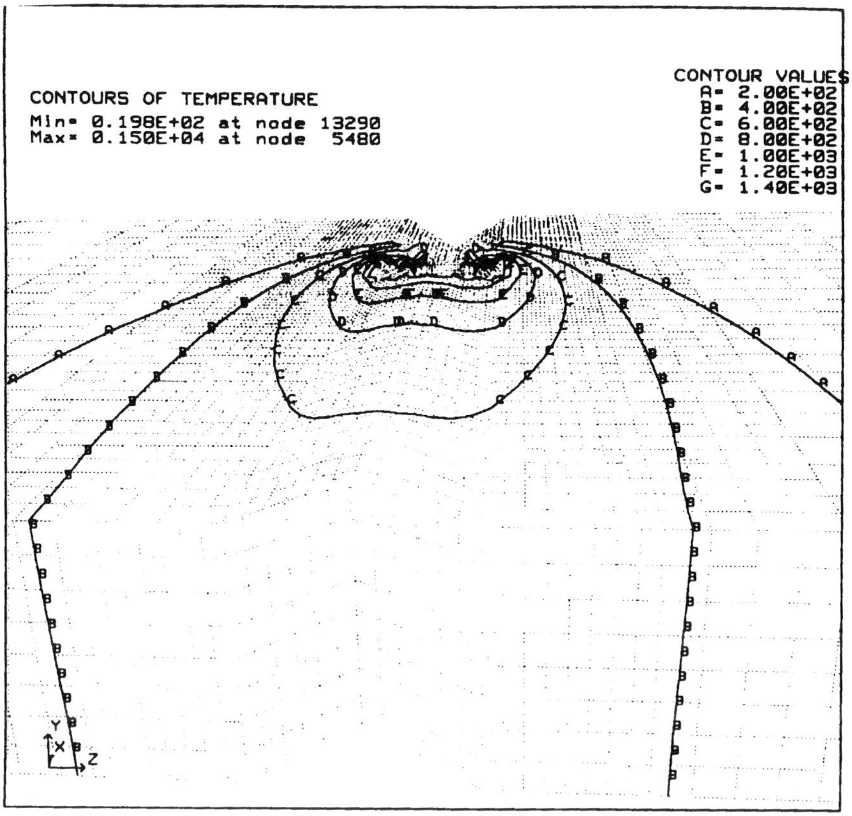

Figure 8. The microstructures in Figs 9 and 10 are computed for the temperature field shown in this figure.

microstructures[8]. some changes have been made in order to satisfy initial conditions in the weld pool or fusion region.

Integration along a flow line gives the thermal history of a material point. In a general FEM mesh with refinement around the weld pool, the creation of flow lines consumes more time than the microstructure calculation itself. The efficiency of the data structure used to organise elements, points and flow lines is critical to the success of the model. The data structure must support efficient ordered traversals of the points on a flow line (starting at the inlet and proceeding downstream to the outlet). It must also allow the element that contains the inlet of a flow line to be found. Loyner used a heap of faces[18], supporting insertion and deletion in $O(\log_2(n))$ operations for n faces. However, given different sized faces this may require a complete search to find the face containing a point. The data structure described in reference 8 meets all the above requirements. Figure 7 shows flow lines in a real discretised domain.

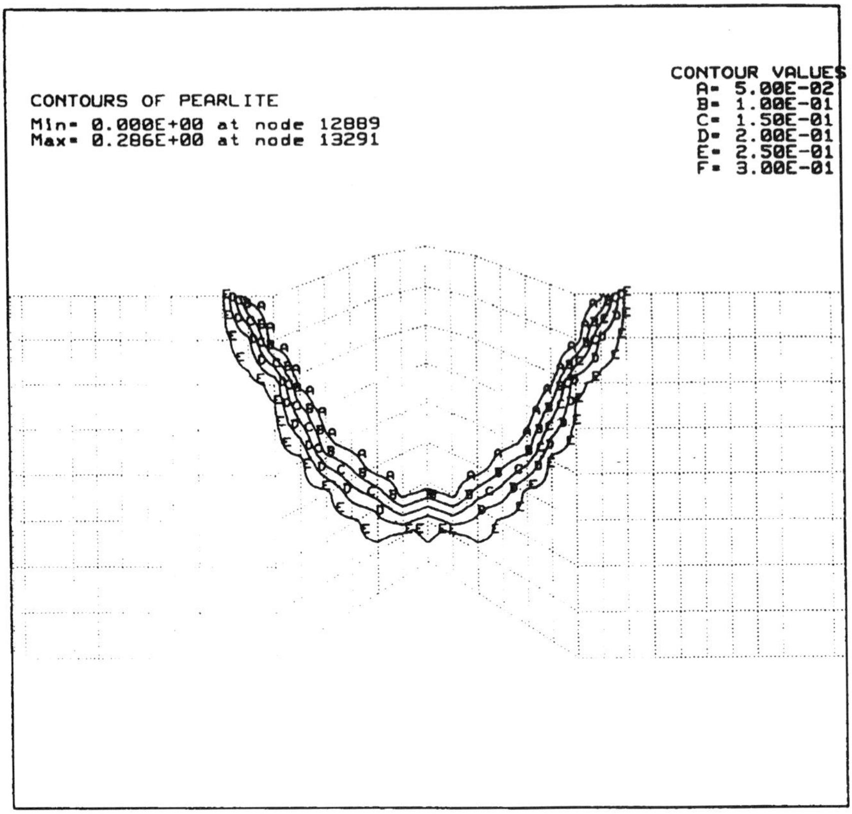

Figure 9. The pearlite contours at the ends of stream lines are shown.

TEST PROBLEMS AND RESULTS

The microstructures were computed first for the temperature field (Fig. 8) obtained from reference 7. The steady state temperature field was computed on a mesh with about 12 000 elements. The welding speed was 1.5mm per second. Weld pool size and shape were measured from an experimental weld. The maximum length was 35 mm, the maximum width 22 mm, and the maximum depth 11 mm. The gross power input in this weld was approximately 1.5 kJ/mm. The test plate size was 150 X 40 X 19 mm. The composition is given in Table 1. Before welding, the initial microstructures was 20% pearlite and 80% ferrite. The initial grain size of austenite was assumed to be 10 μm.

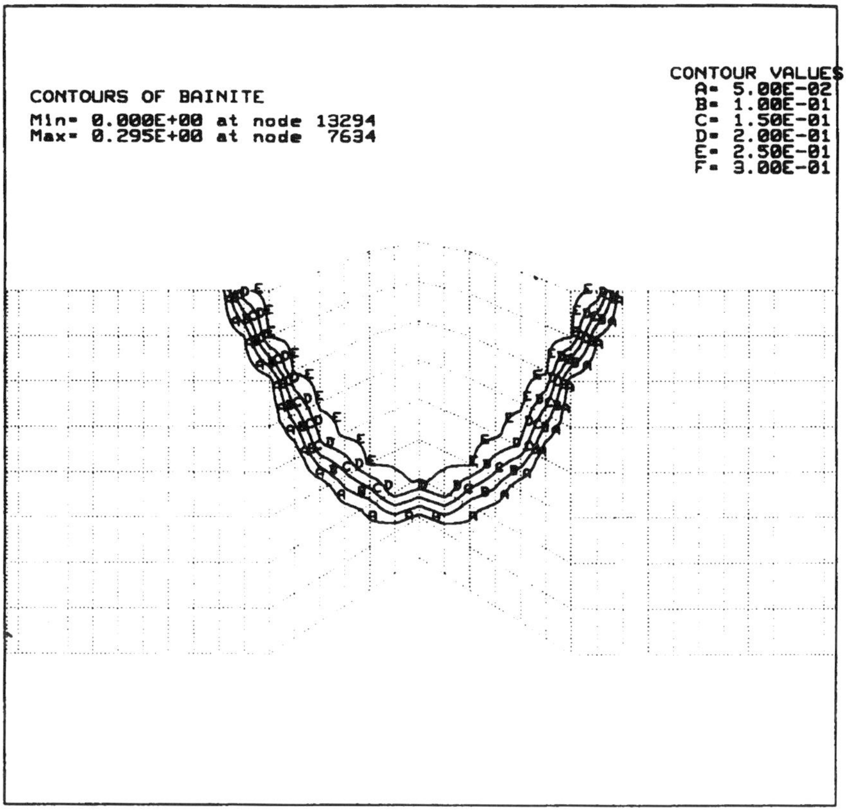

Figure 10. The bainite contours are shown at the end of the flow lines. Contour E is approximately 25% bainite.

Table 1 Chemical composition of the low-alloy steel analysed.

C	Si	Mn	P	S	Al	Nb	N
0.12	0.16	0.91	0.002	0.005	0.04	0.021	0.011

MICROSTRUCTURE EVOLUTION IN A STEADY STATE TEMPERATURE FIELD

The figures in this section show the state at the rear end of the mesh or flow lines. In the regular mesh next to the filler material, each element has a depth of 2.375 mm in the *y*-direction.

In the upstream area, there is no change from the initial microstructures, since the incoming material has not been affected by the heat input from the weld pool. For this temperature field, at the outlet plane, all the austenite has completely decomposed into its transformation products. This full decomposition indicates that the evolution of microstructure in the material flowing out is essentially complete.

Figures 9 and 10 show the iso-percentage contours of pearlite and bainite distributions respectively. After welding, the fraction of ferrite is almost constant at 70%. The major change in the HAZ microstructure is that the formation of bainite reduces the pearlite percentage in the heat-affected zone. in this weld, the cooling rate behind the weld pool is not large enough to form martensite. The microstructure outside the HAZ changes little. Inside the HAZ it is mostly ferrite with a mixture of pearlite and bainite. There is about 20 to 25% bainite near the fusion zone, largely due to the increased austenite grain size. Two to three mm within contour in Figure 8, a layer with the largest grain size and the highest bainite content, is predicted. Since the cooling rate is almost constant for an isothermal contour, variations in microstructure across the HAZ are primarily due to variations in austenite grain size. The cooling rate defined here is dT (°C) / dt (s) between two neighbouring points along a streamline.

CONCLUSIONS

A steady state formulation for computing temperatures and microstructures in arc welds has been reviewed. Relative to the usual Lagrangian formulations, this formulation reduces computing costs and enables higher spatial resolution to be obtained. The model also allows filler material to be added in a natural manner during welding. The test cases indicate that a reasonable accuracy could be achieved with the boundary conditions used. The model can also serve as a good predictor of a full non-linear solver. The modelling of microstructures in a steady state weld analysis has been reviewed. Using this steady state model to compute the evolution of microstructures in the steady state of a long weld reduces computing steps from hundred to one and savings in computing costs are significant. The results of the steady state analysis are also believed to be more reliable.

REFERENCES

1. O. WESTBY: Temperature Distribution in the Workpiece by Welding, Department of Metallurgy and Metals Working, The Technical University, Trondheim, Norway, 1968.

2. Y. UEDA and H. MURAKAWA: Applications of Computer and Numerical Analysis Techniques in Welding Research, *Trans. JWRI*, 1984, **13**, (2) 337–346.
3. D. ROSENTHAL: The Theory of Moving Sources of Heat and Its Application to Metal Treatments, *Trans. ASME*, 1946, **68**, 849–865.
4. R.R. TYKALIN: Energy Sources for Welding, Welding in the world, 1974, Vol. 12, No. 9/10, pp. 227–248 (Houdrement Lecture, International Institute of Welding, London, 1974, p. 1–23).
5. J. GOLDAK, M.J. BIBBY and M. GU: Heat and Fluid Flow in Welds, *Proceedings of the International Institute of Welding Congress on Joining Research*, pp. 69–82, Ed. T.H. North, Chapman and Hall, July 1990.
6. J.M. MCDILL, J.A. GOLDAK, A.S. ODDY and M.J. BIBBY: Isoparametric Quadrilaterals and Hexahedrons for Mesh-Grading Algorithms, *Comm. Appl. Num. Meth.*, **3**, 155–163, 1986.
7. M. GU and J. GOLDAK: Steady State Thermal Analysis of Welds with Filler Metal Addition, *Can. Met. Qtrly*, 1993, **32**, 49–55.
8. M. GU and J. GOLDAK: Modelling the Evolution of Microstructure in the Heat-Affected Zone, *Can. Met. Qtrly*, 1993, **32**, (4) 351–362.
9. M. GU and J. GOLDAK: Steady State Formulation for Stress and Distortion of Welds, *J. Engin. Ind.*, 1994, November, 467–474.
10. K.E. EASTERLING, M.F. ASHBY and LI: A First Report on Diagrams for Grain Growth in Welds, *Acta Metallurgica*, 1982, **30**, 1969–1978.
11. J.C. ION, K.E. EASTERLING and M.F. ASHBY: *Acta Met.*, 1982, **32**, 1949–62.
12. D.F. WATT, L. COON, M. BIBBY, J. GOLDAK and C. HENWOOD: An Algorithm for Modelling Microstructural Development in Weld Heat-Affected Zones (Part A) Reaction Kinetics, *Acta Metall.*, 1988, **36** (11) 3029–3035.
13. C. HENWOOD, M. BIBBY, J. GOLDAK and D. WATT: Coupled Transient Heat Transfer – Microstructure Weld computations (Part B), *Acta Metall.*, 1988, **36**, (11), 3037–3046.
14. PARAMJEET KHORAL: Coupling Microstructure to Heat Transfer computation in Weld Analysis, Master's thesis, Carleton University, Ottawa, Canada, 1989.
15. RAINALD LOHNER: Some Useful Data Structures for the Generation of Unstructured Grids, *Comm. Appl. Numerical Methods*, 1988, **4**, 123–135.
16. J.A. BARLOW: One Weld or Two? The formation of a submerged arc weld pool, *The Welding Institute Research Bulletin*, June 1982, pp. 177–180.

DETERMINATION OF RESIDUAL STRESSES IN SUBMERGED ARC MULTI-PASS WELDS BY MEANS OF NUMERICAL SIMULATION AND COMPARISON WITH EXPERIMENTAL MEASUREMENTS

J.B. ROELENS

Institut de Soudure, 57365 Ennery, France

ABSTRACT

Within the framework of a research programme on the 'Reliability of welded constructions financially supported by CETIM', we have numerically simulated the submerged-arc butt welding of 30 mm thick plates. The modelling was conducted with the help of the Sysweld software and aimed at determining the distribution of the residual stresses throughout the assembly after multi-pass welding. At each stage of the calculation, experimental verifications were made in order to validate the calculation. Comparisons of thermal cycles, sizes of heat affected zones and residual stresses were performed.

DESCRIPTION OF THE CALCULATION

The following numerical simulation is concerned with the modelling of multi-pass submerged-arc butt welding of two unrestrained plates in 30 mm thick A52 steel. The chemical compositions of base metal and filler metal are shown in Tables 1 and 2 respectively. The assembly was produced by depositing sixteen passes in a 60-degree V-groove joint as shown in Fig. 1. Figure 2 shows a cross-section of the welded assembly which permits the determination of the weld metal zone as well as the boundaries of the heat affected zone (HAZ).

Table 1 Chemical composition (wt.%) of the A52 base metal

C	Mn	Si	S	P	Al
0.18	1.39	0.37	0.011	0.022	0.052

Table 2 Chemical composition (wt.%) of the wled metal zone

C	Mn	Si	S	P	Cu	Al
0.073	1.57	0.39	0.009	0.015	0.1	0.009

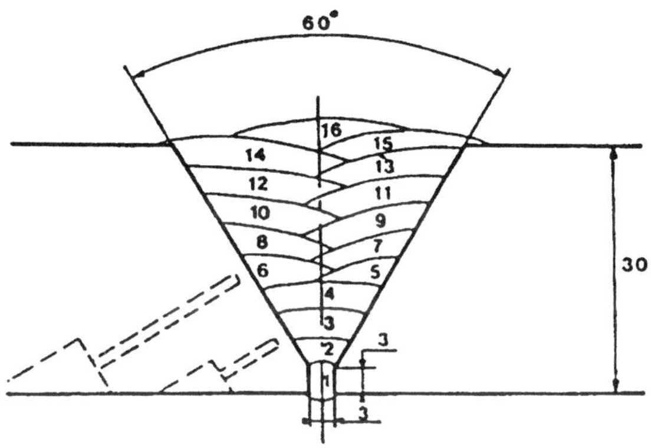

Figure 1. Welding sequence and location of the 16 passes.

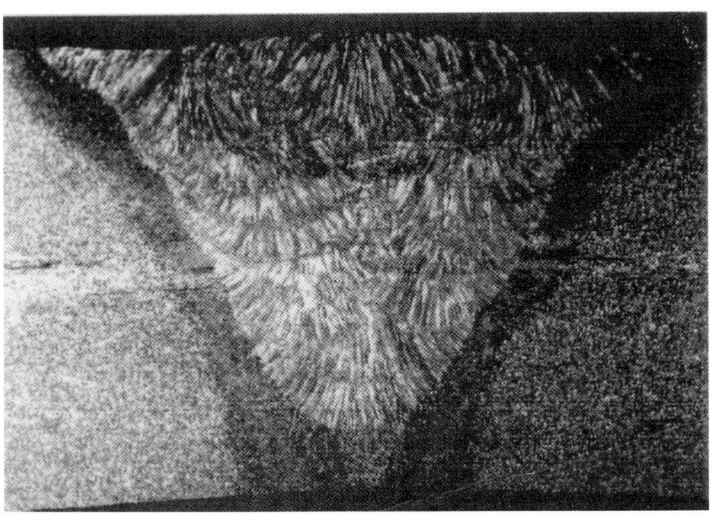

Figure 2. Macrograph of multi-pass weld.

The sixteen passes which form the assembly were sequentially simulated by means of the Sysweld software.[1] This finite element computer code permits the simulation of welding by conducting a thermo-metallurgical calculation and then a thermo-mechanical calculation on the basis of the obtained results. Therefore, the modelling of welding is realistic, because the calculated residual stresses result from a method which accounts for the temperature history of the operation and the metallurgical changes of the materials during heating and cooling. This type of calculation requires the knowledge of the thermal, metallurgical and mechanical properties of the materials of interest, in the various metallurgical states in which they may be found and in the temperature range of 20–1500°C. These data stem from experimental measurements, the open literature and hypotheses.

THERMAL MODELLING

The thermal cycles calculated at any location of the assembly are directly dependent on the modelling of the heat input of the welding process used. We have selected an empirical modelling of this heat input. A temperature rise in the volume of weld bead was imposed until the melting temperature (1500°C) was reached, then this temperature was held in the weld for a period to time corresponding to the solidification of the molten pool and finally the structure was left to cool freely. Figure 3 shows the thermal cycle partly imposed to the successive weld beads during the modelling process. Times t_1 and t_2 of temperature rise and 'molten pool holding' are model variables which may be estimated from the thermal cycles recorded during welding. On the other hand, the values of specific heat C_p and conductivity λ attributed to the material of the weld beads during t_1 and t_2 are model variables which are determined by successive numerical tests. The adjustment of these variables permitted the latent heats of solidification of the weld pool to be taken into consideration. The numerical tests for the 'fitting' of the heat input were made on a simple structure by considering the deposition of a single bead on a plate with the same welding parameters as for multi-pass welding. The discretisation of the structure, illustrated in Fig. 4, contains 169 quadratic elements and 556 nodes. The comparison of the recorded and measured thermal cycles permits the determination by successive approximation of the variables of the thermal model. Then, a thermo-metallurgical calculation of single-pass welding gives the distribution of the metallurgical phases present in the weld. The austenite transformation during the heating of the base material subjected to a temperature in excess of Ac1 leads to the bainite transformation of the material on cooling for the type of material used. Figure 5 shows the bainite distribution in the numerically determined single-pass structure. The

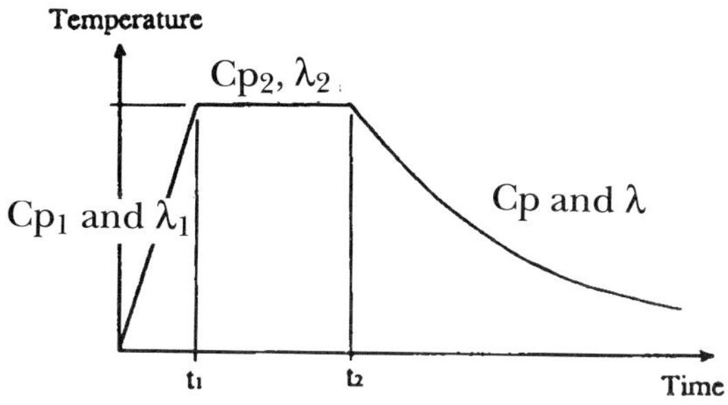

Figure 3. Thermal cycle applied to each of the 16 deposits.

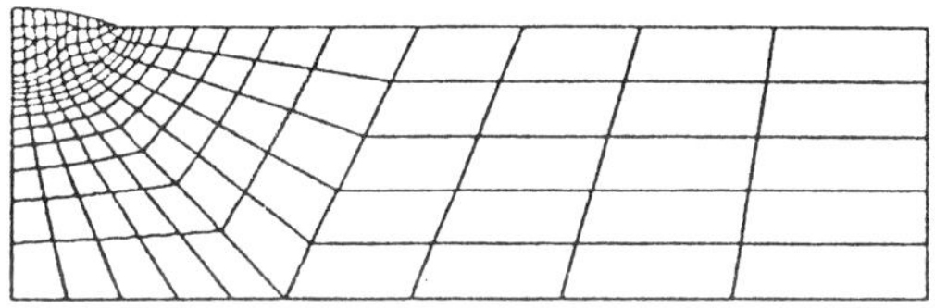

Figure 4. Finite element grid of the single-pass weld.

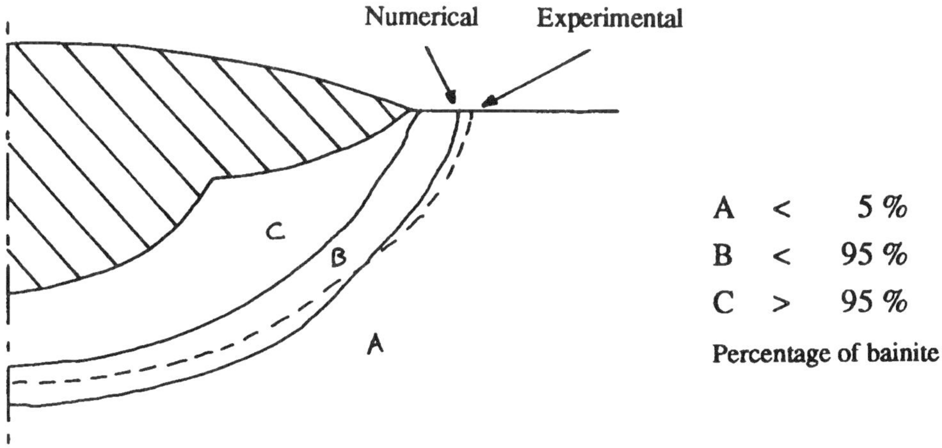

Figure 5. Distribution of the bainite contents in the single-pass weld. Comparison of the experimental and numerical limit of HAZs.

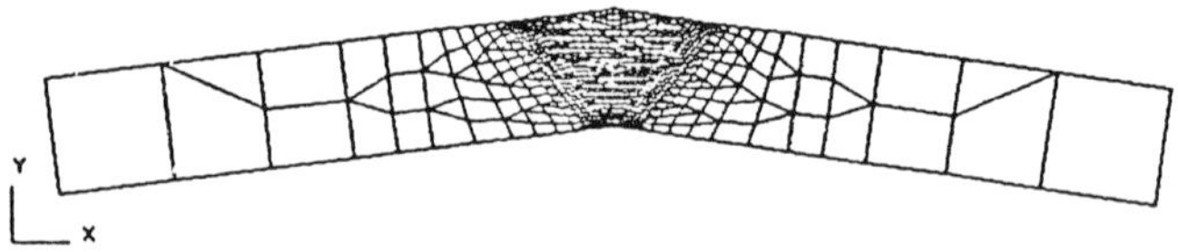

Figure 6. Finite element grid of the multi-pass butt welded assembly.

experimental boundaries of the weld metal zone and HAZ were delineated with a continuous black line. A good agreement between numerical and experimental results may be observed.

RESULTS OF THE MULTI-PASS WELD SIMULATION

The sixteen passes of the welded assembly were sequentially simulated in accordance with the welding procedure. Each pass was the subject of a thermo-metallurgical calculation followed by a thermo-mechanical calculation. The structure discretisation needed for the modelling contained 749 quadratic elements and 2202 nodes. It is shown in Fig. 6. During the execution of the assembly in the shop, thermocouples were welded inside the base metal, in the vicinity of the HAZ, for recording the thermal cycles of some surrounding points during the welding operation. Table 3 compares the temperature measurements of seven thermocouples with the values of temperatures and cooling times calculated at the nodes corresponding to the locations of the thermocouples during the deposition of the sixth bead. The location of this

Table 3 Comparison of thermal results for deposit no. 6

Thermocouple	θ max (°C)		Cooling Time 800–500°C (s)	
	Experimental	Numerical	Experimental	Numerical
R1	648	684	–	–
R2	622	652	–	–
2	>1200	1221	24.5	23.4
4	940	940	24.5	23.3
5	920	863	25.1	24
7	687	674	–	–
8	587	558	–	–

bead within the finite element grid is shown in Figure 7. The numerical results very closely approximate the experimental measurements, thereby validating the thermal modelling. The thermo-metallurgical calculations permitted the determination of the phase proportions for each material. The visualisation of the final metallurgical condition in the weld permitted the calculation to be verified. Figure 8 compares the 'experimental HAZ' and the 'numerical HAZ'. The 'experimental HAZ' which had an average width of 3 mm, was revealed on the cross-section of the assembly by chemical etching; it is clearly visible on the macrograph of Figure 2. The 'numerical HAZ' was defined by a bainite content which was higher than 5%. The right side of Figure 8 shows the distribution of the bainite over the welded assembly; the HAZ was delineated on the left by the black line which represents the limit of the weld metal zone and, on the right, by the dark blue zone which represents the unaffected base metal (bainite content below 5%). The numerical and experimental HAZs were comparable.

Based on the thermo-metallurgical results, two-dimensional thermoplastic calculations were conducted in general plane strain for each of the sixteen beads. For these calculations, the mechanical properties of the materials were dependent on the temperature and the metallurgical phases. Figures 9 and 10 show the evolution of the distribution of the transverse residual stresses σ_{xx} (transverse to welding direction) after each pass at the interpass temperature of 150°C prescribed by the welding procedure. an increase in the levels of the tensile or compressive stresses after each deposit will be noted, while the different zones form and grow larger. For the first deposits, the stresses are low, then increase progressively until peak values are reached after the deposition of the last bead. From the seventh and eighth bead onwards, one clearly sees that the upper part of the bead is in compression and that a zone in tension is forming under the bead. This phenomenon is the same as the one observed during the deposition of a single bead on a plate. The intensity of this zone in tension seems to depend on the thickness of the material under the weld bead. An important zone in tension forms from deposits 10 and 11 onwards, on the lower surface of the weld; this corresponds to a bending stress induced by the straightening of the plates during welding. A wide zone in compression forms right above, for balancing this zone in tension in the material. In the case of an evolution of the longitudinal residual stresses σ_{zz} (direction of welding), one observes the same phenomenon of stress increase after each deposit, except that all the weld metal zone is in tension. Figure 11 shows the distribution of the transverse residual stress σ_{xx} through the thickness of the assembly after cooling to room temperature (20°C); it varies over the thickness from –230 MPa to +320 MPa. Figure 12 shows the distribution of the longitudinal residual stress σ_{zz} through the thickness of the assembly after cooling to room temperature (20°C); it reaches a maximum of 620 MPa. The overall view shows the zones in compression on each side of the assemblies and thus permits the visualisation of the horizontal rotation axis of

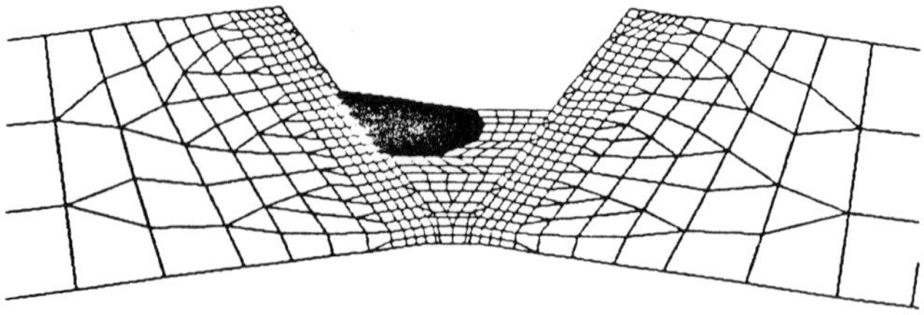

Figure 7. Location of deposit no. 6 within the grid of the multi-pass welded assembly.

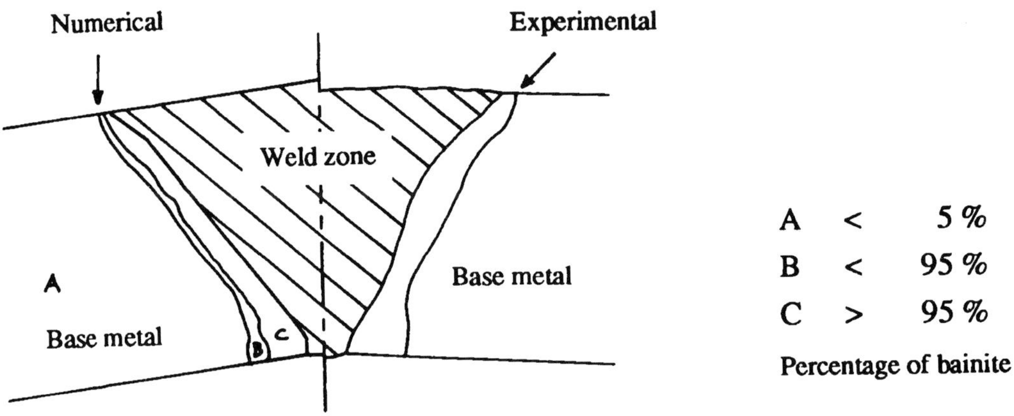

Figure 8. Distribution of the bainite contents in the multi-pass weld. Comparison of the experimental and numerical limit of HAZs.

the cross section (limit between the B and C zones). It was the choice of the two-dimensional calculation in general plane strain which permitted the longitudinal bending effect of the plates to be taken into account.

COMPARISON WITH EXPERIMENTAL MEASUREMENTS

In order to verify the previous mechanical results, we used the X-ray diffraction technique for measuring the residual stresses on the surface of the assembly and the incremental hole drilling technique for measuring these stresses through the weld thickness.[2]

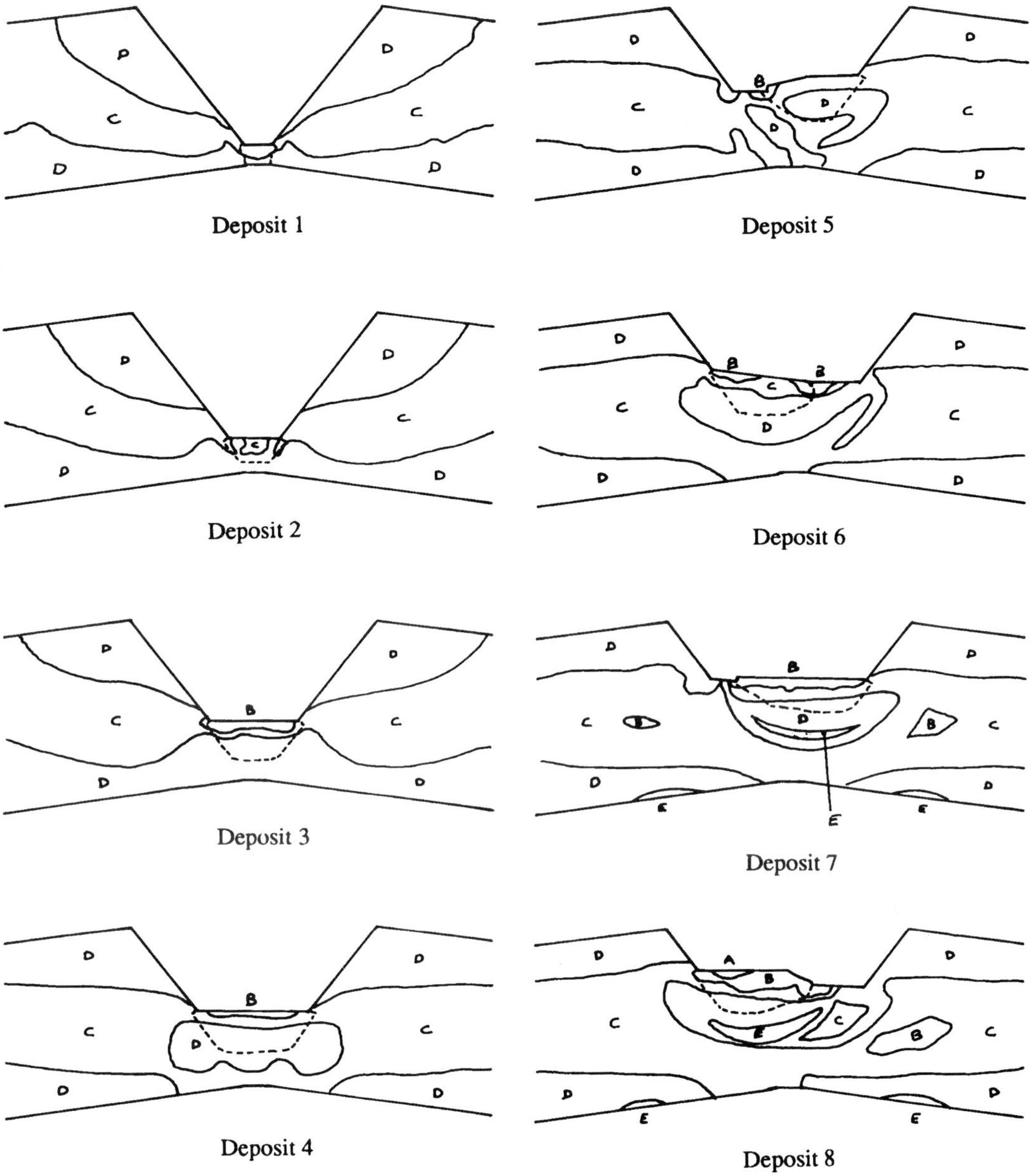

A	<	- 100
B	<	- 50
C	<	0
D	<	50
E	<	150
F	>	150

Stress in MPa

Figure 9. Distribution of the transverse residual stress after each deposit, deposit 1 to 8.

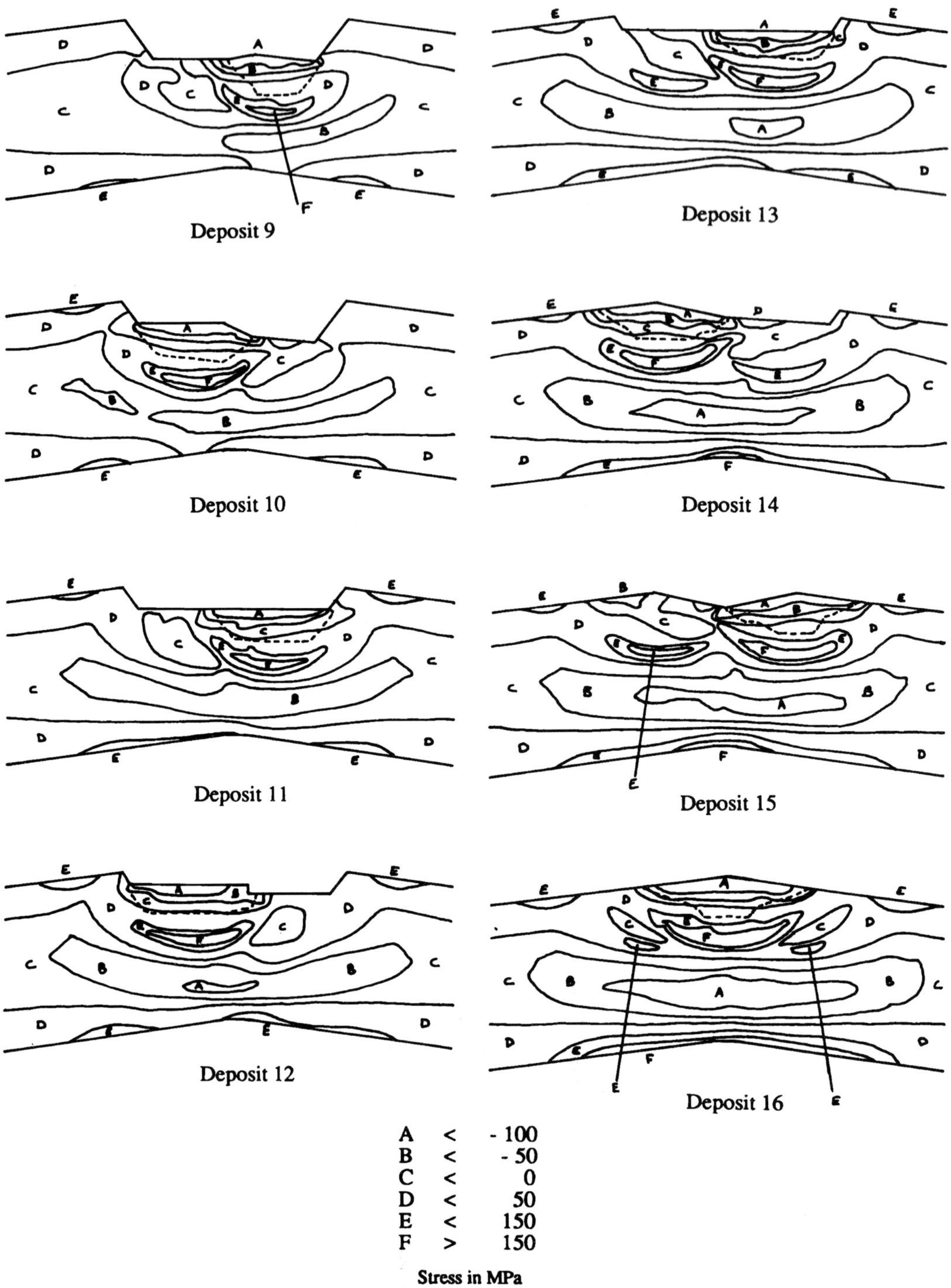

Figure 10. Distribution of the transverse residual stress after each deposit, deposit 9 to 16.

X-RAY DIFFRACTION TECHNIQUE[2]

We performed a series of stress measurements on the upper surface and another series on the lower surface of the assembly in the transverse direction to the weld, at mid-length of the assembly. The measurements of the transverse residual stresses σ_{xx} and longitudinal residual stresses σ_{zz} were made after removal, by electrochemical etching, of a layer of about 0.2 mm of material from the surface. Figures 13 and 14 show the results of the measurements made on the upper surface. For the transverse residual stress, a good agreement between the results of measurements and numerical simulation was observed: the measurement data virtually coincided with the calculated curve. on the other hand, for the longitudinal stress, the shapes of the experimental and numerical curves were almost identical, but the values were very different. This discrepancy between modelling and experiment has been observed in an earlier study[3] and was attributable to the systematic error due to the fact that the three-dimensional effect was not accounted for in the two-dimensional calculation. The modelled cross-section was much more restrained in the simulation than in reality. Under these conditions, the longitudinal stress was thus systematically overestimated. Figures 15 and 16 give the results of measurements made on the lower surface. The observations about the comparison between the measured and calculated residual stresses on the upper surface remain valid in this case.

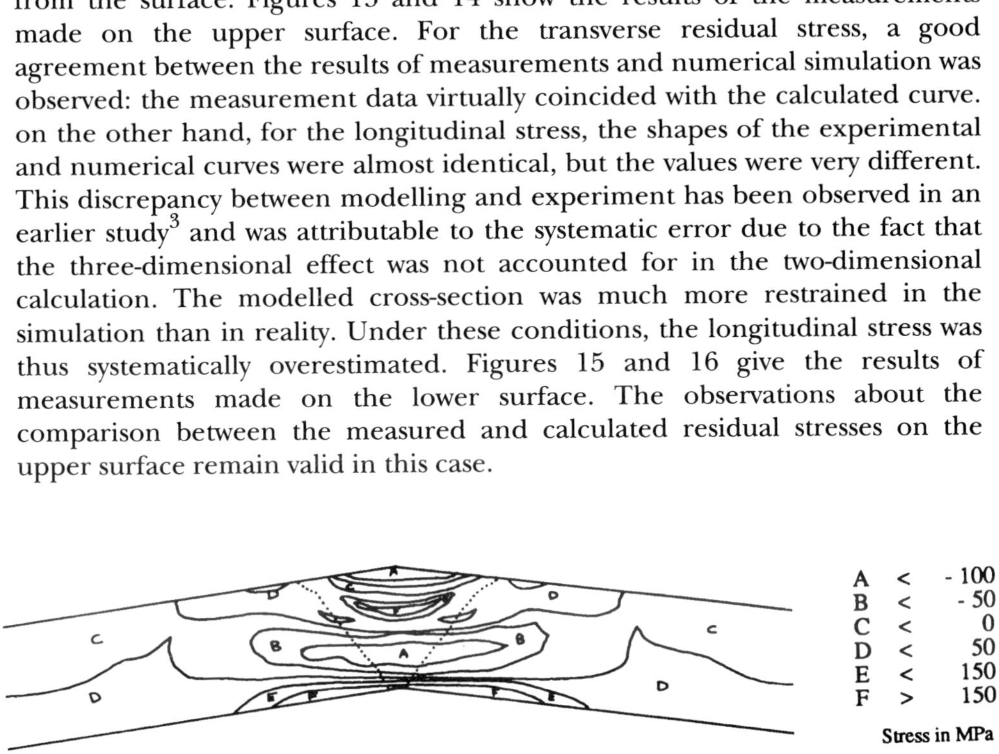

Figure 11. Distribution of the transverse residual stress at 20°C

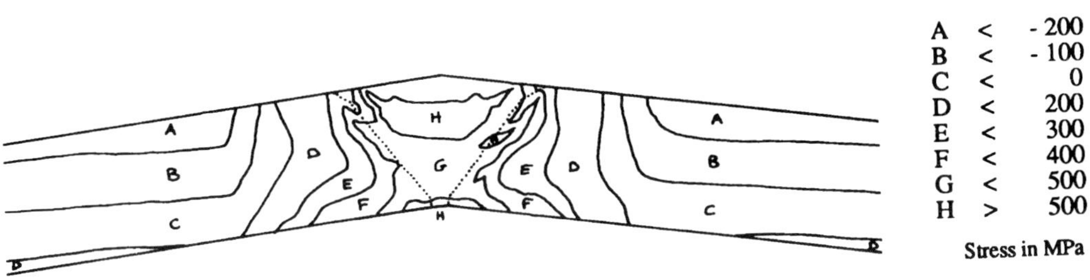

Figure 12. Distribution of the Longitudinal residual stress at 20°C.

THE INCREMENTAL HOLE DRILLING TECHNIQUE[2]

The incremental hole drilling technique consists in analysing the stress relief which takes place after removal of a given volume of metal from a part (calibrated drilling in this case). This technique was applied to the multi-pass butt weld from the upper and lower surface down to the mid-thickness of the weld. Figure 17 is a general view of the implementation of the technique and Fig. 18 shows the locations of the drilled holes required for reaching the residual stresses through the whole thickness of the weld. Small diameter holes (5 mm, 3.5 mm and 2 mm) were drilled on the surfaces and at the bottom of 30 mm diameter holes. The results of measurements made on small diameter holes are shown in Fig. 19 and 20. Figure 19 reveals a good agreement between

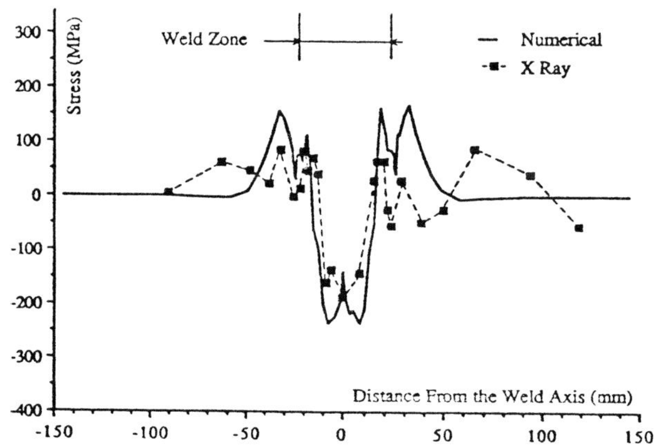

Figure 13. Transverse residual stress (calculated and measured) on the upper surface

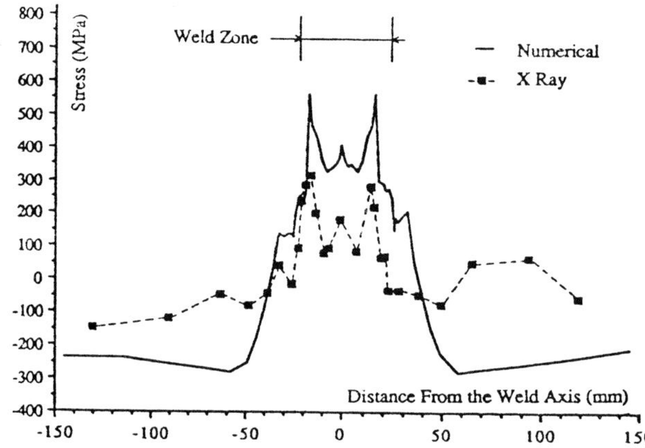

Figure 14. Longitudinal residual stress (calculated and measrued) on the upper surface.

the calculation and the measurements of the transverse residual stress σ_{xx} through the thickness of the weld. The six series of measurements which correspond to the six small-diameter holes reproduce well the stress pattern. Figure 20 compares the calculation and measurement of the longitudinal residual stress σ_{zz} through the thickness of the weld. It evidences an important problem associated with the performance of experiments. As a matter of fact, each drilled hole produces a high stress gradient (up to 600 MPa), and this is not realistic. therefore, a comparison of the results for this stress was not possible. The results of measurements made on 30 mm diameter holes are shown in Figs 21 and 22. Here again, the agreement between measurements and calculations was very good in the case the transverse residual stress, but not exploitable in the case of the longitudinal residual stress.

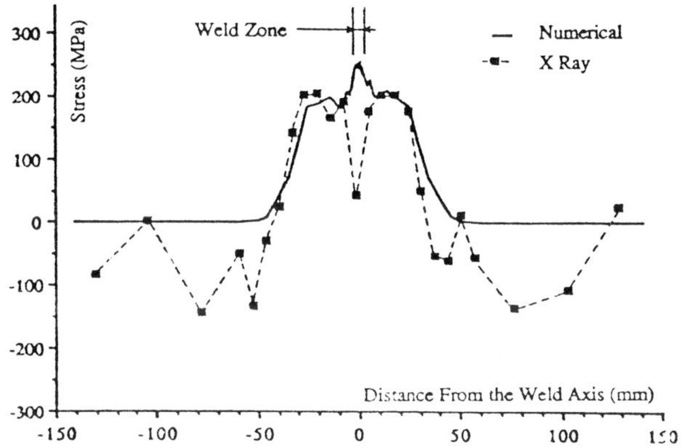

Figure 15. Transverse residual stress (calculated and measured) on the lower surface.

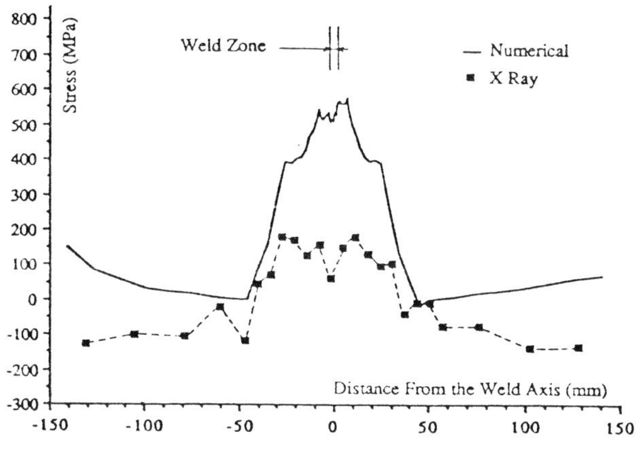

Figure 16. Longitudinal residual stress (calculated and measured) on the lower surface.

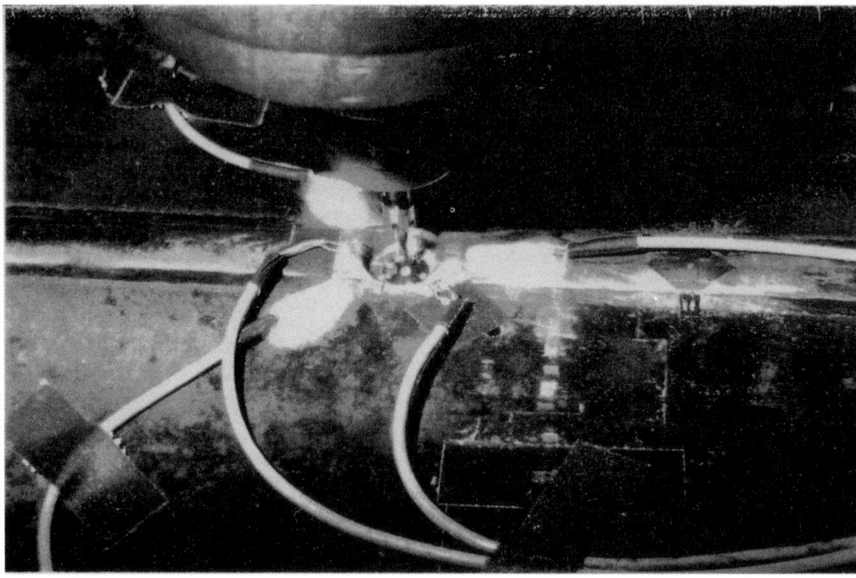

Figure 17. Measurement of the stresses through the weld thickness using the hole drilling technique.

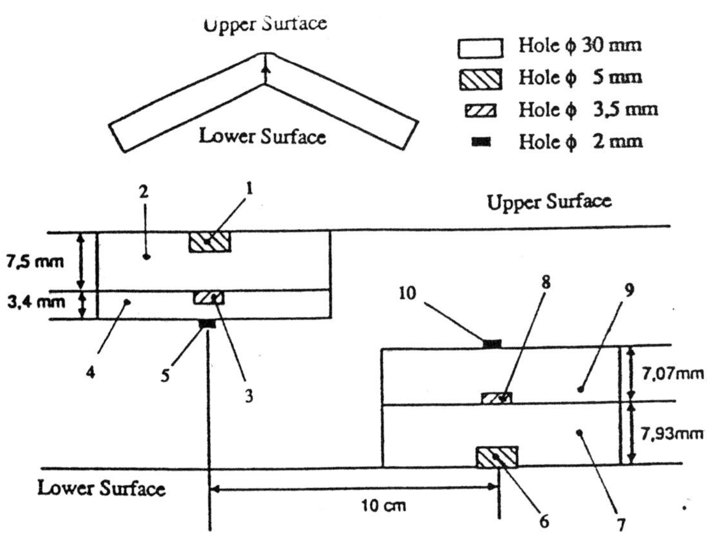

N° Hole	Φ (mm)
1	5
2	30
3	3.5
4	30
5	2
6	5
7	30
8	3.5
9	30
10	2

Figure 18. Location of the drilled holes.

CONCLUSION

The numerical simulation of multi-pass submerged arc butt welding permits a better understanding of the consecutive evolution of the residual stresses in the assembly. Once each phase of the calculation (thermal and metallurgical) had been verified and validated, the weld residual stress state in the whole assembly was obtained. The stress measurements made by using two experimental techniques permitted the validation of this simulation. On the surface, the numerical results were compared with the results of measurements by X-ray diffraction. Through the thickness of the weld, the residual stresses were measured by the incremental hole drilling technique. As a whole, the

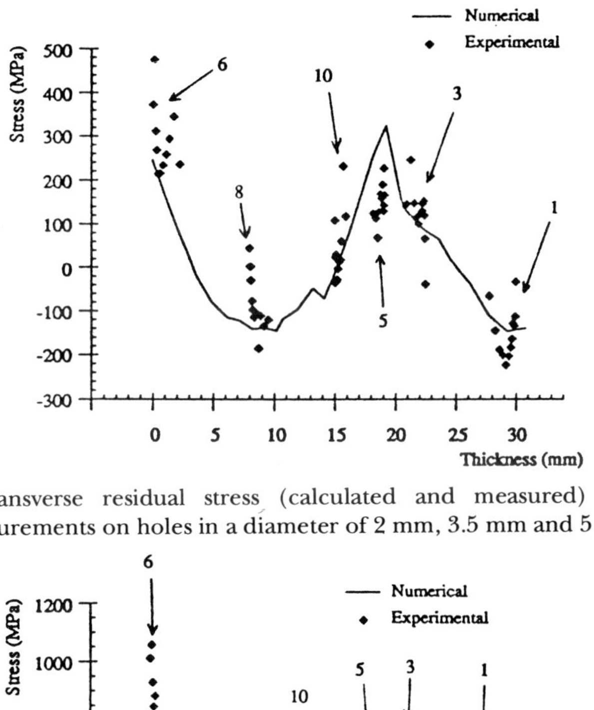

Figure 19. Transverse residual stress (calculated and measured) through the weld thickness. Measurements on holes in a diameter of 2 mm, 3.5 mm and 5 mm.

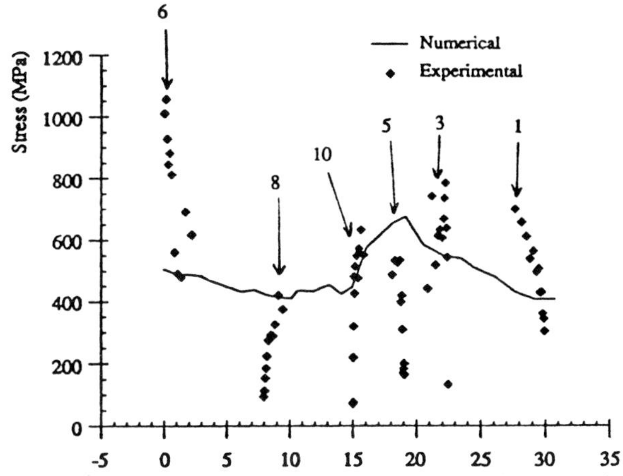

Figure 20. Longitudinal residual stress (calculated and measured) through the weld thickness. Measurements on holes in a diameter of 2 mm, 3.5 mm and 5 mm.

transverse stress was correctly simulated; the comparison of the calculated stresses with the stresses obtained by the two experimental techniques yielded good qualitative and quantitative results. The longitudinal stress, as for it, was more difficult to measure and calculate; the X-ray technique gave satisfactory qualitative results for the stress distribution on the surface, but the incremental hole drilling technique did not permit any significant comparison.

Such a numerical simulation gave satisfactory results in spite of its complicated implementation. As a matter of fact, its adequate application

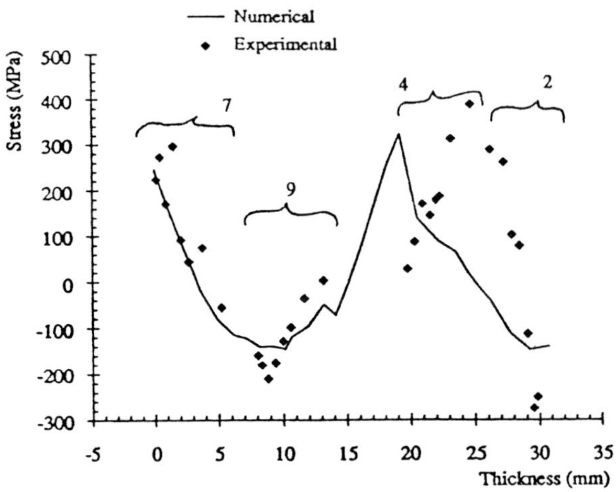

Figure 21. Transverse residual stress (calculated and measured) through the weld thickness. Measurements on holes in a diameter of 30 mm.

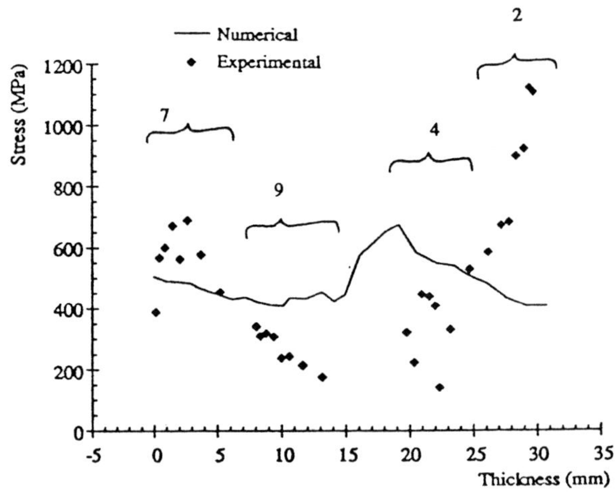

Figure 22. Longitudinal residual stress (calculated and measrued) through the weld thickness. Measurements on holes in a diameter of 30 mm.

requires a great number of material properties to be known. If these properties can be determined more easily in the future, then these complex numerical simulations will be generally used and will obviate the necessity of experimental stress measurements, which are not always reliable.

REFERENCES

1. Sysweld, User manual, Version 231, Framasoft+CSI, (1990).
2. *Etude des contraintes résiduelles de soudage par la méthode du trou incrémental et diffraction des rayons X* (Study of the weld residual stresses by the incremental hole drilling technique and the X-ray diffraction technique), (1992) IIW Doc. X-1257-92, J. LU, Cetim.
3. D. PONT, J.M. BERGHEAU and J.B. ROELENS: *Numerical determination of residual stresses induced by laser surface treatment.* Conférence Européenne sur les Nouvelles Avancées en Calculs de Structures (European Conference on Advances in the Calculation of Structures), (1991). Framasoft+CSI, Institut de Soudure.

SECTION IV
DESIGN ASPECTS

POTENTIAL OF NUMERICAL ANALYSIS OF WELDABILITY IN THE DESIGN PROCESS

DIETER RADAJ
Daimler-Benz AG, Stuttgart, Germany

ABSTRACT

A definition for 'numerical analysis of weldability' is given. Its state of the art is demonstrated by way of weld pool models conceived by W. Sudnik. Its future potential for industrial application by metallurgists, production engineers and design engineers is shown. The demands of industry are specified by way of examples taken from aircraft and automotive engineering. The organisational status of numerical analysis of weldability is clarified considering the new concept of simultaneous engineering within lean production and management.

INTRODUCTION

Numerical weldability analysis is a powerful new research and development tool which is useful for metallurgists, production engineers and design engineers. Whereas the potential for metallurgists and production engineers (especially developers of welding processes) is obvious, the potential of the method in the design process is not so clear.

Numerical analysis of weldability designates the thermodynamical, thermomechanical and microstructural modelling of the welding process including the heat source, the weld pool and the welded structure.

The modelling is performed in respect of the feasibility of the manufacturing process, the suitability of the welded materials and the quality and reliability of the welded structure.

FUNDAMENTALS OF NUMERICAL WELDABILITY ANALYSIS

PRINCIPLES OF MODERN SCIENCE AND TECHNOLOGY

Galileo Galilei (1564–1642), Fig. 1, defined the principles of modern science and technology as:

> 'The book of nature is written in the language of mathematics.'
>
> 'The scientist must decide intuitively which aspects of the phenomena are suitable for mathematical modelling and which ones can be neglected.'
>
> 'Experimental verification is common in those sciences which apply mathematical processes to natural phenomena.'

These definitions constitute the basis of mathematical analysis applied to natural and technical phenomena (weldability included).

Figure 1. Galileo Galilei (1564–1642) at the age of 60; engraving by Ottovio Leoni, 1624.

DEFINITION OF NUMERICAL ANALYSIS OF WELDABILITY

Weldability denotes the possibility to join parts by welding under defined conditions of design, materials and manufacture.

Weldability is conventionally ascertained and further developed by testing (empirical basis).

This process can be enhanced and made more efficient by mathematical modelling and numerical analysis (theoretical basis).

FIELDS OF RESEARCH AND DEVELOPMENT

Simulation of welding processes such as: electric arc welding, resistance spot welding, friction welding, etc., breakdown of the total process into separate physical phenomena.

Welding suitability of materials characterised by suitability indices, temperature dependency of material characteristic values, interpretation of weldability tests, hot and cold cracking susceptibility.

Strength of welded structures comprising residual stresses and distortion, fatigue strength, brittle fracture, buckling strength, crack propagation.

MODELS SIMULATING ASPECTS OF WELDABILITY

Solid state behaviour (welded structure) comprising thermoelastic, elastoplastic and viscoplastic deformation.

Liquid state behaviour (weld pool) comprising laminar or turbulent flow, controlled by electromagnetic, buoyancy and capillary forces.

Gaseous and plasma state behaviour (welding arc) comprising laminar or turbulent flow, controlled by electromagnetic forces.

STATE OF THE ART OF NUMERICAL WELDABILITY ANALYSIS

GENERAL STATE OF THE ART OF WELDABILITY ANALYSIS

The general state of the art of weldability analysis has recently been summarised by Radaj[1] and Buchmayr.[2] Further details may be found in the collections of research papers edited by David and Vitek,[3] Karlsson *et al.*,[4] and Cerjak and Easterling[5] or in the review paper by David and DebRoy.[6] There exists an overwhelming number of approaches to and results from empirically and theoretically based mathematical models of weldability, i.e. of welding processes, of material behaviour in welding and of the strength of welded structures. Substantial progress has been achieved in recent years.

SPECIFIC STATE OF THE ART IN GTA AND GMA WELDING

The more specific state of the art of welding process modelling is considered based on the significant contributions relating to numerical analysis of GTA and GMA welding by Sudnik *et al.*[7, 8] The weld shapes gained from his numerical analysis correlate well with those from real manufacture, Figs 2 and 3 (with the exception of the fusion zone boundary of the GMA weld). The permissible area of current intensity and welding speed for safe GTA welding is at least approximately concurrent with practical experience, Fig. 4.

Numerical analysis is incorporated into 'thermohydrostatic' weld pool modelling comprising conservation of mass and energy, a two-phase boundary zone, a gap with transverse shrinkage, a surface heat source distribution, evaporation, convection and radiation at the surface, the force balance on top-side and bottom-side covering surface tension, hydrostatic pressure and arc pressure. The quasi-stationary state of continuous linear welding in the gravity position is considered. The solid plate surrounding the weld pool is introduced as infinitely extended.

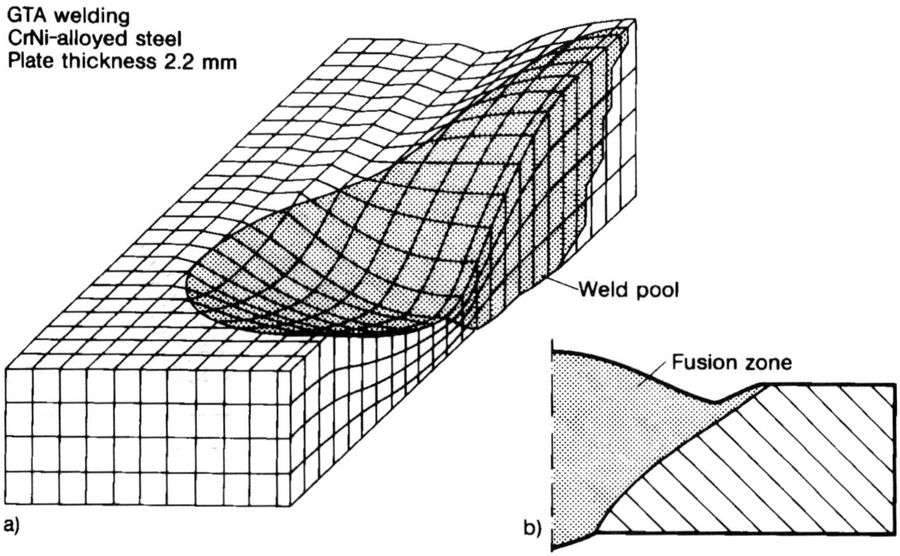

Figure 2. Weld pool (a) and finished weld shape (b). Analytical results based on 'hydrostatic' modelling, gas tungsten arc welding (current intensity $I = 430$ A, welding speed $v = 30$ mm/s, electric arc length $l_a = 1$ mm) of CrNi-alloyed steel (plate thickness 2.2 mm); after Sudnik.[7]

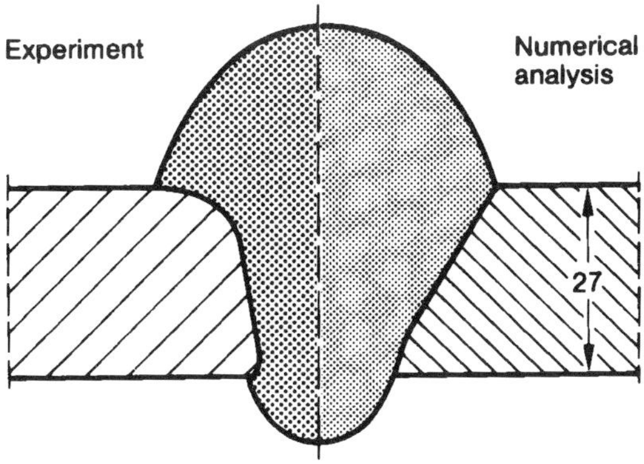

Figure 3. Finished weld shape in gas metal arc welding (current intensity $I = 130$ A, welding speed v = 5 mm/s) of mild steel (plate thickness 2.7 mm, gap width 0.1 mm), analytical result (right-hand side) compared with experimental result (left-hand side); after Sudnik *et. al.*[8]

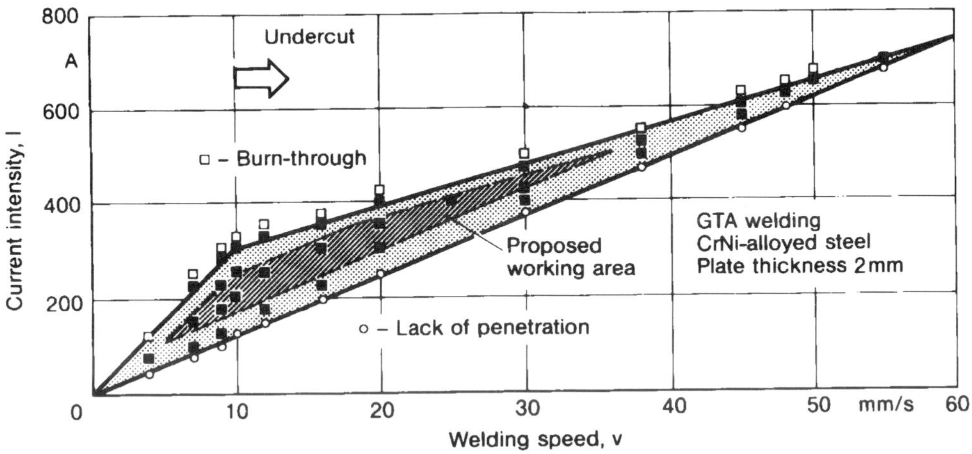

Figure 4. Weld imperfections in gas tungsten arc welding dependent on welding speed and current intensity: CrNi-alloyed steel, plate thickness 2.0 mm, analytical results based on 'hydrostatic' modelling; after Sudnik.[7]

Major uncertainties arise from heat source modelling on the one hand and unknown temperature-dependencies of physical parameters on the other. The uncertainties from heat source modelling are discussed in more detail.

A Gaussian distribution of surface heat sources is applied in the GTA weld model. The heat source density depends on effective arc voltage, effective arc amperage, melting efficiency of the arc, distance from the arc centre and standard deviation of the applied distribution. The voltage is dependent on the amperage via the arc characteristic. The latter depends on the arc length. Approximate empirical correlations have to be introduced.

The GMA weld model on the other hand resolves the arc characteristic into different voltage constituents (anode drop, cathode drop, arc column voltage, ohmic voltage). Effective values are introduced which depend on the operational process characteristics and on the constituents of the inert gas. Arc diameter, drop diameter, drop heat efficiency and gap heat loss must additionally be approximated.

Obviously, empirical knowledge is not substituted by mathematical modelling but only transferred to other areas. The benefit of numerical analysis is not a complete simulation of the welding process but its break-down into the basic aspects of physical behaviour which provides guidelines for successful engineering actions.

INDUSTRIAL POTENTIAL OF NUMERICAL WELDABILITY ANALYSIS

STATE OF INDUSTRIAL APPLICATION

Numerical analysis of weldability has only occasionally been applied to support industrial development projects. There are no routine applications known from industry, but there are pilot applications recorded in the open literature which are based on analysis made in research institutes.

These pilot applications refer to the microstructural state in the heat-affected zone of welded joints and to residual stress, shrinkage and distortion in welded components (e.g. panels or pipes) or in welded specimens used to test weldability (e.g. hot and cold cracking tests).

FUTURE POTENTIAL FOR INDUSTRIAL APPLICATION

Numerical analysis of weldability has a wide potential to support industrial development projects in the future. The potential exists for metallurgists, production engineers and design engineers.

The potential for metallurgists may be found in the following: metal alloy development for welding (base metal and weld metal, fusion zone and heat-affected zone); protective materials development (electrode coatings, shielding gases, welding flux powder); better guidelines for pre-weld and post-weld heat treatment and for multi-pass welding; better control of microstructure, hardness, ductility, hot and cold cracking susceptibility during and after welding; better interpretation of results from hot and cold cracking specimens.

The potential for production engineers may be found in the following: prediction of weld shape (reinforcement, width, undercut, penetration, cross-sectional area) dependent on operational welding parameters; derivation of the limit of operational welding parameters to avoid weld defects or weld imperfections; guidance for enhancing the efficiency of welding processes in a multiparameter environment; derivation of welding sequences with low residual stresses and distortions (or warpage), e.g. welding schedule, multilayer welding; assessment of dimensional and shape derivations after welding; assessment of the effect of mechanical and thermal pre-weld and post-weld treatments on residual stress, distortion, shrinkage, hot and cold cracking susceptibility; prediction of strength reduction during welding; guidance for selecting parameters for sensor-controlled welding; backtracing of the causes of failures of welded structures during manufacture; definition of favourable heating or cooling conditions during welding; definition of favourable clamping conditions (including tack welds) during welding.

The potential for design engineers may be found in the following: assessment of strength of welded structures (brittle fracture, fatigue, corrosion, buckling); assessment of strength improvements by mechanical or thermal pre-weld and post-weld treatments; assessment of strength reductions by tensile residual stresses and distortions; assessment of strength increase by compressive residual stresses; assessment of static, dynamic and vibrational stiffness of welded structures; enhancement of the weldability of components; transfer of results from weld specimens to welded structures; fitness-for-purpose assessment for welded structures with defects or imperfections; material selection for welded structures; backtracing of the causes of failures of welded structures in operational life; assessment of ductility of welded joints.

INDUSTRIAL DEMANDS FOR NUMERICAL WELDABILITY ANALYSIS

OPEN DEMANDS

Urgent demands of industry for numerical weldability analysis have not been met to date. This may be demonstrated by the following three examples taken from aircraft development at Deutsche Aerospace (Airbus Division) and from car development at Mercedes-Benz (Car Division). Deutsche Aerospace (DASA) and Mercedes-Benz are corporate units of Daimler-Benz. Further model and methods development is necessary in response to these demands. Preliminary steps are being taken at present to solve the stated three problems.

LASER BEAM WELDING OF AIRCRAFT SKIN PANELS

Laser beam welding of the longitudinal stiffeners of the skin panels of a commercial aircraft fuselage may reduce the weight of the panels to 80% against the riveted structure now in production, Figs 5 and 6. Welding is performed simultaneously from both sides of the stiffener (vw = 160 mm/s,

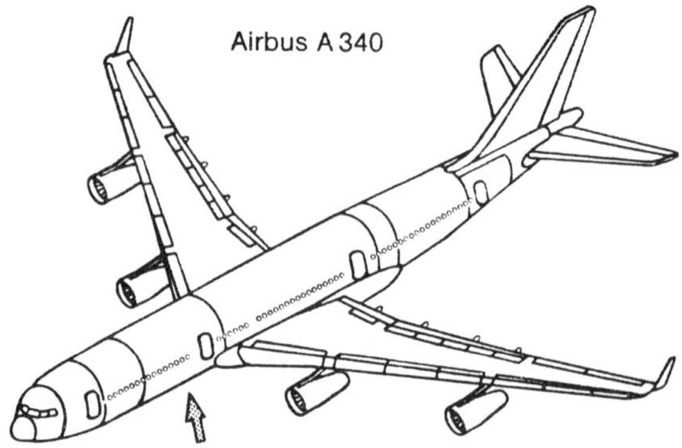

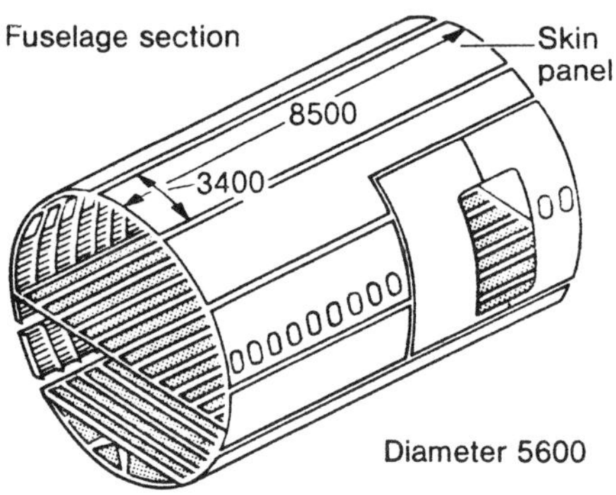

Figure 5. Commercial aircraft 'Airbus A340' with separately manufactured substructures (a) and fuselage section (b) with skin panel to be investigated.

4 kW). The material is the aluminium alloy AA6013. The main problem with welding is to keep the distortion low (especially the transverse deflections) and to reduce the residual stresses (especially the tensile longitudinal stresses).

The following measures are considered for achieving low distortion and residual stresses:

joint type variation,
specific welding sequence,
specific cooling conditions,
specific clamping modes,
welding under longitudinal preload.

A decision on these measures and their possible combination can be based on numerical weldability analysis.

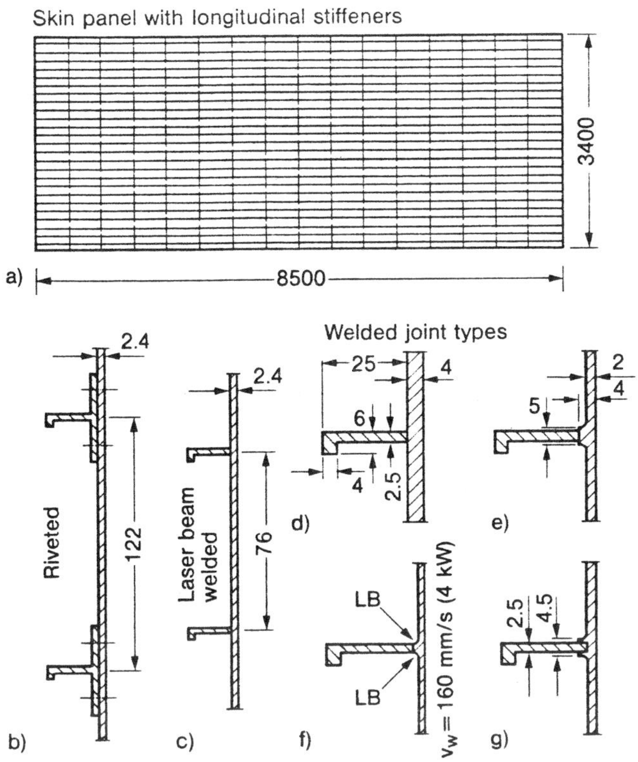

Figure 6. Skin panel with longitudinal stiffeners (a) from fuselage section of the 'Airbus A340', stiffeners riveted (in production) (b) or welded (under consideration) (c), different types of welded joints (d–g).

LASER BEAM WELDING OF THIN-SHEET CAR COMPONENTS

Laser beam welding is under consideration as a promising new method for manufacturing car bodies, truck cabs and undercarriage components from deep-drawn thin sheet material, Fig. 7. It is more expensive than conventional resistance spot welding but it produces lower distortion because of lower and quicker heat input. Lower distortion means higher accuracy of shape and dimensions, i.e. higher quality of manufacture and less post-weld grinding. Accurate assembly and reliable clamping are prerequisites.

The welding process should be optimised to high degree of accuracy of shape and dimensions after welding. The elimination of any problems during welding (e.g. groove movement, gap change, weld defects) should be included. The following measures are considered for achieving these aims.

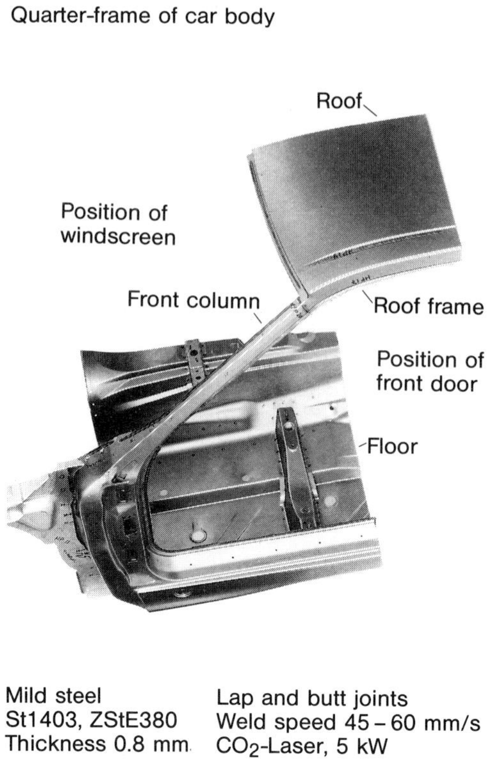

Figure 7. Laser beam welded substructure of a car body (prototype).

Possible design measures are:
type of joints appropriately chosen,
small total weld length,
intermittent welds to be preferred,
subassemblies favourably defined.

Possible manufacturing measures are:

specific welding sequence,
tack welding,
tandem welding,
back-step welding,
simultaneous welding,
simultaneous cooling or heating,
counter-aligning,
preshaping,
bracing,
clamping.

A decision on these measures and their possible combination can be based on numerical weldability analysis.

ELECTRON OR LASER BEAM WELDING OF GEAR SHAFTS

The main shaft of an automatic transmission, Fig. 8, is joined to the sun gear by a circular electron or laser beam weld. The welding process should be optimised for the highest possible flatness of the locating faces near the weld and for the lowest possible axial run-out of the sun gear. Design and process parameters are listed in the figure. The following measures can be taken in order to obtain improved flatness and lower axial run-out:

modification of weld shape,
incomplete through-welding (4.5 mm),
programmed welding speeds,
position, length and number of tack welds,
prewelding with beam splitting,
stress relief measures.

A decision on these measures and their possible combination can be based on numerical weldability analysis.

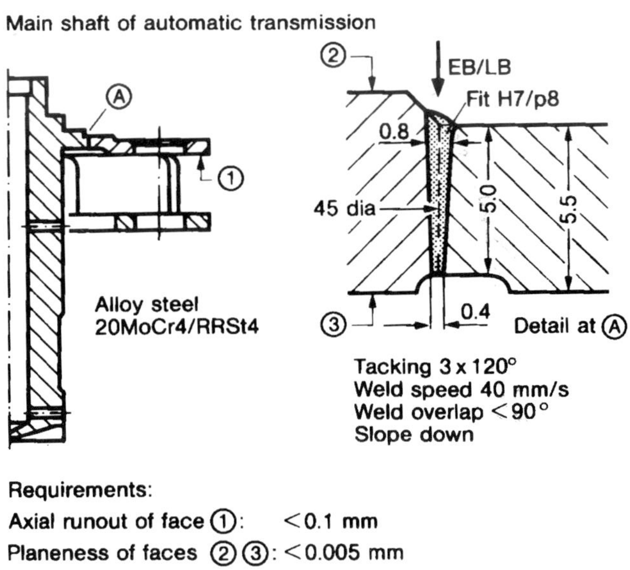

Figure 8. Electron or laser beam welded main shaft of an automatic transmission (in production).

NUMERICAL WELDABILITY ANALYSIS IN THE DESIGN PROCESS

OPERATIONAL STRUCTURE OF THE DESIGN PROCESS

The design process can be subdivided into a sequence of operations according to Fig. 9. Numerical analysis supports the design process. The definition of appropriate manufacturing processes is part of the design process, Fig. 10. Numerical analysis supporting welding process development is specified in Fig. 11.

HISTORY OF NUMERICAL ANALYSIS IN DESIGN

The earliest use of mathematical modelling and numerical analysis is known from structural engineering, especially bridge design (since 1800). Numerical analysis of machine dynamics has been performed since 1930. Aircraft

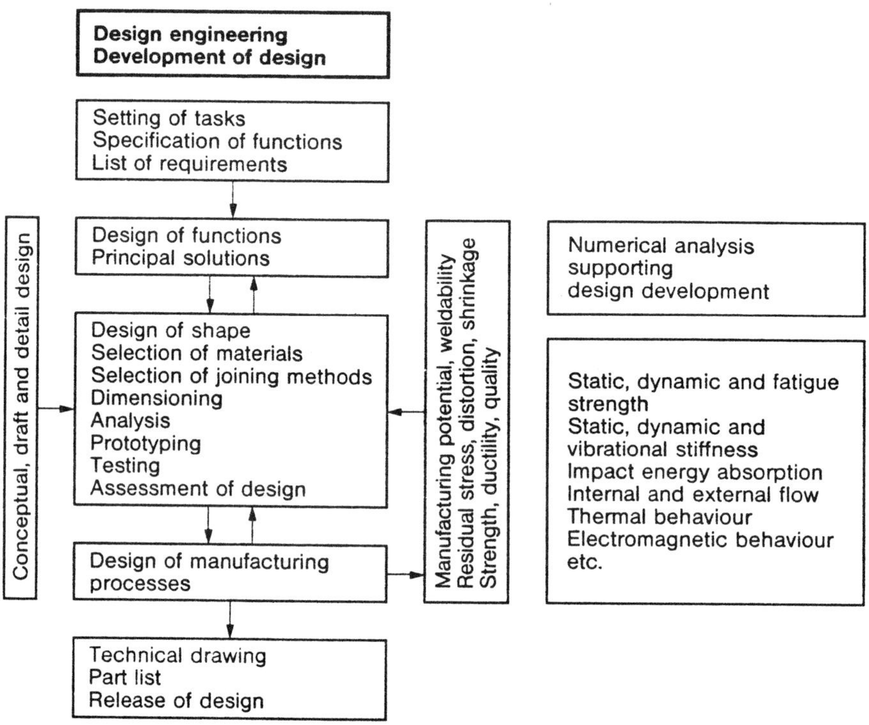

Figure 9. Sequence of operations in the design process which is supported by numerical analysis.

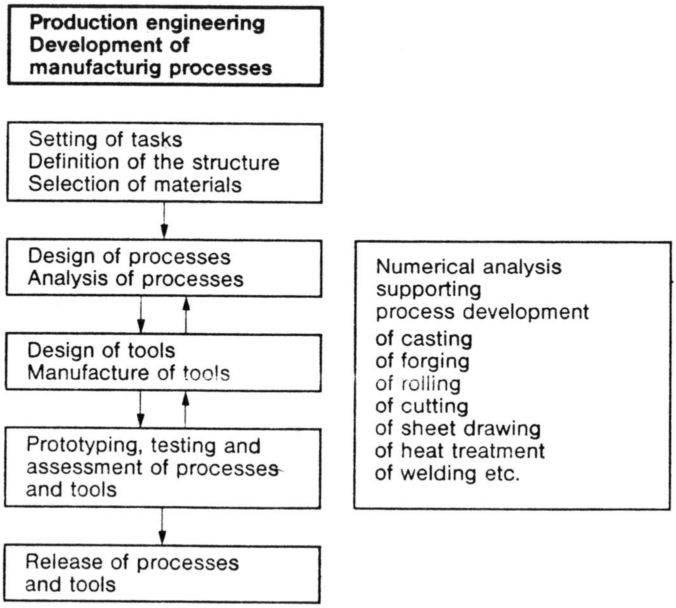

Figure 10. Definition of appropriate manufacturing processes within the design process.

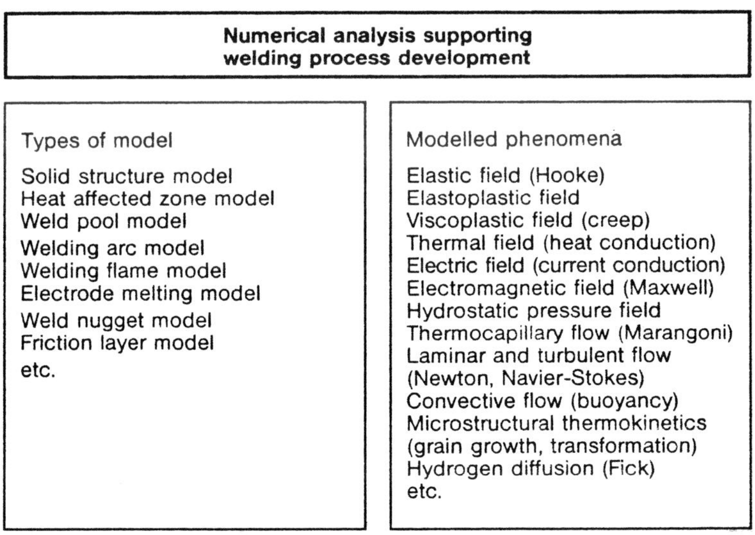

Figure 11. Numerical analysis supporting welding process development, types of model and modelled phenomena.

structures have been numerically analysed in respect of strength and aerodynamics since the same time, with much progress since 1950, when the first computers became available. Car bodies followed in 1970 with increasing computer performance. The arrival of supercomputers in 1980 opened a new area of highly non-linear and time-dependent models which became application-relevant in 1990 as in the numerical analysis of structural crash, turbulent flow, metal forming, heat treatment, and welding.

ORGANISATIONAL STATUS OF NUMERICAL ANALYSIS IN INDUSTRY

The original function of numerical analysis in the design process was a rational support of design decisions which were mainly based on empirical knowledge, ingenuity and on trial and error. An exception was the situation in structural engineering where the theory of statics was constitutive from the very beginning.

Numerical analysis in recent years moved from a supporting role to a central position in nearly all fields of engineering. Rational decision making based on computer output has been widely accepted among personnel within the companies, between cooperating companies (e.g. aerospace industry) and between supplier and customer. This is true despite the well founded objections which can be raised against any modelling method based on theory.

The common product data base comprising especially geometrical data which describe shape and dimensions has become the central reference quantity of the development process for the groups mentioned above. The 'computerised' analyst is therefore on the way to take up the leadership in the design process.

SIMULTANEOUS ENGINEERING WITHIN 'LEAN PRODUCTION AND MANAGEMENT'

The manufacturing industry, especially the automotive industry, is going through a process of fundamental organisational change at present. The new concept of 'lean production' is substituting the conventional concept of mass production based on the subdivision of the work content into small fractions ('rations') according to O. Taylor ('rationalization'). The aim of 'lean production' is to reduce the organisational overhead (therefore better termed 'lean management').

These are the main characteristics of lean production and management according to the relevant MIT Report[10, 11]:

production teams with a maximum amount of tasks and responsibility,
specific systems of error detection and error elimination,
intermediate storage ('zero-buffering', 'just-in-time delivery'),

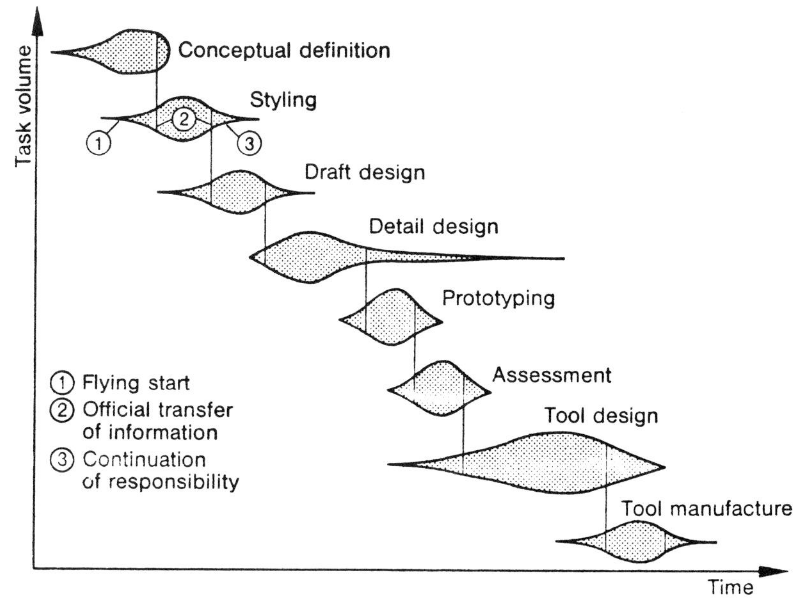

Figure 12. Chronological order of styling, design engineering and production engineering with only minor overlapping; after Whitney.[9]

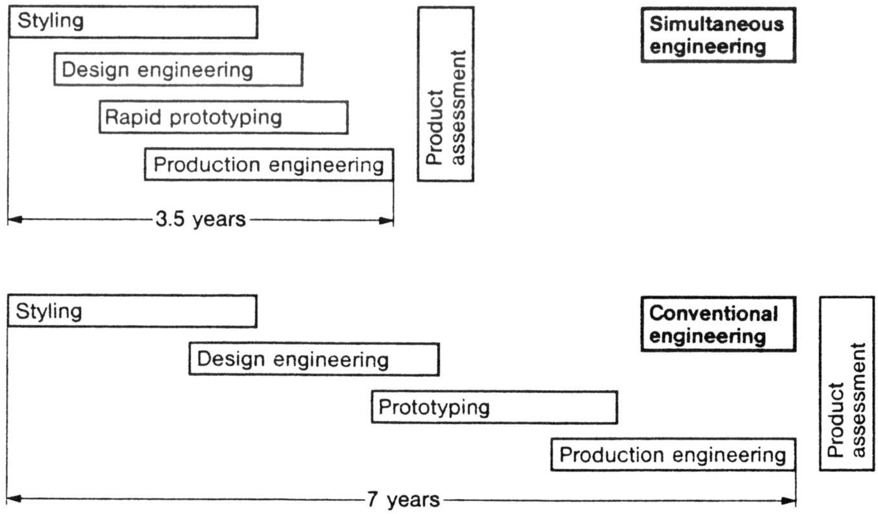

Figure 13. Simultaneous engineering compared with conventional engineering, reduction of total development time of car bodies by increased overlapping of operations.

hierarchic structure of deliveries by sub-contractors,
temporary development teams instead of permanent development sections,
customer demands directly coupled with supplier activities,
balancing of varying customer demand by globalised business,
company groups with close capital links.

The temporary development teams have to handle a specified development task within a definite time limit. This brings about simultaneous engineering activities instead of the traditional sequential procedure, Figs. 12 and 13. The main feature of simultaneous task handling is the common product data base used in design and production engineering instead of individual task-oriented data bases, Fig. 14.

INTEGRATION OF NUMERICAL ANALYSIS OF WELDABILITY

It is straightforward to couple numerical weldability analysis to the common data base of simultaneous engineering mentioned above, especially to its geometrical contents (CAD data).

The complete set of geometrical data is needed both for numerical control and numerical analysis, e.g. laser beam welding of thin sheet car body components. The CAD/CAM welding data will supplement the CAD data of

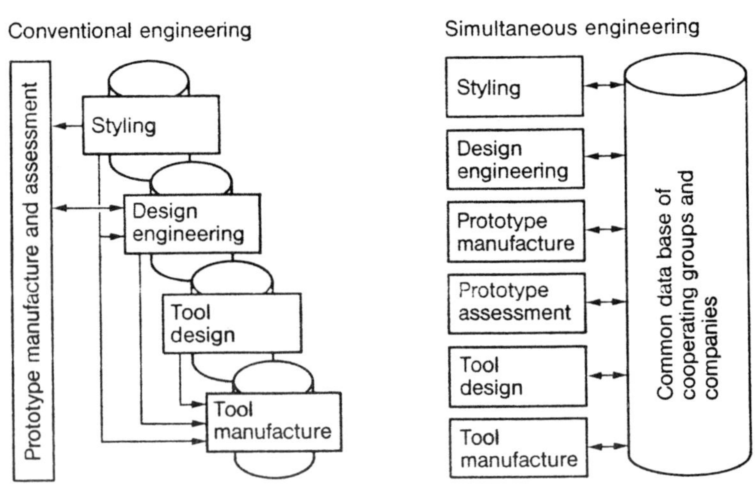

Figure 14. Common product data base in sumultaneous engineering instead of individual task-oriented data bases in conventional engineering.

styling and body engineering, and the CAD/CAM data for moulding, casting and milling the press dies and for cutting the sheet material before and after deep-drawing.

The task to be solved, before numerical analysis of residual stresses and distortions caused by welding can be assigned to the designer's responsibility to assess service strength and manufacturing quality, is the availability of equivalent moving heat source distributions for the analysis model. These distributions will depend on the welding process, the operational parameters, the welded materials, the component thickness and on further parameters. They are not yet available.

The software available at present for supporting the design process in respect of welding and weldability has been reviewed by Buchmayr (paper in reference 5). The review covers the software specifying possible design variants, software assessing the strength according to different codes, software specifying and qualifying welding procedures and software determining the weldability of base and filler metals.

DESIGN OF WELDING MACHINERY

Numerical weldability analysis may be applied advantageously to the development of new welding processes and welding tools or to the better adaptation of old ones. The developers of welding machinery should be aware of this potential of numerical weldability analysis.

CONCLUSIONS

The following conclusions are drawn from the contents of this paper:

Numerical weldability analysis based on intelligent modelling and empirically defined correction factors may provide guidelines for successful engineering actions in design, production and metallurgy.

Further model and methods development is necessary in response to industrial demands.

Simultaneous engineering within the concept of lean production facilitates the integration of numerical weldability analysis into the development process.

Equivalent moving heat source distributions must be made available as a basis of numerical weldability analysis in the design process. Numerical weldability analysis may be applied advantageously to the design of welding machinery.

REFERENCES

1. D. RADAJ: *Heat Effects of Welding – Temperature Field, Residual Stress, Distortion.* Springer-Verlag (1992).
2. B. BUCHMAYR: *Computer in der Werkstoff- und Schweißtechnik – Anwendung von mathematischen Modellen.* DVS-Verlag, Düsseldorf (1991).
3. S.A. DAVID and J.M. VITEK (Eds.): *Recent Trends in Welding Science and Technology.* Materials Park, OH., ASM International (1990).
4. L. KARLSSON, L.-E. LINDGREN and M. JONSSON, (Eds.): *Mechanical Effects of Welding.* Berlin: Springer-Verlag (1992).
5. H. CERJAK, and K.E. EASTERLING (Eds.): *Mathematical Modelling of Weld Phenomena.* London: The Institute of Materials (1993).
6. S.A. DAVID and T. DEBROY: Current issues and problems in welding science. *Science*, 1992, **257**, 497–502.
7. W. SUDNIK: Untersuchung von Schmelzschweißtechnologien anhand physikalisch-mathematischer Modelle. *Schweißen u. Schneiden*, 1991, **43**, (10), 588–590.
8. Y. DILTHEY, G. HABEDANK, T. REICHEL, W. SUDNIK and A. IWANOW: Numerische Simulation des Metall-Aktivgasschweißprozesses. *Schweißen u. Schneiden*, 1993, **45**, (3), 148–153.
9. D.E. WHITNEY: *State of the art in Japanese CAD methodologies for mechanical products.* Charles Stark Draper Laboratory, Inc.; CSDL-P-3126, Cambridge, MA 02139, 1991.
10. P.J. WOMACK, D.T. JONES and D. ROOS: *The Machine That Changed the World.* New York: Rawson Ass. et. al. 1990; German edition: *Die zweite industrielle Revolution in der Autoindustrie.* Frankfurt: Campus-Verlag (1991).
11. G. LAY: Schlankheitskur mit Schönheitsfehlern. *Technische Rundschau* (Bern) **85**, 1993 (5), 36–39.

SECTION V

PC SOFTWARE PACKAGES

PC-BASED PREDICTION OF WELDABILITY TO ASSURE WELDMENT QUALITY

B. BUCHMAYR

Institute for Materials Science and Welding Technology, Graz University of Technology, Kopernikusgasse 24, A-8010 Graz, Austria

ABSTRACT

Based on the analysis of influencing factors and their effect on the quality-determining properties of a weldment, weak points can be revealed and avoided by appropriate counter-measures. The quantitative prediction of the effects of various influencing parameters promotes goal-oriented planning and an improvement of the weldment quality. In most cases statistical methods, primarily using multiple regression analysis are employed for this purpose. Although such approaches may give useful results, they are not able to describe effects on a mechanistic basis. Causal correct interpretations are only possible by fundamental models.

In this article the principles of the different approaches applicable to weldability problems are considered regarding their accuracy, applicability and their potential for materials and process development. By considering the cold cracking problem, the strong and weak points of these approaches will be highlighted. Furthermore, software packages which support the quality assurance of weldments are also considered.

INTRODUCTION

In order to assure the quality of weldments, factors determining quality have to be taken into consideration as have possible defects and their prevention in the planning stage. Starting with an analysis of the goal parameters and any influencing parameters, quantitative correlationships or software packages are required for the prediction of the presumed behaviour of a weldment. Due to the complex processes involved during welding, regression analysis is mainly performed, in which the most significant factors are considered with respect to one particular aim (strength, hardness, toughness, crack sensitivity etc.) in one single equation. Very often linear multiple regression analysis or polynomial regression is used, without consideration of the relevant mechanistic phenomena. In contrast to this procedure, there are complex numerical models which take into account governing mechanisms and the sequential effect on the desired property.

The principles of these approaches will be discussed and available software packages, which support the assessment of weldability, will be considered.

ANALYSIS OF THE REQUIREMENTS

As a first step, the customer-specific requirements must be examined and appropriate measures derived in order to meet them. This goal-oriented procedure is the first principle to ensure excellent quality assurance.

Secondly, there should be a quantitative analysis of the significant parameters, which have to be adapted to obtain the minimum requirements.

Thirdly, scheduled goals are checked to verify they have been reached, i.e. quality control and weldment testing.

The final documentation demonstrates the quality assurance and stores this valuable information for the next job.

METHODS TO ASSURE WELDMENT QUALITY

For the planning stage, several methods of quality assurance have been developed, such as Quality Function Deployment (QFD) and Failure Mode and Effect Analysis (FMEA). For example, the 'House of Quality', in which the customer's requirements are directed to characteristic demands regarding the product itself, which may be a weldment. Figure 1 shows a simplified correlation matrix in order to find significant parameters and to set priorities for the QA-work. A similar principle is applied using the FMEA, where potential failures (defects) are considered and an analysis of their causes is performed. Via a ranking of risk factors, the most likely or most important failures should be avoided.

The main purpose of these approaches is to recognise effective measures for sufficient quality in a comprehensive and objective manner. The assessment is however only enumerative (a list of failure causes) and may be weighted by a simple scale (e.g. from 1 to 7). A quantitative prediction of the effect of influencing factors on the desired properties is however not included.

In order to perform *quantitative* predictions of the effect of influencing parameters, it is usual to use empirical equations which take into account the

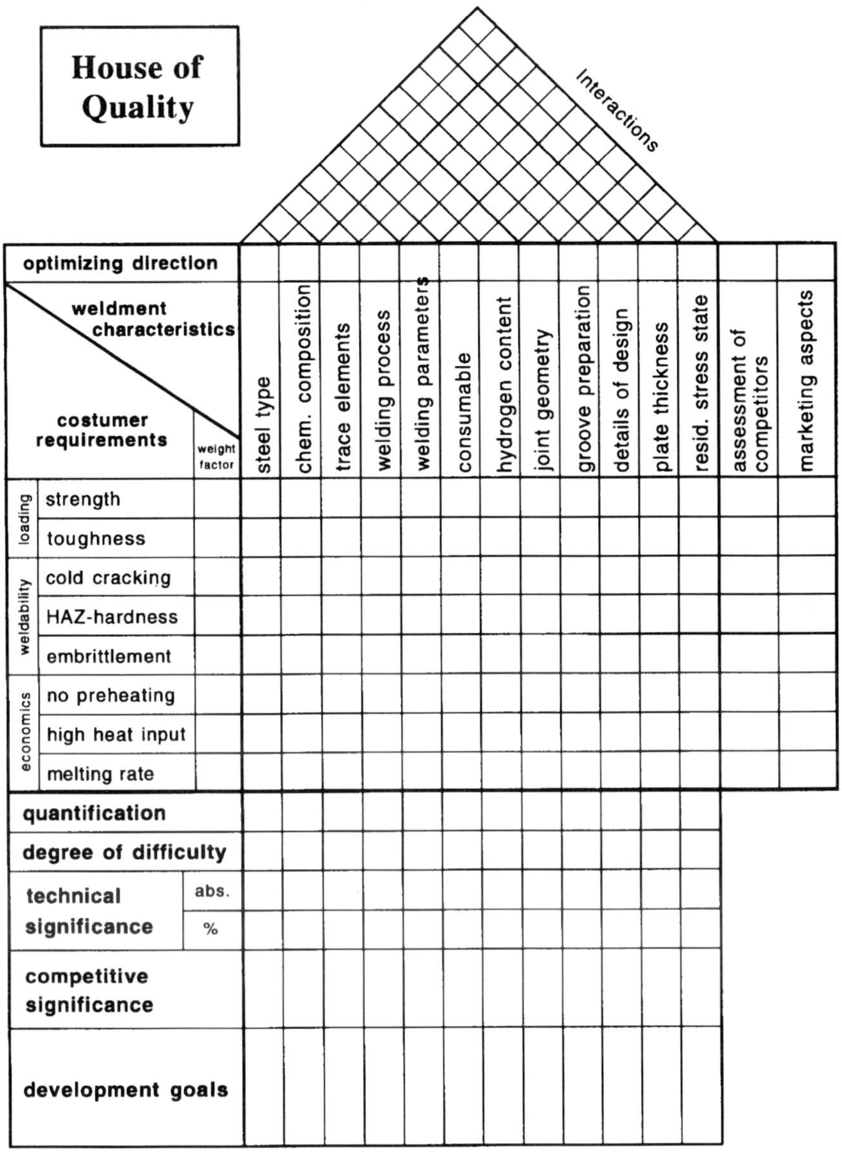

Figure 1. House of Quality (QFD-method) to analyse the user requirements and the measures to reach these goals.

most significant parameters. Using a large set of experimental data and by multiple regression analysis, predictive equations can be derived. For this task, several commercial programs are available, like the statistics package 'Statgraf'. The key element for a successful application is however the choice of the equation type. Without knowledge of the dominant mechanisms involved, many engineers use additive, linear equations, which only give moderate correlation coefficients. A better predictive capability provides such equations, which take into account the physical basis of a phenomena. For example, the analytical solution of the temperature field during welding can be a good approximation (using some simplifications, like a point heat source, constant thermophysical properties, neglect of the heat loss), and by the implementation of adjustable parameters the real behaviour can be sufficiently described. Additional problems arise when there are mutual influences between the parameters. In these cases, reliable equations cannot be found and a strictly forward approach is not applicable. A typical example for such a case is the prediction of the distortion of a weldment.

In order to solve such complex and multiphased problems, more and more workers now try to simulate dominant processes on a metallurgical basis. This approach allows a correct chaining of lumped phenomena and their interactions as well as the prediction of the multiple effect on the influencing factors.

Therefore, we can distinguish between different levels (depths) of consideration required for quality determining factors such as:

- an enumerative approach (e.g. Ishikawa-diagram);
- a weighted approach (e.g. House of Quality, FMEA);
- a statistical, empirical approach (regression analysis);
- a physical based approach (= mathematical models),

whereby all have their eligibility and do not exclude each other.

In order to demonstrate the typical potential of these approaches, the problem of cold cracking is considered in the following chapter.

ASSESSMENT OF COLD CRACKING SENSITIVITY OF MODERN STRUCTURAL STEELS

The fact that there are many different experimental procedures to test the sensitivity for cold cracking may demonstrate the difficulties involved with this problem.

The main factor in cold cracking is the critical combination of the following three properties:

1. hardenability of the base material or weld metal;
2. hydrogen content in the weldment;
3. level of residual stresses.

These factors are thus dependent on other secondary and interacting factors. For a quantitative assessment, there are several approaches in use, which provide a measure for the sensitivity, such as:

- carbon equivalent;
- critical HAZ-hardness;

- critical cooling time;
- critical implant stress;
- critical plate thickness;
- minimum preheating temperature and others.

Taking into consideration the necessary preheating temperature, there are several empirical equations in use,[1–5] which describe the most important influencing factors in different ways.

It is a question of a sound experimental basis, careful assessment and the selection of a proper analytical function, which determine the range of applicability and ability for extrapolations. In the following, some principal statements will be made concerning a successful assessment of weldability aspects.

DERIVATION OF EMPIRICAL EQUATIONS

First of all, it is necessary to plan a well-balanced set of experiments regarding the various influencing factors so that the effect of a single influencing factor (by keeping others constant) can be assessed. If this is not done, a robust equation type cannot be found, and only linear multiple regression analysis can be done. This results in a limited applicability. A better procedure would be to study the effect of single influencing factors and to add the non-linear effects of all significant parameters. By assuming an additive superposition, better predictions are given in many cases. Indicators for the goodness of fit are not only multiple regression coefficients but also the so-called *t*-values, which measure the significance of the influencing parameters.

Such a procedure has been applied[6] to determine an equation for the preheating temperature. The data set used was based on results of Tekken- and CTS-Tests, whereby the chemical composition, the plate thickness, the hydrogen content and the heat input was varied carefully. For the choice of proper equations, the individual effects of single parameters were studied. The adjustable parameters were then found by a non-linear regression analysis. Starting values for the variations of the most significant parameters were chosen as follows: thickness = 30 mm, hydrogen content = HD4, heat input = 1 kJ/mm.

Figure 2 shows schematically the effect of the single parameters and non-linearities and mutual interactions can be observed.

The equation found by Uwer *et al.*[6,7] has the form:

$$T_0 = 700{*}\mathrm{CET} + 160{*}\tan h(d/35) + 62{*}\mathrm{HD}^{0.35} + (53{*}\mathrm{CET}\text{-}32){*}Q - 330$$

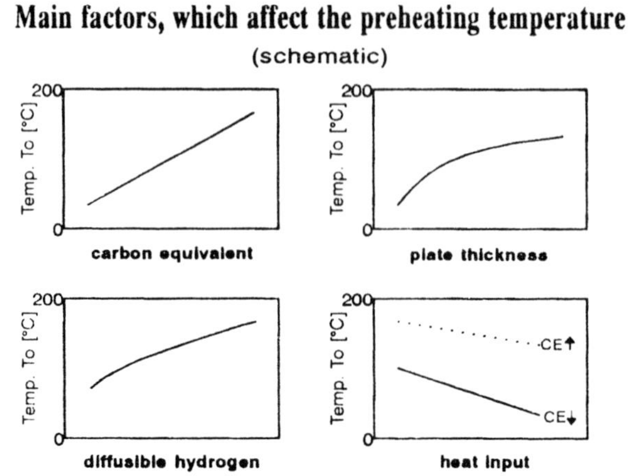

Figure 2. Parameters which determine the necessary preheating temperature to avoid cold cracking.[6]

where CET = C + (Mn + Mo)/10 + (Cr + Cu)/20 + Ni/40, d is the plate thickness in mm, HD is the diffusible hydrogen content in $cm^3/100$ g, Q is the heat input in kJ/mm.

Due to this well-considered procedure, it is obvious that the equation has a better applicability and accuracy compared with just linear approaches.

It should also be noted that the application of any regression equation requires the validity range to be taken into account concerning chemical composition, welding parameters etc.

Furthermore, it is not surprising that the different regression based equations (e.g. for the prediction of the preheating temperature) give quite different results. This is caused by different sets of empirical data with specific peculiarities (i.e. different ranges of chemical composition, different ways of steel making, different steel types etc.) and also by differences in the assessment. Under given conditions (steel type, plate thickness, hydrogen content), a comparison of the results using well-known predictive equations[1–5] revealed differences in the predicted preheating temperature of more than 100°C!

MODELLING OF PROPERTY-DETERMINING PROCESSES

As described[8], fundamental metallurgical calculations of phenomena which have significant influence on the properties can provide detailed causalities. Nevertheless, it was also shown[8] that some weldability problems cannot be solved, like the sensitivity for cold cracking, because a failure criterion or a critical constellation of microstructure, residual stress state and hydrogen content is not yet known.

Further details about the usage of numerical models in the field of welding technology can be found[9].

A COMPREHENSIVE ASSESSMENT OF WELDABILITY ASPECTS

In order to assess the weldability of a particular steel type in a comprehensive manner, the approaches for the description of the cold cracking sensitivity are not sufficient. For instance, an increase of the heat input would reduce the likelihood for cold cracking but on the other hand important properties like strength and toughness would deteriorate. This example should demonstrate that various considerations are needed to find optimal welding conditions. In many cases, the final selection of appropriate welding parameters results from a compromise of diverging influences.

To support a comprehensive evaluation of various properties of a weldment, the software package 'HAZ-Calculator' was developed. This program provides about 50 evaluation routines, which assesses the effect of various parameters, like chemical composition, welding parameters, plate thickness etc. An overview of the routines implemented can be obtained from the main menu in Fig. 3. A detailed description is given elsewhere.[10, 11]

A CASE STUDY USING THE HAZ-CALCULATOR

Consider the case where for a given structural steel type (StE355), the welding parameters have to be chosen so that for an application in a sour gas environment the HAZ-hardness should not exceed the value of 250 HV (to avoid stress-induced corrosion). Due to economical reasons, there should be no preheating and no PWHT.

Using routine 8 in the main menu, the HAZ-hardness and the amount of martensite can be predicted as a function of the cooling time $t_{8/5}$. Figure 4 shows the hardness curve for the considered steel grade with the given chemical composition. As a result, it may be seen from the diagram that for the specified requirements the cooling time should be longer than 35 s. In a next step by calling routine 4, the user can determine the welding parameters for a given plate thickness, joint type etc., which fulfil the lower limit regarding heat input. Afterwards, one can also check the need for preheating using routine 12, which also provides information regarding the selection of a proper electrode type with a limited hydrogen content. Finally, the mechanical properties of the heat-affected zone can be predicted as a function of the heat input.

With this multi-step procedure, the welding engineer has obtained all the information he needs to specify proper welding conditions. By relying on this pre-assessment, welding trials can be significantly reduced. It could also be possible to consider scatters in the chemical composition and welding parameters in order to see their effect on the required weldment properties.

OTHER SOFTWARE PACKAGES FOR WELDABILITY ASSESSMENT

A tabular overview of various computer programs which are available for the assessment of weldability is given in Table 1. All of them are based on equations which have been derived statistically from a large set of experimental results.

Some of them cover only a particular problem (like weld thermal cycle prediction or dilution calculation), whereas others are more comprehensive. In any case it is always necessary that the ranges of application are defined, but unfortunately this information is not always provided.

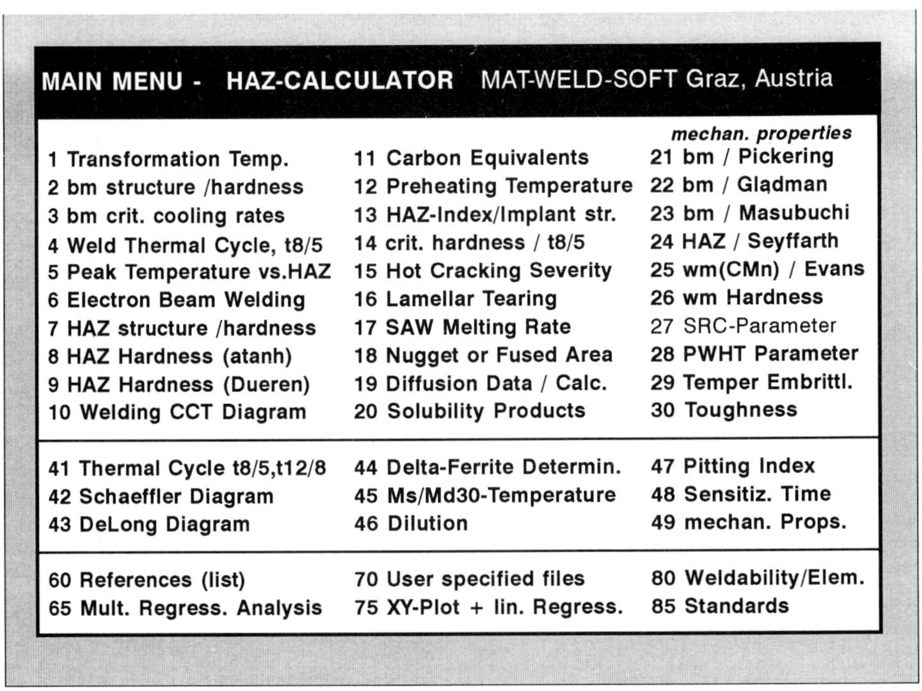

Figure 3. Main menu of the software package 'HAZ-Calculator'.

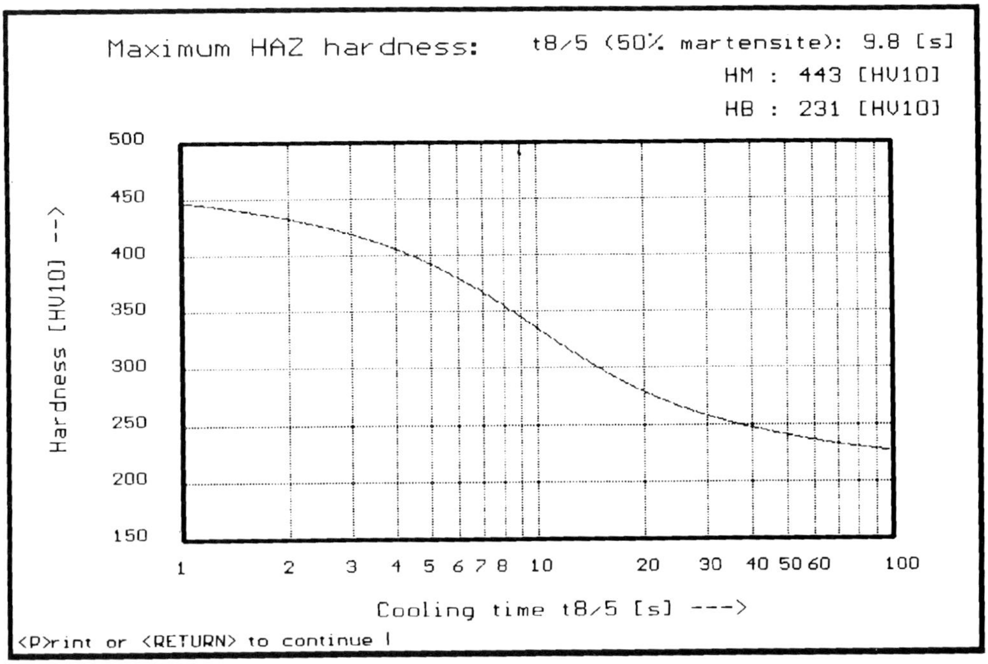

Figure 4. Calculated HAZ-hardness as a function of cooling time for the steel grade StE355 (using HAZ-Calculator / routine 8).

Table 1 List of available software packages, which support the assessment of weldability

Name	Supplier	Description
Ferrite predictor	AWI, USA	Prediction of delta-ferrite
HAZ-Calculator	MAT WELD SOFT Graz, A	50 metallurgical-based approaches
N.N.	Frank, Germany	Several routines in BASIC-code
N.N.	Thier, Germany	Weld thermal cycle calculations
N.N.	Thier, Germany	Calculation of welding CCT diagram
PREHEAT	TWI, UK	Calculation of preheating temperature according to BS5153
Schaeffler-DeLong	ESAB AB, Sweden	Prediction of delta-ferrite and graphics
WELDY	Gut, CH	Calculation routines for low-alloy steels
WELDWARE	Seyffarth, Germany	Assessment of weldability of low-alloy-steels

THE USE OF KNOWLEDGE-BASED (EXPERT) SYSTEMS

Based upon the complexity and experience-based procedures applied in welding technology, expert systems or knowledge-based systems seem destined for this purpose. In welding technology, typical applications for knowledge-based systems are:

- diagnosis (e.g. for welding or machine defects);
- selection problems (consumables, welding parameters, processes);
- scheduling (groove form, welding specifications, PWHT);
- process control (robot control or online process control);
- training of personnel.

Some examples of powerful expert systems in the field of welding technology are MAGXPERT, MIG EXPERT, Weld Defect, EUROSEL, WELDEX.

For the development of any expert system, a recent evaluation of a DVS-working group suggested NEXPERT OBJECT as an appropriate XPS-shell for welding applications.

THE QA-COMPONENT DOCUMENTATION

To ensure welding quality, not only the aforementioned causal software packages are useful, there is also a considerable input by programs for the administration of WPS and PQRs. In references 12 and 13, several computer programs have been listed and described for this purpose. The rapid access to this valuable information (which proved to be an excellent choice) is as important as any calculation routine.

CONCLUSIONS REGARDING THE CHOICE OF A SUITABLE METHOD

Depending on the objectives, the welding engineer can select from various approaches because each method has its typical application field and capability to support a particular welding problem. Table 2 summarises the characteristics of the various methods and their capability.

The approaches can be distinguished by their applied depth of knowledge and to the number of aspects considered. Besides statistical QA-methods, which are mainly used by QA-managers, there are specific semi-empirical programs which assist the welding engineer and special programs which cover small but very detailed problems. In Fig. 5, different methodologies and typical user groups were assessed for current situations.

SUMMARY

The scheduling of high quality weldments can be supported by several computer-based methods or software packages. In the first planning phase, methods like FMEA, QFD or diagnostic and scheduling expert systems can provide useful information. For quantitative assessments either empirical-based regression equations or numerical models can be used. These methods demonstrate the effect of influencing factors in order to optimise welding conditions. Their correct use however very much depends on the range of application. In the future, further improvements of both materials and welding processes will be based more and more on mathematical models, and so it seems indispensable that R&D departments take advantage of these powerful tools.

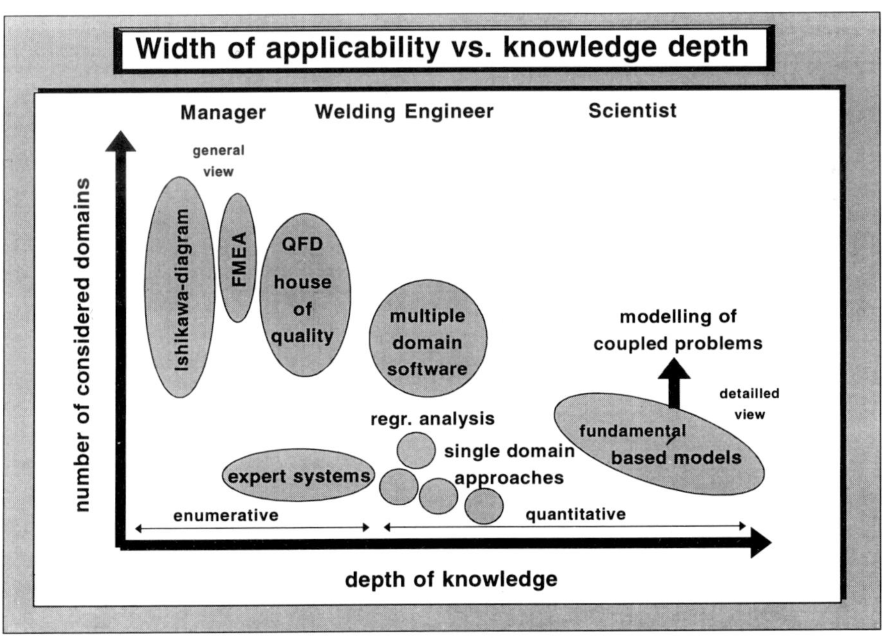

Figure 5. Classification of principal methods which support the assessment of weldability, regarding their applicability and applied depth of knowledge.

Table 2 Comparison of principal methods applicable for the assurance of weld quality

Method	Advantages	Disadvantages	Typical applications
Ishigawa diagram	Simple, easy to understand and clearly arranged Partitioned into primary, secondary and tertiary influencing factors	No quantitative evaluation No order of the effect of parameters	System analysis
House of Quality FMEA	Definition and analysis of the requirements Assessment of the influencing factors regarding their relevance Coupling of various characteristic properties Identification of development priorities Comprehensive consideration of various aspects (also non-numerical items)	No quantitative prediction of the effect of influencing factors concerning a particular property Only an approximative order of the influencing factors	Strategic scheduling and quality assurance applications
Expert systems	Knowledge-based, logical support of decisions Handling of vague information	Mostly non-quantitative statements Only narrow problem fields	Failure diagnosis scheduling tool
Empirical equations (regression analysis)	Simple to derive → cost-saving Wide applicability Goal-oriented approach Determination of the significance and the effect of the influencing factors Short computation time Ä applicable for online process control Balanced data set Ä good prediction accuracy	Requires a well-balanced empirical data set Just one goal value can be predicted Validity only in a limited parameter range Microstructural causalities are taken into account insufficiently Non-linear influences are seldom considered	Property prediction and process control
Mathematical modelling	Mechanism-based description of the effect of parameters by coupling of relevant phenomena Local and sequential correct simulation of the property-determining processes Leads to improved understanding Identification of further development potentials	Time and cost consuming Simulation of processes kinetics requires adjustment to experimental data Correlationship between microstructure and mechanical properties are insufficient understood Long response times	Tool for research and development

REFERENCES

1. C. DÜREN: *3R-int.*, 1985, **24**, 3–8.
2. Y. ITO and K. BESSYO: *J. JWES*, 1968, **37**, 9, 983–991.
3. M. INAGAKI *et al.*: IIW-Doc. IX-1412-86.
4. H. SUZUKI: IIW-Doc. IX-1074-78.
5. O.W. BLODGETT: *Welding Journal*, 1979, **58**, 12, 26–42.
6. D. UWER, H. HÖHNE: Schweissen und Schneiden, 1991, **43**, H(5), 282–286.
7. J. DEGENKOLBE and D. UWER: *DVS-Berichte*, 1992, **140**, 43–51.
8. B. BUCHMAYR: this book, 120–137.
9. B. BUCHMAYR: *Computer in der Werkstoff- und Schweißtechnik – Anwendung von mathematischen Modellen*, Fachbuchreihe Schweißtechnik, 1991, **112**, DVS-Verlag Düsseldorf.
10. B. BUCHMAYR: *Schweißen und Schneiden*, 1989, **41**, (2), 69–75.
11. B. BUCHMAYR: *DVS-Berichte*, 1991, **133**, 56–60.
12. B. BUCHMAYR: *Swiss Materials*, 1992, **4**, (5), 5–15.
13. P. DREWS (ed.): Proc. IIW-Workshop, *Computer-Aided Welding*, Madrid, Sept. 92.

INDEX

UNIV DUNE

DURHAM UNIVERSITY LIBRARY
3 0104 00866887 7